Achieving sustainable cultivation of sorghum

Volume 2: Sorghum utilization around the world

It is widely recognised that agriculture is a significant contributor to global warming and climate change. Agriculture needs to reduce its environmental impact and adapt to current climate change whilst still feeding a growing population, i.e. become more 'climate-smart'. Burleigh Dodds Science Publishing is playing its part in achieving this by bringing together key research on making the production of the world's most important crops and livestock products more sustainable.

Based on extensive research, our publications specifically target the challenge of climate-smart agriculture. In this way we are using 'smart publishing' to help achieve climate-smart agriculture.

Burleigh Dodds Science Publishing is an independent and innovative publisher delivering high quality customer-focused agricultural science content in both print and online formats for the academic and research communities. Our aim is to build a foundation of knowledge on which researchers can build to meet the challenge of climate-smart agriculture.

For more information about Burleigh Dodds Science Publishing simply call us on +44 (0) 1223 839365, email info@bdspublishing.com or alternatively please visit our website at www.bdspublishing.com.

Related titles:

Achieving sustainable cultivation of sorghum Volume 1:
Genetics, breeding and production techniques
Print (ISBN 978-1-78676-120-0); Online (ISBN 978-1-78676-122-4, 978-1-78676-123-1)

Achieving sustainable cultivation of maize Volume 1: From improved varieties to local applications
Print (ISBN 978-1-78676-008-1); Online (ISBN 978-1-78676-010-4, 978-1-78676-011-1)

Achieving sustainable cultivation of maize Volume 2: Cultivation techniques, pest and disease control
Print (ISBN 978-1-78676-012-8); Online (ISBN 978-1-78676-014-2, 978-1-78676-015-9)

Achieving sustainable cultivation of wheat Volume 1: Breeding, quality traits, pests and diseases
Print (ISBN 978-1-78676-016-6); Online (ISBN 978-1-78676-018-0, 978-1-78676-019-7)

Achieving sustainable cultivation of wheat Volume 2: Cultivation techniques
Print (ISBN 978-1-78676-020-3); Online (ISBN 978-1-78676-022-7, 978-1-78676-023-4)

Chapters are available individually from our online bookshop: https://shop.bdspublishing.com

climate•SMART•publishing
IN AGRICULTURAL SCIENCE

BURLEIGH DODDS SERIES IN AGRICULTURAL SCIENCE

NUMBER 32

Achieving sustainable cultivation of sorghum

Volume 2: Sorghum utilization around the world

Edited by Professor William Rooney
Texas A&M University, USA

burleigh dodds
SCIENCE PUBLISHING

Published by Burleigh Dodds Science Publishing Limited
82 High Street, Sawston, Cambridge CB22 3HJ, UK
www.bdspublishing.com

Burleigh Dodds Science Publishing, 1518 Walnut Street, Suite 900, Philadelphia, PA 19102-3406, USA

First published 2018 by Burleigh Dodds Science Publishing Limited

Library of Congress Control Number: 2017956487

British Library Cataloguing in Publication Data
A catalogue record for this book is available from the British Library

ISBN 978-1-78676-124-8 (print)
ISBN 978-1-78676-126-2 (online)
ISBN 978-1-78676-127-9 (online)
ISSN 2059-6936 (print)
ISSN 2059-6944 (online)

Typeset by Deanta Global Publishing Services, Chennai, India
Printed by Lightning Source

Contents

Series list

Title	Series number
Achieving sustainable cultivation of maize - Vol 1 From improved varieties to local applications Edited by: Dr Dave Watson, CGIAR Maize Research Program Manager, CIMMYT, Mexico	001
Achieving sustainable cultivation of maize - Vol 2 Cultivation techniques, pest and disease control Edited by: Dr Dave Watson, CGIAR Maize Research Program Manager, CIMMYT, Mexico	002
Achieving sustainable cultivation of rice - Vol 1 Breeding for higher yield and quality Edited by: Prof. Takuji Sasaki, Tokyo University of Agriculture, Japan	003
Achieving sustainable cultivation of rice - Vol 2 Cultivation, pest and disease management Edited by: Prof. Takuji Sasaki, Tokyo University of Agriculture, Japan	004
Achieving sustainable cultivation of wheat - Vol 1 Breeding, quality traits, pests and diseases Edited by: Prof. Peter Langridge, The University of Adelaide, Australia	005
Achieving sustainable cultivation of wheat - Vol 2 Cultivation techniques Edited by: Prof. Peter Langridge, The University of Adelaide, Australia	006
Achieving sustainable cultivation of tomatoes Edited by: Dr Autar Mattoo, USDA-ARS, USA & Prof. Avtar Handa, Purdue University, USA	007
Achieving sustainable production of milk - Vol 1 Milk composition, genetics and breeding Edited by: Dr Nico van Belzen, International Dairy Federation (IDF), Belgium	008
Achieving sustainable production of milk - Vol 2 Safety, quality and sustainability Edited by: Dr Nico van Belzen, International Dairy Federation (IDF), Belgium	009
Achieving sustainable production of milk - Vol 3 Dairy herd management and welfare Edited by: Prof. John Webster, University of Bristol, UK	010
Ensuring safety and quality in the production of beef - Vol 1 Safety Edited by: Prof. Gary Acuff, Texas A&M University, USA & Prof.James Dickson, Iowa State University, USA	011
Ensuring safety and quality in the production of beef - Vol 2 Quality Edited by: Prof. Michael Dikeman, Kansas State University, USA	012
Achieving sustainable production of poultry meat - Vol 1 Safety, quality and sustainability Edited by: Prof. Steven C. Ricke, University of Arkansas, USA	013
Achieving sustainable production of poultry meat - Vol 2 Breeding and nutrition Edited by: Prof. Todd Applegate, University of Georgia, USA	014
Achieving sustainable production of poultry meat - Vol 3 Health and welfare Edited by: Prof. Todd Applegate, University of Georgia, USA	015

Managing soil health for sustainable agriculture - Vol 2 049
Monitoring and management
Edited by: Dr Don Reicosky, USDA-ARS, USA

Rice insect pests and their management 050
E. A. Heinrichs, Francis E. Nwilene, Michael J. Stout, Buyung A. R. Hadi & Thaís Freitas

Improving grassland and pasture management in temperate agriculture 051
Edited by: Prof. Athole Marshall & Dr Rosemary Collins, University of Aberystwyth, UK

Precision agriculture for sustainability 052
Edited by: Dr John Stafford, Silsoe Solutions, UK

Achieving sustainable cultivation of temperate zone tree fruit and berries – Vol 1 053
Physiology, genetics and cultivation
Edited by: Prof. Gregory Lang, Michigan State University, USA

Achieving sustainable cultivation of temperate zone tree fruit and berries – Vol 2 054
Case studies
Edited by: Prof. Gregory Lang, Michigan State University, USA

Agroforestry for sustainable agriculture 055
Edited by: Prof. María Mosquera-Losada, University of Santiago de Compostela, Spain & Dr Ravi Prabhu, World Agroforestry Centre (ICRAF), Kenya

Achieving sustainable cultivation of tree nuts 056
Edited by: Prof. Ümit Serdar, Ondokuz Mayis University, Turkey & Emeritus Prof. Dennis Fulbright, Michigan State University, USA

Assessing the environmental impact of sustainable agriculture 057
Edited by: Prof. Bo P. Weidema, Aalborg University/2.-0 LCA Consultants, Denmark

Critical issues in plant health: 50 years of research in African agriculture 058
Edited by: Dr. Peter Neuenschwander, IITA & Dr. Manuele Tamò, IITA

Achieving sustainable cultivation of vegetables – Vol 1 059
Physiology, breeding, cultivation and quality
Edited by: Emeritus Prof. George Hochmuth, University of Florida, USA

Achieving sustainable cultivation of vegetables – Vol 2 060
Case studies
Edited by: Emeritus Prof. George Hochmuth, University of Florida, USA

Advances in Conservation Agriculture – Vol 1 061
Systems and science
Edited by: Prof. Amir Kassam, University of Reading, UK

Advances in Conservation Agriculture – Vol 2 062
Practice and benefits
Edited by: Prof. Amir Kassam, University of Reading, UK

Achieving sustainable greenhouse cultivation 063
Edited by: Prof. Leo Marcelis and Dr Ep Heuvelink, Wageningen University, The Netherlands

Achieving carbon-negative bioenergy systems from plant materials 064
Edited by: Dr Chris Saffron, Michigan State University, USA

Achieving sustainable cultivation of citrus and other tropical and subtropical fruits – Vol 1 065
Citrus fruits
Edited by: Prof. Elhadi Yahia, Universidad Autonoma de Queretaro, Mexico

Achieving sustainable cultivation of citrus and other tropical and subtropical fruits – Vol 2 066
Other tropical and subtropical fruits
Edited by: Prof. Elhadi Yahia, Universidad Autonoma de Queretaro, Mexico

Pesticides and agriculture: Profit, politics and policy 067
Edited by: Dr Dave Watson, formerly CIMMYT, Mexico

Acknowledgements

We wish to acknowledge the following for their help in reviewing particular chapters:

- Chapter 1: Dr Sergio Román Othón Serna Saldívar, ITESM, Mexico; Dr Robert Moreau, USDA-ARS, USA
- Chapter 2: Dr Jeff Dahlberg, University of California, USA
- Chapter 3: Dr Scott Finlayson, Texas A&M University, USA
- Chapter 4: Prof. Gyebi Duodu, University of Pretoria, South Africa
- Chapter 5: Dr Scott Sattler, USDA-ARS, USA; Dr Yi-Hong Wang, University of Louisiana at Lafayette, USA; Prof. Juan Landivar, Texas A&M University, USA
- Chapter 6: Dr Cary W. Herndon, Jr., Mississippi State University, USA
- Chapter 7: Dr Claudia Guimaraes, EMBRAPA, Brazil
- Chapter 8: Prof. Guoqing Liu, Heibei Academy of Agricultural and Forestry Sciences, China
- Chapter 9: Dr Benôit Clerget, CIRAD, France; Dr Henry Ojulong, ICRISAT, Kenya

Introduction

Sorghum is one of the world's major cereals, cultivated in the semi-arid tropics for a growing range of uses. Like other crops it faces the need to meet rising demand whilst reducing its environmental impact and adapting to the challenges of climate change. These challenges are addressed in the two volumes of *Achieving sustainable cultivation of sorghum*:

- Volume 1 Genetics, breeding and production techniques
- Volume 2 Sorghum utilization around the world

The chapters in this volume (Volume 2) discuss the varied uses of sorghum as a food grain, as well as a forage and energy crop. The volume also includes detailed case studies of key challenges facing sorghum cultivation in regions such as Asia, Africa and South America.

Part 1 Sorghum utilization

Chapter 1 addresses the structure and chemistry of sorghum grain Sorghum is grown around the world under diverse environmental conditions. Combined with the high degree of genetic diversity present in sorghum, this can result in substantial variability in grain composition and grain quality. While similar to other cereal grains such as maize and the millets in some aspects, sorghum has many unique compositional traits which impact grain quality and utilization. The chapter reviews the composition of sorghum grain, including its physical structure, composition of bio-macromolecules and micronutrients such as starch, proteins, lipids, phenolic compounds and vitamins and minerals. The chapter also includes a summary of research showing how various factors can impact sorghum grain composition.

Following from the first chapter's focus on the chemistry and physiology of sorghum, Chapter 2 reviews the domestication, spread and uses of sorghum as a crop. Although only a minimal component of human food in developed countries, sorghum remains the fifth most important cereal crop in terms of production and acreage This is due to its robustness as a cereal crop, biotic and abiotic tolerances, and diversity of both varieties and uses. The chapter describes the diversity of sorghum and related species, sorghum's anthropocentric origins, spread and domestication as well as reviewing its utility as both a source of nutrition for humans and domestic animals. The chapter also examines the population genetics of global sorghum and the crop's relationship to its wild relatives. The chapter covers early evidence of sorghum use and introduces the sorghum conversion (SC) program, as well as current techniques such as introgression of distant wild crop relatives and whole genome sequencing.

Returning to the theme of the structure of sorghum, Chapter 3 focuses on sorghum crop physiology and development. A thorough understanding of how grain sorghum develops is essential in determining how best to manage this crop. Grain sorghum has three predictable phases of growth: vegetative (planting to panicle initiation); reproductive (panicle initiation to flowering); and grain filling (after flowering to maturity). The chapter describes each of the phases in detail, observing how each stage is affected by factors such as genotype, temperature and photoperiod. Climatic conditions for optimal growth are discussed, as well as the effects and physiological basis of drought, high temperature and elevated carbon dioxide tolerance. Due to climate change, sorghum-producing areas are facing increased exposure to water deficits and extreme temperatures, making understanding of plant response to environmental stresses even more important. The chapter concludes that the use of high-yielding varieties and proper management practices will be essential to the future of this crop.

Chapter 4 begins to analyse the uses of sorghum, considering sorghum's role as a food grain. Grain sorghum is the fifth most important cereal crop in the world, exceeded only by rice, wheat, corn and barley, and it is a dietary staple for more than half a billion people in more than 30 countries. Sorghum is a principal food in areas of Africa and Asia as well as being used as a raw material for alcoholic drinks and beverages. The chapter describes the protein and starch makeup of sorghum and the effects of cooking and processing sorghum on its digestibility and nutrient availability. The chapter assesses the protein quality of sorghum for human use, describes the sensory and functional characteristics of food products containing sorghum and outlines the health benefits of sorghum.

Continuing the theme of the uses of sorghum, Chapter 5 addresses the role of sorghum as a forage and energy crop. Sorghum is an important source of grain and fodder, forage, and biomass throughout the world. In the United States, grain sorghum is the prominent crop; however, non-grain sorghum plays an important role as a feedstock globally. The chapter reviews the different types of sorghum and their uses. The chapter begins by considering the types of sorghum used for forage and biomass, and then offer a detailed discussion of the use of forage sorghum as feed for ruminant animals. The chapter reviews the advantages of using sorghum as a dedicated energy crop, and finally considers the composition and uses of sweet sorghum cultivars.

Part 2 Sorghum production and improvement across the world

The second part of the book discusses ways of improving cultivation in regions such as South America, Europe, Asia and Africa. Chapter 6 examines the introduction of new technologies and market strategies for sorghum producers in developing countries, using the experience of the Sahel as a case study. Sorghum produced principally for home consumption is typically characterized by low yields, input use and market prices. These are interrelated but must be addressed using different strategies. The chapter reports on the field research of a program combining both agronomic and economic components. The program initially included four Sahelian countries and then focused on Mali as it moved from the pilot stage to implementation. The chapter focuses particularly on the introduction of new sorghum technologies and the improvement of marketing strategies, as well as the ability of the system to respond to emerging problems. The chapter examines sorghum performance and potential in Mali and Burkina Faso before considering strategies for breaking farmers out of subsistence. It outlines the success of the pilot project and methods for scaling up improvements, before considering second generation problems and the importance of institutional innovation in the form of farmers' associations as marketing cooperatives. Finally, the chapter considers the growth of a secondary market for sorghum.

Building on the themes of Chapter 6 but moving from Africa to South America, Chapter 7 examines the improvement of sorghum cultivation in South American countries. Sorghum production in Brazil has expanded significantly in recent years, with a current planted area of close to one and a half million hectares. Sorghum provides an excellent source of grain to feed cattle, poultry and swine, as well as forage for cattle, especially in situations of drought and conditions of low soil fertility. It can also be used in bioenergy production. The chapter reviews the history of sorghum production in Brazil, and then considers the improvements necessary to ensure than sorghum production can continue to meet demand, including improving sorghum tolerance for aluminium and lack of phosphorus. The chapter emphasizes the need to develop adapted sorghum cultivars which will grow well in tropical soils and produce crops suited to their particular uses. It addresses genetic and molecular approaches to studying biomass production and sorghum quality and examines the challenge of improving disease resistance, as well as outlining the pests of sorghum in the region.

Moving from South America to Asia, Chapter 8 considers the improvement of sorghum production and utilization in Asian countries. Sorghum was once a major crop in dry regions. However, its production area has decreased drastically, especially in Asia. In order to revitalise sorghum production, sorghum research must be reoriented to resolve current production constraints and improve productivity. The chapter reviews potential measures to improve sorghum cultivation in Asia, including genetic improvement, improved processing technologies, marketing strategies and policy measures.

The volume's final chapter, Chapter 9, returns to Africa to consider sorghum cultivation and improvement in West and Central Africa (WCA). The diversity of sorghum cultivated in Africa attests to the African origin of this crop. Farmers in West and Central Africa (WCA) have developed strategies of using varietal diversity to minimize risk and maximize productivity in the context of complex and diverse adaptive challenges. The chapter describes the major sorghum production systems, options for intensification and sustainable production, genetic diversity and opportunities to breed for enhancing sorghum crop value, as well as opportunities for progress in strengthening seed systems.

Part 1

Sorghum utilization

Chapter 1

Structure and chemistry of sorghum grain

S. R. Bean, B. P. Ioerger, J. D. Wilson, M. Tilley, D. Rhodes and T. J. Herald, USDA-ARS, USA

1 Introduction

Sorghum, almost always introduced as the fifth most important grain in the world, is known for its high degree of genetic diversity, with just over 43 000 accessions held in the USDA germplasm collection alone (Kimber et al., 2013). This diversity is very relevant to the discussion of sorghum grain composition since it results in phenotypes that can include virtually all aspects of grain composition and end-use quality for a wide range of uses spanning food, feed and fuel. The genetic diversity of sorghum may harbour numerous useful genes related to grain composition that, once identified, could be moved into elite adapted lines.

Sorghum is grown around the world in diverse environments, and often under harsh conditions (extreme heat, drought and low soil fertility). Therefore sorghum is exposed to diverse climatic conditions, pests, soil types and management/cultural practices, all of which can influence grain composition, properties and end-use quality.

Summarizing the grain composition of sorghum can be challenging due to the full range of possible genetic x environmental x management interactions ('G x E x M') that must be considered when attempting to compare grain compositional studies. Added to the G x E x M variability that impacts grain composition, differences in analytical methods used

http://dx.doi.org/10.19103/AS.2017.0015.21

to measure grain composition must be considered. This is especially true when comparing data generated across time, since methods and techniques change as new instruments become available with improved sensitivity and resolution, costs of tests decrease and new assays are developed. Thus comparing results of grain composition studies can be challenging as samples may have been grown in widely different environments and analysed by a wide range of methods. Recent studies using large, genetically diverse sample sets grown over multiple locations all analysed with a common technique can be useful for gauging the extent of variability in composition present in sorghum grain.

2 Physical grain properties

Physical grain structure and composition

Grain composition is closely intertwined with the physical structure of the grain at several levels. At a simple physical level, the sorghum grain can be considered to be composed of the outer layers (i.e. pericarp), the endosperm and the germ (Rooney and Miller, 1982; Evers and Millar, 2002). The major structures of the grain are shown in Fig. 1. As these different components vary in their chemical composition, their relative proportions in a given grain will influence the overall composition of the grain. Keeping all the factors that can influence grain composition in mind, on average sorghum is often reported to be

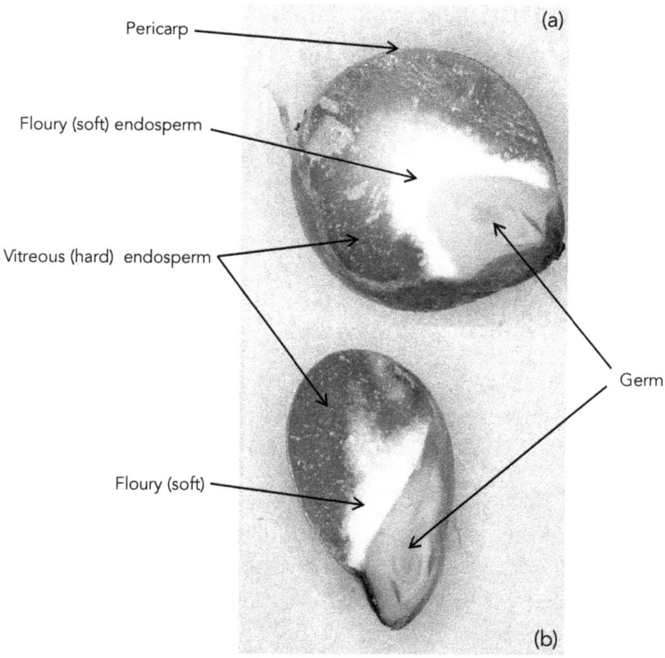

Figure 1 Major structures of the sorghum grain through (a) cross section and (b) longitudinal section of a sorghum kernel.

made up of 3–6% pericarp, 84–90% endosperm and 5–10% germ (Hubbard et al., 1950; Rooney and Miller, 1982; Rooney and Serna-Saldivar, 2000).

The outer layer of the grain, the pericarp, is composed of multiple sublayers including the epicarp, mesocarp and endocarp (Waniksa and Rooney, 2000) The thickness of the pericarp can vary widely not only across genotypes but across individual grains (Earp et al., 2004a). The sorghum pericarp is unique among the cereal grains in that it contains starch, and the amount of starch found in the pericarp is related to the thickness of the outer layers (Earp et al., 2004a). Pericarp protein content can vary from ~4 to 9%, lipid content from ~3 to 6% and starch content in the 30% range (Waniska and Rooney, 2000). Pericarp thickness has been positively correlated to polyphenol content in sorghum (Beta et al., 1999).

The majority of the fibre in sorghum grain is found in the pericarp, with some contained in cell walls throughout the grain (Waniska and Rooney, 2000). Fibre is composed of several components (cellulose, hemicellulose and lignin) and thus can be reported in various ways. Waniska and Rooney (2000) compiled the fibre content of sorghum using multiple categories. Total dietary fibre has been found to range from ~6 to 16% depending on the methodology used (Waniska and Rooney, 2000). Likewise, soluble fibre was reported to range from 0.4 to 1.6% and insoluble fibre from 5.4 to 15.1% (Waniska and Rooney, 2000).

The endosperm forms the majority of the sorghum and is composed primarily of the outer hard (vitreous or corneous) fraction and the inner soft (opaque or floury) fraction (Waniska, 2000). The outer region of the endosperm is covered by the aleurone layer, which in sorghum is composed of a single layer of cells (Ogbonna, 2011). The aleurone layer contains proteins, lipids, trace elements and inclusion bodies – organelles which may contain phytic acid (phytin) (Zeleznak and Varriano-Marston, 1982; Serna-Saldivar and Rooney, 1995; Shapter et al., 2010). Beneath the aleurone layer is the peripheral endosperm that has a dense layer of cells with increased protein concentration and small starch granules of 1–2.5 μm in diameter (Zeleznak and Varriano-Marston, 1982; Serna-Saldivar and Rooney, 1995). Physically, waxy sorghum types were found to differ from normal sorghum types by having less peripheral endosperm, which was postulated to be one reason why waxy sorghum types have improved starch digestibility (Sullins and Rooney, 1974).

The vitreous outer layer of the endosperm is composed of densely packed cells that have higher protein content and smaller starch granules than the inner floury endosperm (Zeleznak and Varriano-Marston, 1982; Serna-Saldivar and Rooney, 1995). The dense nature of the vitreous endosperm causes the protein bodies to be physically indented into the starch granules. The starch granules themselves are polygonal, and both starch granules and protein bodies are covered in protein matrix (Seckinger and Wolf, 1973; Hoseney et al., 1974). The inner floury endosperm contains a more open structure where the starch granules are not packed tightly and still retain a spherical shape. Likewise, the protein bodies are not pushed into the starch granules and the protein matrix is not a continuous film-like cover in the floury endosperm (Seckinger and Wolf, 1973; Hoseney et al., 1974). The relative amounts of vitreous to floury endosperm may vary across genotypes. Based on weight, the outer vitreous layer has been reported to make up 54–90% of the endosperm (Watterson et al., 1993; Ioerger et al., 2007) and by area 9–80% (Kirleis et al., 1984; Shull et al., 1990; Watterson et al., 1993). Overall grain hardness has been correlated to the ratio of vitreous to floury endosperm of the grain (Hallgren and Murty, 1983).

The final component of the grain to discuss is the germ, which is composed of the embryonic axis and scutellum (Waniska and Rooney, 2000). Physically, the germ constitutes only 5–10% of the total grain; however, it accounts for the majority of all the lipids, vitamins and minerals (Waniska and Rooney, 2000). The germ also accounts for ~15% of the total protein in the grain (Waniska and Rooney, 2000), and while this is a relatively small portion, the protein quality of the germ is higher, having substantially greater levels of lysine than that found in the proteins of the endosperm (Serna-Saldivar and Rooney, 1995).

Grain size and hardness

Compared to maize, sorghum grains are relatively small and typically round with one slightly flattened side (Reichert et al., 1988). Sorghum grain varies widely in both grain size (diameter) and weight. Thousand kernel weight has been reported to range from 30 to 80 g in sorghum (Rooney and Serna-Saldivar, 2000; Chiremba et al., 2012). Accordingly, average individual kernel weight also varies from 3 to 80 mg (Serna-Saldivar and Rooney, 1995; Bean et al., 2006; Sukumaran et al., 2012). Individual kernel size and weight are highly correlated, as is expected (e.g. Bean et al., 2006). While the averages reported above reflect differences among genotypes, kernel size and weight can vary within a given lot of grain from the same genotype and vary within a given panicle of sorghum (Heiniger et al., 1993; Lee et al., 2002; Gambin and Borras, 2005).

Several methods have been used to evaluate grain hardness in sorghum including abrasive decortication (Reichert et al., 1981; Lawton and Faubion, 1989), particle size index, Brabender hardness tester, Stenvert micro-hammermill test (Pomeranz, 1986) and the single kernel characterization system (SKCS) (Pedersen et al., 1996; Bean et al., 2006). As each of these methods measure grain hardness in different ways, they may only be weakly correlated to each other. For example, SKCS hardness was found to be only weakly correlated to the abrasive hardness index (AHI; $r = 0.61$; Bean et al., 2006). However, all methods reveal that sorghum grains can vary substantially in hardness. Reichert et al. (1981) utilized the tangential abrasive dehulling device to calculate an AHI on 31 sorghum samples and reported AHI values ranging from 5 to 12.8. Bean et al. (2006) reported AHI values ranging from 6 to 22 for a diverse set of 101 sorghum genotypes. This same set of samples ranged from ~36 to 78 for the SKCS hardness index. SKCS grain hardness in a subset of the genetically and phenotypically diverse sorghum association panel (SAP) of Casa et al. (2008) was reported to have an average of 78.15 with a standard deviation of 18.97 (Sukumaran et al., 2012).

Grain colour

The last element of the physical structure to consider is that of grain colour. As with all the physical traits discussed so far, sorghum grain colour can vary substantially. Sorghum grain can appear white, yellow, red, brown or black (Rooney and Miller, 1982). Within each colour category, various shades and hues may exist as well creating a wide variety of colours. Grain colour in sorghum is controlled by the colour of the pericarp, but is also influenced by pericarp thickness, endosperm colour and the testa layer if present (Rooney and Miller, 1982; Rooney, 2000). The genetic basis for sorghum grain colour is known, with pericarp colour controlled by the genes R and Y. The combination R_Y_ gives a red colour and rrY_ a yellow colour. White pericarp is determined by the gene combinations RRyy and rryy. These primary pericarp colours are in turn modified by other genes including the intensifier (I) which intensifies colour in red and yellow sorghum (Rooney and Miller, 1982; Rooney, 2000).

As previously mentioned, the thickness of the layers in the pericarp can modify overall grain colour. The mesocarp, one of the internal layers of the pericarp, influences grain colour by its thickness and is controlled by the Z gene. A thick mesocarp (zz) gives a 'chalky' look to grain by influencing the colour of the pericarp and endosperm, and a thin mesocarp (genetically ZZ or Zz) gives a 'translucent' look to grain, allowing the colour of the endosperm beneath the mesocarp to be easily seen (Rooney, 2000). An additional modifier to grain colour is the presence of a pigmented testa layer, determined by the B1 and B2 genes. A sorghum genotype having dominant forms of these genes (B1_ B2_) will have a pigmented testa. In sorghum genotypes where one of these genes is in the recessive form (i.e. b1B2, B1b2 or b1b2), the grain will not have a pigmented testa (Earp et al., 1983, 2004b). The pigmented testa layer, when present, can modulate grain colour in sorghum genotypes that have a thin mesocarp and white pericarp, giving the grain a brown colour. Likewise, the S gene controls the presence of brown pigment in the pericarp (Rooney and Miller, 1982).

Finally, grain colour can be influenced by the presence of carotenoids in the endosperm (Rooney and Miller, 1982). Carotenoids give the endosperm a yellow colour, which in turn can influence overall grain colour in genotypes that have a thin outer layer and lack a pigmented testa layer (Serna-Saldivar and Rooney, 1995).

3 Chemical composition of grain: starch and proteins

Starch composition

As with all cereals, starch is the major component of the sorghum grain and total starch levels in the grain can vary widely. Starch is also the major carbohydrate fraction of the grain (Table 1). A survey of 50 sorghum genotypes selected for use in foods found starch levels ranging from 59 to 69% (Dicko et al., 2006). The starch content of the SAP (Casa et al., 2008) from two years grown in two environments (Kansas and Indiana) was reported to have an average of 69.3% with a standard deviation of 2.44% (n = 299) (Sukumaran et al., 2012). A similar sample set with more samples (n = 390) was reported to have starch content of ~62–71% by Rhodes et al., (2016) when grown over three years in South Carolina. Murray et al., (2008) reported starch content ranging from 45 to 69% in a recombinant inbred population grown over two years at two locations in Texas. These studies illustrate both the genetic potential for variability and the impact of the environment on starch content in

Table 1 General carbohydrate composition of sorghum grain

Component	Content (%)
Free sugars	1.3–5.2
Total starch	45–71
Crude fibre	1.2–6.6
Total dietary fibre	5.8–16.1

Adapted from Waniska and Rooney, 2000; Dicko et al., 2006; Murray et al., 2008; Sukumaran et al., 2012; Rhodes et al., 2016.

sorghum. Starch content and composition can be impacted by the environment through several different pathways (Thitisaksakul et al., 2012; Beckles and Thitisaksakul, 2014) and thus it is not surprising to see a high degree of variability across studies.

Starch structure and composition can be discussed at several levels. At the macromolecular level, starch is organized into granules, which can vary in size and shape. Sorghum starch granules vary in their shape and can be spherical or polygonal (Benmoussa et al., 2006; Sang et al., 2008). Location in the grain can impact the appearance of starch granules: those found in the outer vitreous (hard) endosperm are tightly packed and have a polygonal shape, while those in the inner floury (soft) endosperm are not densely packed together and retain a more spherical shape. Several studies have reported a wide range of starch granule size in sorghum with values ranging from 4 to 35 μm (reviewed in Zhu, 2014). Sorghum starch granules contain channels visible on the surface, and which enter the surface of the granule and run towards the hilum (Huber and BeMiller, 2000).

At a molecular level, starch exists as two polymers: amylose and amylopectin. Amylopectin forms the bulk of sorghum starch with 'normal' sorghum genotypes typically containing 70–77% amylopectin (Waniska and Rooney, 2000). Amylopectin is a highly branched polymer formed of linear glucose units branched at α-(1-6) linkages. While starch is difficult to accurately isolate and characterize (e.g. Syahariza et al., 2010), the overall size of sorghum amylopectin has been reported to be ~9000–13 000 degrees of polymerization (DP_n) (Gaffa et al., 2004) with a molecular weight of $2.8–3.2 \times 10^8$ Da (Matalanis et al., 2009).

In addition to the overall size of amylopectin, it is possible to study the fine structure of amylopectin through the branch chain-length distribution, which is related to the crystalline structure of the starch (Hizukuri, 1985). Amylopectin branch chain length varies widely across various sorghum sources, due to either broad genetic diversity or significant environmental effects (Zhu, 2014).

Starch is also composed of amylose; for the most part, a linear glucose polymer connected through α-(1-4) linkages. As with other starch traits, amylose content can vary widely in sorghum. In 'normal' sorghum genotypes, amylose content has been reported to range from 23 to 30% (Waniska and Rooney, 2000). Amylose varies in its properties with reported molecular weight varying between 6.4×10^6 and 1.1×10^7 Da and DP_n of 1150–1390 (Matalanis et al., 2009; Gaffa et al., 2004).

As with other cereal grains, 'waxy' germplasm exists in sorghum in which little to no amylose is produced (<5% amylose). Amylose content in cereal grains is controlled by the enzyme granule-bound starch synthase (GBSS). For sorghum, four allelic variants of the GBSS gene have been identified which lead to the development of the waxy trait in sorghum grain (Pedersen et al., 2005; Hamblin et al., 2007; McIntyre et al., 2008; Sattler et al., 2009; Lu et al., 2013). Heterowaxy sorghum lines exist in which the waxy trait is expressed within a panicle in a 1:3 ratio (waxy to non-waxy) (Lichtenwalner et al., 1978).

Despite being the major component in the grain, sorghum starch has received less research attention compared to other components such as proteins and phenolic compounds. Combined with technical difficulties in isolating and characterizing starch accurately, there is relatively little detailed characterization of sorghum starch. Zhu (2014) listed several areas of research needed with regard to sorghum starch, including screening of additional genetic resources to identify germplasm with new and unique starch properties, conducting additional research on starch modification and structure, identifying relationships between starch properties and product quality, and standardization of methods used in characterizing sorghum starch.

Protein content and composition

Generally, there is a wide range of protein content for sorghum with 7–15% being a common range reported when summarized across studies (Waniska and Rooney, 2000). As mentioned in the introduction, the protein variation between studies could be due to differences in genotypes, environments, crop management practices and analytical methods. Protein content in the SAP (Casa et al., 2008) has been measured in multiple studies, providing information on the variability of protein content across a large and diverse sample set in multiple years and locations. Sukumaran et al. (2012) reported an average total protein content of 13.85% with a standard deviation of 1.67 in the SAP (n = 299) grown in Kansas and Indiana. Rhodes et al. (2016) reported total protein ranging ~8–19% (n = 390) in the SAP grown over three years in South Carolina. Murray et al. (2008) reported protein content ranging from 9 to 19% in two generations of recombinant inbred lines (n = 176), again demonstrating the wide range of protein content that can be present in sorghum grain.

Of the major physical grain components, the endosperm contains the majority (~80%) of the total protein found in the grain, with the germ containing roughly 15% of overall protein (Serna-Saldivar and Rooney, 1995). The proteins of the endosperm are primarily prolamins and as such are high in proline and glutamine. While high in the branched chain amino acids, the endosperm proteins are very low in the key amino acid, lysine (Table 2). Conversely, the proteins of the germ contain much higher levels of lysine.

In addition to their physical locations in the grain, cereal proteins are classified based on a number of other criteria, namely by their division into fractions or classes based on solubility (Osborne, 1907, 1924). Significant information has been added to solubility-based classifications, and various methods have been used to define the solubility classes. This has resulted in several nomenclature schemes and extraction procedures existing in the literature (e.g. Landry and Moureaux, 1970; Jambunathan et al., 1975; Taylor et al., 1984a; Hamaker et al., 1995; Landry, 1997). While there can be some confusion over

Table 2 Essential amino acid content of sorghum grain[a]

	g/100 g protein[b]								
Reference	Ile	Leu	Lys	Met	Phe	Thr	Trp	Val	His
Khalil et al., 1984	4	13.7	2.6	1.4	5.2	3.2	0.9	4.8	1.8
Pontieri et al., 2010	2.06	13.3	2.67	3.62	6.37	3.34	n/a	3.61	2.44
Ejeta et al., 1987	3.8	13.1	2.5	1.7	5.2	3.3	0.9	5.1	2.2
Ebadi et al., 2005	4.01	12.53	1.79	1.43	4.68	3.22	2.24	5.28	2.27
Ahmed et al., 1996	4.1	13.5	2.4	1.9	4.7	4.3	n/a	7.6	3.1
Subramanian et al., 1990	4.28	14.61	2.44	0.86	4.5	3.32	n/a	5.77	2.07
Solsuski and Imafidon, 1990	2.26	9	1.32	9	3.02	1.96	0.75	3.64	1.38
da Silva et al., 2011	3.75	14.8	2.08	1.65	4.74	2.73	n/a	4.79	1.95
Youssef, 1998	3.65	13.2	2.11	1.36	5.62	3	n/a	4.55	1.82

[a] From Bean and Ioerger 2015 with permission; data adapted from listed references.
[b] Values represent averages when study used multiple samples.

classification and nomenclature, inclusion of additional biochemical data, gene sequence homology and protein functionality have improved the current classification systems.

On a broad scale, the major cereal protein classes can be considered to consist of the albumins, globulins, prolamins and glutelins (Branlard and Bancel, 2006). The sorghum albumins (water-soluble proteins) and globulins (salt-soluble proteins) are located mainly in the outer layers of the grain and inside the germ, and overall represent some 10–30% of the total grain protein (Taylor et al., 1984a; FAO, 1995; Wall and Paulis, 1978). While not as well characterized as the prolamins, a recent proteomics study identified heat shock proteins, enzymes, metabolic proteins and chaperone proteins in the water- and salt-soluble fractions (Cremer et al., 2014).

The majority of sorghum grain proteins are the prolamins, specifically known as the kafirins in sorghum. The kafirins are aqueous-alcohol-soluble proteins and as prolamins have high levels of proline and glutamine (over 30% of the total amino acid residues in the kafirins) (Taylor, 1983; Belton et al., 2006). Reports on the kafirin levels in sorghum vary widely based on the methodology and samples used, but in general comprise ~50–70% of total protein (Taylor et al., 1984a; Hamaker et al., 1995).

Prolamins are not composed of a single, uniform protein, but rather a mixture of various related proteins that can exist as monomers or cross-linked into polymers. In sorghum, the kafirins have been divided into classes or families in a similar manner as that of maize prolamins, the zeins (Esen, 1987; Shull et al., 1991), with which the kafirins share a high degree of sequence similarity and chemical properties (Shull et al., 1991, 1992; Mazhar et al., 1993; DeRose et al., 1989; Belton et al., 2006; Xu and Messing, 2009).

The most abundant sub-family of kafirins are the α-kafirins which comprise ~70% of the total prolamins (Watterson et al., 1993). When separated by SDS-PAGE according to their molecular weight, the α-kafirins generally migrate as two distinct bands of molecular weight – 23kDa and 25kDa (Shull et al., 1991) or 22kDa and 28kDa (Mazhar et al., 1993). Predictions of molecular weight from gene sequences reveal primary molecular weights of 19kDa and 22kDa (DeRose et al., 1989; Belton et al., 2006; Holding, 2014).

At the phylogenetic level, the α-kafirins have been placed in the 'Group I' prolamin family with one group belonging to the K1α19 sub-family (the '19-kD' group) and the K1α22 sub-family (the '22-kD' group) (Xu and Messing, 2009). Interestingly, while SDS-PAGE reveals only two major α-kafirin bands, there may be as many as 19 α-kafirin genes expressed (Xu and Messing, 2008). Separation methods that do not rely on separation by molecular weight, such as capillary electrophoresis, isoelectric focusing and reversed-phase HPLC, reveal a greater degree of heterogeneity within the α-kafirins (Taylor and Schüssler, 1984; Sastry et al., 1986; Smith, 1994; Bean et al., 2000, 2011; Blackwell and Bean, 2011), which may reflect the larger numbers of genes expressed than what protein separations via SDS-PAGE suggests (Fig. 2).

The next abundant class of kafirins are the γ-kafirins; members of the phylogenetic groups II γ1 k2γ27 and k2γ50 (Xu and Messing, 2009). The amount of γ-kafirins varies by location in the grain and makes up ~5–12% of the prolamins in the outer vitreous (hard) endosperm and ~11–21% in the inner floury (soft) endosperm (Watterson et al., 1993; Ioerger et al., 2007). While present at substantially lower levels than α-kafirin, γ-kafirin may play an important role in the structure of protein bodies and functionality of kafirins due to the high levels of cysteine present (7 mol%) (Duodu et al., 2003; Belton et al., 2006).

Interestingly, the most well-known γ-kafirin migrates on SDS-PAGE as the largest of the kafirins with an apparent molecular weight of ~28 kDa, but the amino acid sequence reveals a molecular mass of only ~20 kDa (Belton et al., 2006). For the '28kDa' γ-kafirin,

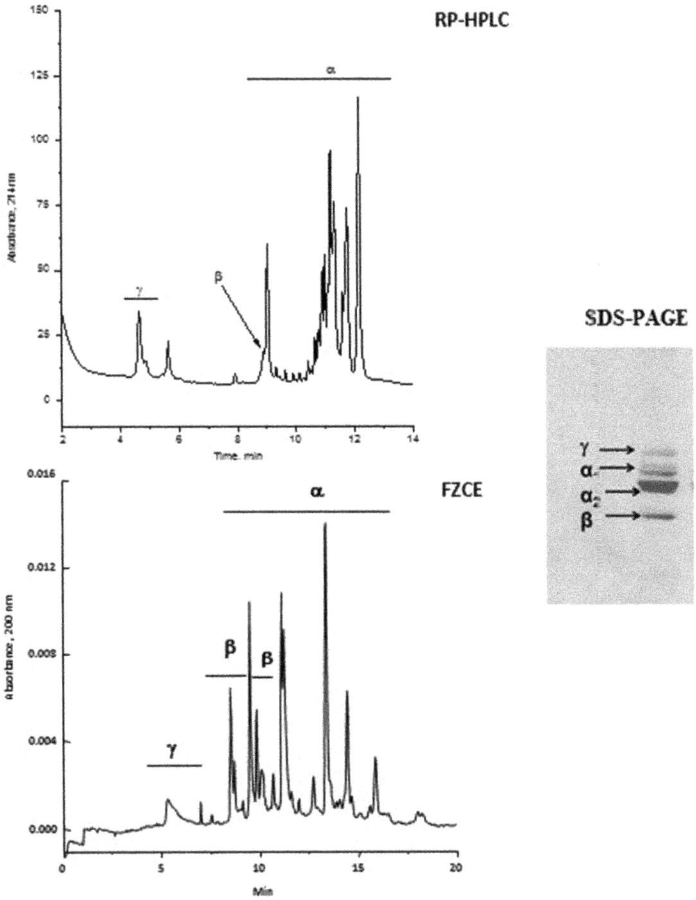

Figure 2 Separation of kafirins by reversed-phase HPLC (RP-HPLC), free zone capillary electrophoresis (FZCE) and SDS-PAGE. Adapted from Bean et al. (2000, 2011) and Bean (2003).

only one γ-kafirin gene is expressed in sorghum, with five allelic variants having been identified (Xu and Messing, 2008; Laidlaw et al., 2010).

While not well characterized, and possibly often misidentified as a dimer, an additional high-molecular-weight γ-kafirin has been identified (Evans et al., 1987; Belton et al., 2006; Cremer et al., 2014). In a recent study, sequence data for this high-molecular-weight γ-kafirin revealed homology to high-molecular-weight γ-prolamins in maize and other species (Fig. 3). The high-molecular-weight γ-kafirin has been reported to migrate on SDS-PAGE ~49 kDa (e.g. Evans et al., 1987); however, sequence information predicted the protein to be ~36 kDa (Cremer et al., 2014). The lower-molecular-weight γ-kafirin also migrates in SDS-PAGE separations larger than its predicted molecular weight (Belton et al., 2006). In maize, a high-molecular-weight γ-zein has been identified at the surface of protein bodies (Woo et al., 2001) and thus this protein could have a role in protein cross-linking and protein body structure. More research is needed to characterize the

Figure 3 Phylogenetic relationships of high-molecular-weight gamma kafirin and other prolamins. Adapted from Cremer et al. 2014, with permission.

high-molecular-weight γ-kafirin in sorghum and determine what role it may play, if any, in protein body structure and formation.

The last of the well-characterized kafirins are the β-kafirins. β-kafirin content varies by location in the grain, with β-kafirin making up 7–8% of total protein in outer hard (vitreous) endosperm and 10–13% in the soft (floury) inner endosperm (Watterson et al., 1993). The β-kafirins belong to the 'Group II' prolamin phylogenetic family and there is only one β-kafirin gene expressed (Xu and Messing, 2008) with four allelic variants identified (Laidlaw et al., 2010).

Characterization of a β-kafirin gene revealed it coded for a protein with molecular weight of ~18–19 kDa (Chamba et al., 2005). Like the γ-kafirin, β-kafirin contains high levels of cysteine (10 cys residues and 5.8 mol%) and may play a role in protein cross-linking and the digestion-resistant structure of protein bodies (El Nour et al., 1998; Belton et al., 2006). Interestingly, natural β-kafirin null mutant lines did not show unusual *in vitro* protein digestibility (raw or cooked) compared to other sorghum lines (Cremer et al., 2014).

The δ-kafirins make up the last of the kafirin groups and have had little research conducted on them. Izquierdo and Godwin (2005) reported δ-kafirin as a protein with 147 amino acids and a molecular weight of ~16 kDa. Belton et al. (2006) reported two δ-kafirin DNA sequences that have significant homology with the 14 kDa δ-zein of maize. δ-kafirin has been reported to make up less than 1% of total prolamin (Laidlaw et al., 2010).

The prolamins described above are found primarily in discrete, spherical protein bodies in both the vitreous and floury endosperm. The protein bodies are made up of prolamins, with the interior containing mostly α-kafirins and small amounts of β- and γ-kafirin scattered throughout (Shull et al., 1992). The surface of the protein bodies contains mostly β- and γ-kafirins thought to be disulphide cross-linked together (Belton et al., 2006). The protein bodies in the outer vitreous endosperm were reported by Shull et al. (1992) to exist as 0.3–1.5 μm 'spheroids'. Those protein bodies found in the central floury endosperm were similar to those in the vitreous endosperm, but smaller in size and more irregular in shape.

Sorghum is known to be unique in that protein digestibility decreases when wet cooked (Duodu et al., 2003). When sorghum is cooked with a reducing agent, protein digestibility is restored to close to that of the raw grain, demonstrating that the decrease in digestibility is related to the formation of disulphide bond cross-links (Hamaker et al., 1986). This cross-linking is visible in cooked materials as a highly linked web (Hamaker and Buguso, 2000).

Related to the reduced protein digestibility in cooked sorghum, mutant sorghum lines with irregularly shaped protein bodies have been identified (Weaver et al., 1998; Oria et al., 2000). These mutants tend to have higher protein digestibility than wild-type sorghum and most importantly, retain high levels of digestibility after being cooked. Oria et al. (2000) reported that protein bodies in the mutant sorghum had γ-kafirins located primarily in the bottom of creases, rather than consistently around the edges as in the wild-type sorghum lines.

The last major class of sorghum proteins is the glutelins which have been poorly characterized. These proteins are soluble in acid or alkali pH, and have been reported to make up anywhere from ~25 to 50% of total protein (Beckwith, 1972; Landry and Moureaux, 1970; Taylor et al., 1984b; Subramanian et al., 1990; Nucere and Sumrell, 1979). As these proteins are generally extracted after all other classes, one reason for the wide variability in glutelin content may be due to differences in protein extraction methods and incomplete extraction of previous protein classes (Taylor et al., 1984a). Very little research has been conducted to identify the proteins that make up the glutelin fraction. Recently, several alkali-soluble proteins were identified to be heat shock proteins, chaperonins, luminal binding proteins, enzymes and vicilin- and legumin-like storage proteins (Cremer et al., 2014).

4 Chemical composition of grain: lipids and phenolic compounds

Lipid content and composition

Total lipid content in sorghum grain is low, averaging only 3–4%, but ranging from ~0.5 to 5% (Rooney, 1978; Waniska and Rooney, 2000; Sukumaran et al., 2012; Moreau et al., 2016; Rhodes et al., 2017). Like most cereal grains, lipids are found primarily in the germ and contain various classes of lipids including non-polar lipids, glycolipids and phospholipids (Taylor and Belton, 2002). As with other compounds found in grains, the methodology used to extract lipids influences the composition reported, but sorghum has been reported to have ~80% neutral lipids, ~18 phospholipids, ~2.5% glycolipids and low levels of sterols (Taylor and Belton, 2002). Sterols present have been found to be in the form of free phytosterols and fatty acyl phytosterol esters (Singh et al., 2003). Sorghum grain has also been found to contain long-chained lipids of which ~37–44% were reported to be policosanols (Hwang et al., 2004).

The major component of sorghum lipids are triacylglycerols (~80–90% of total lipid), with linoleic acid as the major fatty acid, followed by oleic acid (Waniska and Rooney,

2000; Taylor and Belton, 2002). Sorghum genotypes can vary in lipolytic enzymes, which may impact end-use quality in food and bio-fuel markets (Moreau et al., 2016).

Phenolic content and composition

Phenolic compounds in sorghum are located primarily in the outer layers of the grain, as is the case for most cereal grains. Researchers have reported that sorghum and more specifically sorghum bran contains significant levels of phenolic compounds that offer health-promoting bioactive properties (Dykes et al., 2005; Gu et al., 2007). Phenolic compounds found in sorghum grain are primarily flavonoids and phenolic acids (Awika and Rooney, 2004; Dykes and Rooney, 2006). Such phenolic compounds are bioactive phytonutrients and are secondary plant metabolites and are often expressed by the plant in response to abiotic stress (climate, disease, UV light, etc.).

The measurement of these compounds and their antioxidant capacity is an important tool to screen and compare the phytochemical composition across sources. Chemical assays such as oxygen radical absorption capacity (ORAC), 2,2-diphenyl-1-picrylhydrazyl (DPPH) and ferric reducing antioxidant power have been employed to measure antioxidant capacity in many types of materials including sorghum and are well documented in the literature (e.g. Awika et al., 2003; Thaipong et al., 2006). Even though these are widely employed there are some limitations to the assays. The chemical antioxidant capacity assays do not determine the bioavailability of the phytochemicals (Blasa et al., 2011), thus providing a weak relationship to *in vivo* assays. Wolf and Liu (2007) developed a cell culture model to measure antioxidant capacity in a range of fruits noted to possess phenolic compounds. Cell culture models can determine phytochemical uptake by the cell, and thus circumvent the cost, expedit the process and address some of the concerns of uptake, distribution and metabolism of the current chemical antioxidant capacity assays.

Sorghum contains several different forms of flavonoids (reviewed in Awika and Rooney, 2004). The most widely studied group of sorghum flavonoids is the condensed tannins (proanthyocyanidins). Awika and Rooney (2004) recently reviewed tannin content in sorghum and reported tannin content ranging from 10 to 68 (mg/g sorghum). Rhodes et al. (2014, 2016) reported similar ranges in the SAP panel.

Tannins are water soluble, with molecular weight ranging from 500 to 3000kDa. Condensed tannins are secondary metabolic products that accumulate exclusively in the seed testa underneath the pericarp. Condensed tannins are flavan-3-ol unit oligomers or polymers (epicatechin and catechin) with hydrolysis-insensitive carbon–carbon bonds (Kaluza et al., 1980).

Condensed tannins in sorghum have been intensely researched due to their negative impact on the nutritional quality of sorghum, namely reduced protein digestibility and reduction in micronutrient absorption (Awika and Rooney, 2004). The presence of tannins in sorghum grain has been a concern through much of the twentieth century because they bind nutrients, thus limiting weight gain and absorption for humans and ruminants. Consequently, commercial sorghum in the United States is virtually all tannin-free.

While condensed tannins do have a negative impact on nutritional aspects of sorghum, these compounds are potent antioxidants and in this role, may have unique health benefits for humans (Awika and Rooney, 2004). Hagerman et al. (1998) showed that sorghum tannins were 15–30 times more effective than simple phenolics in radical scavenging ability. Awika and Rooney (2004) reported that bran from high-tannin sorghums had an antioxidant capacity nearly 30 times higher than grapes.

The renewed interest in tannin sorghum has awakened old testing concerns. The methods historically used to qualitatively determine the presence of tannins include the scratch and bleach tests (Waniska et al., 1992). These are simple and rapid (30 min) to conduct but are not quantitative. The scratch test requires scrapping off the pericarp, examining the grain for a brown pigmented testa which signifies the presence of condensed tannin. This is not always as straightforward as it sounds, as the testa layer is not always easily identified and it is difficult to check large number of breeding materials in the field/lab. The bleach test has its own set of challenges that include bleach concentration, temperature of the solution, time of exposure and amount of agitation. Furthermore, the bleach test is not as accurate with nursery or small sample size (too much variation in results) and any physically damaged kernels or the presence of mould can give false positives (Awika et al., 2001).

The HCl-vanillin method is one of the most frequently cited quantitative for determining tannin content in sorghum, is well established and carries validity (Maxson and Rooney 1972; Price et al., 1978; Awika and Rooney 2004). Because of the frequent citation of the HCl-vanillin assay in the literature the method is one of choice for researchers to compare tannin data across laboratories. That does not mean the HCl-vanillin assay is without drawbacks. Limiting factors to the HCl-vanillin assay are the relatively long assay time (8 h), labour intensive, low throughput, cost of disposable supplies, the need for a highly trained personnel for reproducibility and not conducive for screening large sample sets.

A rapid and reliable method to determine tannin content will be essential if large sample sets of sorghum are to be screened and marketed for health-promoting compounds. Herald et al. (2012, 2014) developed high-throughput 96-well micro-titre assays for total polyphenol content and 2,2-diphenyl-1-picrylhydrazyl (DPPH) scavenging capacity for sorghum that offered good accuracy and precision.

Another class of flavonoid in sorghum that is gaining research interest is the 3-deoxyanthocyanins. 3-deoxyanthocyanins for the most part provide the red, and in some cases, the black colour, to sorghum grain. The 3-deoxyanthocyanins of sorghum have a unique chemical structure that is only rarely found in other plant-based foods and interest in these compounds is due to their potential health benefits (Awika and Rooney, 2004; Awika et al., 2004, 2005; Rhodes and Kresovich 2016).

In addition to the range of flavonoids present in sorghum, sorghum grain also contains a wide range of phenolic acids, with ferulic acid as the major component (Subba Rao and Muralikrisnha, 2002; Awika and Rooney, 2004; Awika et al., 2004; Dykes and Rooney, 2006; Chrandasekara and Shahidi, 2011). Phenolic acids, especially ferulic acid, have been related to grain hardness in sorghum, potentially due to cross-linking of phenolic acids to cell walls in the pericarp and aleurone layers of the kernel (Chiremba et al., 2012). Like other compounds in the sorghum grain, total phenolic acid content can vary widely (Dykes et al., 2005).

5 Chemical composition of grain: vitamins and minerals

Sorghum, like the majority of cereals, contains numerous vitamins and minerals. Sorghum is often referred to as a 'good source' of B vitamins with the exception of B-12 (Waniska and Rooney, 2000). Sorghum is often cited as a 'good source' of potassium and an 'adequate' source of iron, zinc, copper and magnesium (Waniska and Rooney, 2000). Vitamin and mineral content of sorghum as listed in the USDA National Nutrient Database is shown in Table 3.

Table 3 General vitamin and mineral content of sorghum[a]

Vitamins	Amount per 100 g
Vitamin C (mg)	0
Thiamin (mg)	0.332
Riboflavin (mg)	0.096
Niacin (mg)	3.688
Pantothenic acid (mg)	0.367
Vitamin B-6 (mg)	0.443
Vitamin B-12 (ug)	0
Vitamin A (IU)	0
Vitamin E (mg)	0.5
Minerals	
Ca (mg)	13
Fe (mg)	3.36
Mg (mg)	165
P (mg)	289
K (mg)	363
Na (mg)	2
Zn (mg)	1.67
Cu (mg)	0.284
Mn (mg)	1.605
Se (ug)	12.2

[a] From the USDA food composition database (https://ndb.nal.usda.gov/ndb/).

A recent study investigated the mineral content of the Casa et al. (2008) SAP grown over multiple years at one location in Mexico (Puerto Vallarta) and two locations in the United States (Texas and South Carolina) (Shakoor et al., 2016). This study provided an interesting look at both the genetic variability present in sorghum for a range of minerals as well as how the environment can influence the mineral composition of sorghum. Shakoor et al. (2016) reported a high degree of variability for Na, Ca, Al and Ni with substantially less variability found for P, K, S and Mg. Gerrano et al. (2016) evaluated mineral content of 22 genotypes grown over two years in South Africa and also reported substantial variability in mineral content across the samples.

Of the vitamins and minerals found in cereal grains, vitamin A, iron and zinc are especially important micronutrients for populations around the world where deficiencies exist (Zimmermann and Hurell, 2007; Khush et al., 2012). Studies measuring iron and zinc in sorghum lines in regions where sorghum is a staple food have found varying levels of iron and zinc. For example, Kayodé et al. (2006) reported Fe and Zn levels in farmers' samples grown in Benin ranging from 3.0 to 11.3 mg/100g of Fe and 1.1 to 4.4 mg/100g of Zn. Kumar et al. (2012) identified 29 genotypes that were high in Zn and Fe content from a pool of over 2200 lines and compared the iron and zinc values to commercial sorghum lines. Fe values varied from 2.6 to 6.0 mg/100g and Zn ranged from 2.1 to 5.7 mg/100g in this study. Gerrano et al. (2016) reported a significant relationship between Fe and Zn content in 22 genotypes. Several studies have addressed the possibility of breeding sorghum for increased levels of both Fe and Zn and suggest this as an important area of future research (Kumar et al., 2013; Gerrano et al., 2016; Shakoor et al., 2016).

Several studies have investigated the levels of carotenoids and their genetic control in sorghum. Carotenoids are a class of lipophilic (fat-soluble) pigments several of which have important implications for human nutrition (Rao and Rao, 2007). For sorghum, interest has focused on specific carotenoids which are vitamin A precursors (Taylor et al., 2014). Vitamin A deficiency is estimated to impact ~250 million pre-school aged children and can result in blindness and increases in mortality of pregnant women and infants (Khush et al., 2012). Vitamin A deficiency is high in regions of the world where sorghum is often grown. In Ethiopia, for example, vitamin A deficiency has been labelled as 'a serious public health problem' (Demissie et al., 2010). Carotenoid levels of sorghum are therefore of interest for combating vitamin A deficiency in those regions of the world where sorghum is a food staple and maize cannot be grown (Fernandez et al., 2008). Reports of total carotenoid content in sorghum grain vary widely ranging from ~0.17 to 30 ppm (Waniska and Rooney, 2000; Salas Fernandez et al., 2008, 2009; Kean et al., 2011; Moreau et al., 2016). This large range may be due to genetic and environmental differences among samples used as well as differences in extraction and analytical methodologies. Carotenoids can be difficult to extract and analyse and precautions such as extracting under yellow light, storing ground material under nitrogen atmosphere and so on are often recommended (Rodriguez-Amaya and Kimura, 2004).

The major carotenoids of sorghum are typically reported to be zeaxanthin and lutein (Waniska and Rooney, 2000) while a recent analysis of commercial US sorghum hybrids identified some genotypes with β-carotene as the major carotenoid (Moreau et al., 2016). Quantitative trait loci have been identified for various carotenoids in sorghum grain (Fernandez et al., 2008) and transgenic sorghum having increased levels of carotenoids have been developed (Lipkie et al., 2013). Thus the possibility to increase levels of carotenoids in sorghum grain, either through biotechnology or through traditional breeding exists.

Like all cereal grains, sorghum also contains phytic acid. Phytic acid is the main phosphate storage vehicle in seeds (Kumar et al., 2010). Phytic acid is an important factor related to nutritional quality of foods as it can interfere with bioavailability of minerals, impact protein digestibility and reduce lipid bioavailability (Duodu et al., 2003; Kumar et al., 2010). Ironically, several potential therapeutic benefits of phytic acid have also been reported (Urbano et al., 2000; Kumar et al., 2010).

Levels of phytic acid have been reported to range from 0.3 to 3.5% (Garcia-Estepa et al., 1999; Elkhalil et al., 2001; Osman, 2004; Kayodé et al., 2006). Because of the anti-nutritional impact of phytic acid on cereal-based foods, the effect of pre-treatment and food processing methods to reduce/degrade phytic acid in sorghum have been reported by several research groups. Methods investigated to reduce phytic acid levels in sorghum include fermentation (Mahgoub and Elhag, 1998; Makokha et al., 2002; Osman, 2004; Towo et al., 2006); malting and germination (Mahgoub and Elhag, 1998, Elkhalil et al., 2001; Makokha et al., 2002; Afify et al., 2011); soaking, milling and heat treatment (Mahgoub and Elhag, 1998; Afify et al., 2011); and enzymatic treatment with phytase (Towo et al., 2004).

6 Factors that affect grain composition

As indicated several times throughout this chapter, grain composition can be influenced by various factors. It is well known that environmental factors can impact grain composition (Wang and Frei, 2011). There is a wide range of environmental factors that can affect crops, including carbon dioxide levels, temperature, moisture, soil chemistry, crop management practices and pest infestations (Annicchiarico, 2002).

Research has shown that sorghum grain composition can vary across environments and/or years. For example, Griess et al. (2010) and Yaseen et al. (2014) reported that environmental differences accounted for most of the variation observed in kernel hardness and starch properties for sorghum grown at various locations in Nebraska. Pedersen and Kofoid (2003) reported that starch content and seed size varied across growing years for 16 sorghum lines grown in Nebraska. Zhan et al. (2003) also reported that starch content varied across different locations for nine sorghum hybrids and, accordingly, ethanol and lactic acid yields produced from these sorghum lines also varied by location. Beta and Corke (2001) observed across-environmental changes in sorghum starch properties such as amylose content, pasting properties and thermal properties. Kaufman et al. (2013) reported that tannin content and chemistry varied between years for select sorghum lines grown at the same location over two years. Taleon et al. (2014) observed significant genetic and environmental interactions impacting content of several sorghum flavonoids. Zn and Fe content have been reported to vary across environments, and at least one study has indicated that variability may be due more from environmental influences rather than genetics (Kayode et al., 2006). The presence of arbuscular mycorrhizal fungi and their interaction with sorghum genotypes has also been reported to influence mineral content in sorghum (Cobb et al., 2016).

The above studies demonstrate that, as expected, the environment can influence sorghum grain composition, and thus grain quality. There has, however, been less research conducted to examine how specific types of environmental variation (i.e. abiotic stress) influence sorghum grain composition. Of specific importance in this area is how drought and heat stress, two of the major environmental factors facing sorghum worldwide (Belton and Taylor, 2004; Ananda et al., 2011; Assefa et al., 2014), impact sorghum grain composition.

Ananda et al. (2011) found no effect of drought stress on ethanol production from sorghum. As starch is a major component of ethanol production, this research suggests that drought stress may have little impact on sorghum starch synthesis. Conversely, Liu et al. (2013) reported that sorghum grown under non-irrigated conditions had less starch than sorghum grown under irrigated conditions and that the starch functional properties differed between irrigated and non-irrigated samples, suggesting that drought stress does affect sorghum starch properties as it does in other cereal grains. The degree to which drought stress will affect sorghum starch properties and how the timing of drought stress will affect sorghum starch are unknown. Given that sorghum is one of the most drought-tolerant cereal grains, this is an important area of research that needs additional attention.

Li et al. (2013) reported that starch from sorghum exposed to high temperature had reduced starch content (starch weight per grain) and smaller granule size distribution, less branching and greater content of long to short amylopectin branches. Amylose content was not changed by exposure to high temperatures, however (Li et al., 2013). This work clearly illustrates that heat stress will impact sorghum starch properties; however, more work is needed to elucidate the mechanisms behind this in order to alleviate detrimental effects of heat stress on sorghum starch properties.

Wu et al. (2016, 2017) reported that abiotic stress had a significant impact on the polyphenol content, and subsequently antioxidant capacity, as well as physical properties of sorghum. In this research, samples grown under deficit water conditions had higher polyphenol content and greater antioxidant capacity.

Likewise, water stress was also reported to impact mineral content of sorghum. Paiva et al. (2017) examined the effect of water stress on mineral content of 100 sorghum genotypes. Water stress was found to decrease the level of Mn across all the samples and decrease the levels of P and Mg across the majority of the samples. Sulphur content was decreased in just over half of the samples tested. For K, Ca, Cu, Fe and Zn the authors concluded that 'genotypes and other factors' contributed more to levels of minerals than did water stress.

7 Summary

Sorghum is well known for its genetic diversity. This diversity is reflected in the wide-ranging variability present for virtually all sorghum grain components. Grain composition in sorghum is not only due to genetic variability however, as research has demonstrated that environmental factors can significantly impact grain composition. Research on sorghum grain composition is increasingly focused on improving the nutritional quality of sorghum, especially regarding vitamin and mineral content. Ongoing screening of multiple sorghum populations developed to identify the natural variation of and underlying genetic contributions to grain composition is accelerating crop improvement. As screening of diverse populations of sorghum continues, new sources of germplasm may be identified and genetic markers may be developed for the sorghum breeding community to use in improving sorghum grain composition. The statement of Taylor and Belton (2002) when discussing the genetic diversity of sorghum is very relevant in this regard: 'It would appear that there are all kinds of novel types of sorghum just waiting to be exploited…'. Screening of diverse sorghum populations for grain composition and quality traits to discover 'novel types' of sorghum and incorporate them into breeding programmes is an important area of future research.

8 Future trends

To take advantage of the variability in sorghum grain composition, some thoughts on avenues for future research are as follows: 1) continued screening of diverse germplasm collections for grain composition to take advantage of the genetic diversity of sorghum with regard to both improving grain composition and identifying germplasm with desirable traits for new uses of sorghum grain; 2) identification of genes and genetic markers controlling variation of grain composition traits; 3) development of new and modification/improvement of existing methods for measuring grain composition in sorghum, especially high-throughput methods that will contribute to the improvement of genetic analysis that require large sample sets; 4) focus on the inter-related effects of G x E x M on grain composition, including all types of environments where sorghum can be grown and with regard to sustainable agricultural practices; and 5), consideration of the impact of grain processing on the constituents of the grain, or in effect 'G x E x M x P' which reflects the reality of how grain is used (i.e. all uses beyond utilization of the raw grain).

9 Where to look for further information

Some references that readers may wish to consult related to sorghum grain composition and the factors that can influence grain chemistry are listed below:

1 Hulse, J. H., Laing, E. M. and Pearson, O. E. 1980. *Sorghum and the Millets: Their Composition and Nutritive Value.* Academic Press, London. This book (997 pages) is an amazingly thorough review of early work on sorghum and millet grain composition as it relates to human nutrition. The authors reviewed over 1700 papers and made the effort to put grain composition data in common units where possible and contacted authors of publications to clarify details as needed. This is a thorough resource and overview of older research on sorghum grain composition.

2 Waniska, R. D. and Rooney, L. W. 2000. Structure and chemistry of the sorghum caryopsis. In: C. W. Smith and R. A. Frederiksen (Eds), *Sorghum: Origin, History, Technology, and Production.* Wiley, New York. pp. 649–88. Another extremely thorough review of sorghum grain composition including several very useful tables of summarized compositional data.

3 Chapters 7 'Sorghum', 8 'Sorghum: Subsistence Types', 9 'Sorghum: Commercial Types', 10 'Specialty Types' and 11 'Sorghum: Fuel and Utility Types' in National Research Council. 1996. *Lost Crops of Africa. Volume 1: Grains.* National Academy Press, Washington, D.C. These chapters provide a unique view of sorghum grain composition and how it relates to different uses.

4 The following recent review articles emphasizing various constituents of the sorghum grain are excellent starting points for detailed information on sorghum grain composition:
   a Belton, P. S., Delgadillo, I., Halford, N. G. and Shewry, P. R. 2006. Kafirin structure and functionality. *J. Cereal Sci.* 44:272–86.
   b Awika, J. M. and Rooney, L. W. 2004. Sorghum phytochemicals and their potential impact on human health. *Phytochemistry* 65:1199–1221.
   c Zhu, F. 2014. Structure, physicochemical properties, modifications, and uses of sorghum starch. *Compr. Rev. Food Sci. Food Saf.* 13:597–610.

5 The following reviews offer insights related to grain composition of sorghum and how it relates to grain quality, utilization and nutritional properties:
   a Duodu, K. G., Taylor, J. R. N., Belton, P. S. and Hamaker, B. R. 2003. Factors affecting sorghum protein digestibility. *J. Cereal Sci.* 38:117–31.
   b Taylor, J., Anyango, J. O. and Taylor, J. R. N. 2013. Developments in the science of zein, kafirin, and gluten protein bioplastic materials. *Cereal Chem.* 90:344–57.
   c Taylor, J., Schober, T. and Bean, S. R. 2006. Non-traditional uses of sorghum and pearl millet. *J. Cereal Sci.* 44:252–71.
   d Taylor, J. R. N., Belton, P. S., Beta, T. and Duodu, K. G. 2014. Increasing the utilization of sorghum, millets and pseudocereals: Developments in the science of their phenolic phytochemicals, biofortification and protein functionality. *J. Cereal Sci.* 59:257–75.
   e Wang, D., Bean, S., McClaren, J., Seib, P., Tuinstra, M., Lenz, M., Wu, X. and Zhao, R. 2008. Grain sorghum is a viable feedstock for ethanol production. *J. Industrial. Microbio. and Biotechnol.* 35:313–20.
   f Taylor, J. R. N and Emmambux, M. N. 2010. Review: Developments in our understanding of sorghum polysaccharides and their health benefits. *Cereal Chem.* 87:263–71.

10 Acknowledgements

It is impossible to review the literature with regard to sorghum grain composition without reading the amazing depth of research conducted in this area by Dr. Lloyd Rooney's lab at Texas A&M and Dr. John Taylor's lab at the University of Pretoria. The research of Dr. Rooney and Dr. Taylor on sorghum grain composition covers basic grain chemistry, composition and structure through detailed research on various biomolecules in sorghum and has provided much needed information on sorghum grain structure, composition and chemistry. Their research expands greatly beyond grain composition into various aspects of nutrition and many types of utilization as well. The authors acknowledge the vast amount of research in this area done by Dr. Rooney and Dr. Taylor. Names are necessary to report factually on available data; however, the U.S. Department of Agriculture neither guarantees nor warrants the standard of the product, and use of the name by the U.S. Department of Agriculture implies no approval of the product to the exclusion of others that may also be suitable.

11 References

Afify, A. E.-M. M. R., El-Beltagi, H. S., Abd El-Salam, S. M. and Omran, A. A. 2011. Bioavailability of iron, zinc, phytate and phytase activity during soaking and germination of white sorghum varieties. *PLoS ONE* 6:e25512. doi:10.1371/journal.pone.0025512.

Ananda, N., Vadlani, P. and Prasad. P. V. V. 2011. Evaluation of drought and heat stressed grain sorghum (Sorghum bicolor) for ethanol production. *Ind. Crops Prod.* 33:779–82.

Annicchiarico, P. 2002. Genotype x environment interaction: Challenges and opportunities for plant breeding and cultivar recommendations. FAO Plant production and protection paper 174.

Assefa, Y., Roozeboom, K., Thompson, C., Schlegel, A., Stone, L. and Lingenfelser, J. 2014. *Corn and Grain Sorghum Comparison: All Things Considered*. Amsterdam: Academic Press.

Awika, J. M., McDonough, C. M., Rooney, L. W. and Waniska, R. D. 2001. False positives for tannin sorghum in non-tannin sorghum using the bleach test. *International Sorghum and Millets Newsletter* 42:58–62.

Awika, J. M., Rooney, L. W., Wu, X., Prior, R. L. and Cisneros-Zevallos, L. 2003. Screening methods to measure antioxidant activity of sorghum (Sorghum bicolor) and sorghum products. *J. Agric. Food Chem.* 51:6657–62.

Awika, J. M. and Rooney, L. W. 2004. Sorghum phytochemicals and their potential impact on human health. *Phytochemistry* 65:1199–221.

Awika, J. M. Rooney, L. W. and Waniska, R. D. 2004. Properties of 3-deoxyanthocyanins from sorghum. *J. Agric. Food Chem.* 52: 4388–94.

Awika, J. M., McDonough, C. M. and Rooney, L. W. 2005. Decorticating sorghum to concentrate healthy phytochemicals. *J. Agric. Food Chem.* 53:6230–4. https://doi.org/10.1021/jf0510384.

Bean, S. R. 2003. Evaluation of novel pre-cast gradient gels for the separation of sorghum proteins. *Cereal Chem.* 80:500–4. https://doi.org/10.1094/cchem.2003.80.5.500.

Bean, S. R., Lookhart, G. L. and Bietz, J. A. 2000. Acetonitrile as a buffer additive for the separation of maize (*Zea mays* L.) and sorghum (*Sorghum bicolor* L.) storage proteins by HPCE. *J. Agric. Food Chem.* 48:318–27. https://doi.org/10.1021/jf990786o.

Bean, S. R., Chung, O. K., Tuinstra, J. F. and Erpelding, J. 2006. Evaluation of the single kernel characterization system (SKCS) for measurement of sorghum grain attributes. *Cereal Chem.* 83:108–13. https://doi.org/10.1094/cc-83-0108.

Bean, S. R., Ioerger, B. P. and Blackwell, D. L. 2011. Separation of kafirins on surface porous reversed phase-high performance liquid chromatography columns. *J. Agric. Food Chem.* 59:85–91. https://doi.org/10.1021/jf1036195.

Bean, S. R. and Ioerger, B. P. 2015. Sorghum and millet proteins. In Ustunol, Z. (Ed.) *Applied Food Protein Chemistry*. John Wiley & Sons, Oxford, UK, pp. 323 –59.
Beckles, D. M. and Thitisaksakul, M. 2014. How environmental stress affects starch composition and functionality in cereal endosperm. *Starch* 66:58–71. https://doi.org/10.1002/star.201300212.
Beckwith, A. C. 1972. Grain sorghum glutellin: Isolation and characterization. *J. Agric. Food Chem.* 20:761–4. https://doi.org/10.1021/jf60182a020.
Belton, P. S. and Taylor, J. R. N. 2004 Sorghum and millets: Protein sources for Africa. *Trends Food Sci. Technol.* 15:94–8. https://doi.org/10.1016/j.tifs.2003.09.002.
Belton, P. S., Delgadillo, I., Halford, N. G. and Shewry, P. R. 2006. Kafirin structure and functionality. *J. Cereal Sci.* 44:272–86. https://doi.org/10.1016/j.jcs.2006.05.004.
Benmoussa, M., Suhendra, B., Aboubacar, A. and Hamaker, B. R. 2006. Distinctive sorghum starch granule morphologies appear to improve raw starch digestibility. *Starch/Starke* 58:92–9. https://doi.org/10.1002/star.200400344.
Beta, T., Rooney, L. W., Marovatsanga, L. T. and Taylor, J. R. N. 1999. Phenolic compounds and kernel characteristics of Zimbabwean sorghums. *J. Sci. Food Agric.* 79:1003–10. https://doi.org/10.1002/(sici)1097-0010(19990515)79:7%3C1003::aid-jsfa317%3E3.0.co;2-f.
Beta, T. and Corke, H. 2001. Genetic and environmental variation in sorghum starch properties. *J. Cereal Sci.* 34:261–8. https://doi.org/10.1006/jcrs.2000.0379.
Blackwell, D. L. and Bean, S. R. 2011. Separation of alcohol soluble sorghum proteins using non-porous cation-exchange columns. *J. Chromatog. A.* 1230:48–53. https://doi.org/10.1016/j.chroma.2012.01.063.
Blasa, M., Angelino, D., Gennari, L. and Ninfali, P. 2011. The cellular antioxidant activity in red blood cells (CAA-RBC): A new approach to bioavailability and synergy of phytochemicals and botanical extracts. *Food Chem.* 125:685–91. https://doi.org/10.1016/j.foodchem.2010.09.065.
Branlard, G. and Bancel, E. 2006. Protein extraction from cereal seeds. In Thiellment, H., Zivy, M., Damerval, C. and M'echin, V. (Eds) *Methods in Molecular Biology, Vol. 335: Plant Proteomics: Methods and Protocols*. Totowa, NJ: Humana Press Inc, pp. 15–25.
Casa, A. M., Pressoir, G., Brown, P. J., Mitchell, S. E., Rooney, W. L., Tuinstra, M. R., Franks, C. D. and Kresovitch, S. 2008. Community resources and strategies for association mapping in sorghum. *Crop Sci.* 48:30–40. https://doi.org/10.2135/cropsci2007.02.0080.
Chamba, E. B., Halford, N. G., Forsyth, J., Wilkinson, M. and Shewry, P. R. 2005. Molecular cloning of β-kafirin, a methionine-rich protein of sorghum grain. *J. Cereal Sci.* 41:381–3.
Chandrasekara, A. and Shahidi, F. 2011. Determination of antioxidant activity in free and hydrolyzed fractions of millet grains and characterization of their phenolic profiles by HPLC-DAD-ESI-MSn. *J. Funct. Foods* 3:144–58. https://doi.org/10.1016/j.jff.2011.03.007.
Chiremba, C., Taylor, J. R. N., Rooney, L. W. and Beta, T. 2012. Phenolic acid content of sorghum and maize cultivars varying in hardness. *Food Chem.* 134:81–8. https://doi.org/10.1016/j.foodchem.2012.02.067.
Cobb, A., Wilson, G. W., Goad, C. L. Bean, S. R., Kaufman, R. C., Herald, T. J. and Wilson, J. D. 2016. The role of arbuscular mycorrhizal fungi in grain production and nutrition of sorghum genotypes: Enhancing sustainability through plant-microbial partnership. *Agric. Ecosyst. Environ.* 233:432–40.
Cremer, J. E., Bean, S. R., Tilley, M., Ioerger, B. P., Ohm, J.-B., Kaufmann, R. C., Wilson, J. D., Innes, D. D., Gilding, E. K. and Godwin, I. D. 2014. Grain sorghum proteomics: An integrated approach towards characterization of seed storage proteins in kafirin allelic variants. *J. Agric. Food Chem.* 62:9819–31.
Demissie, T., Ali, A., Mekonen, Y., Haider, J. and Umeta, M. 2010. Magnitude and distribution of vitamin A deficiency in Ethiopia. *Food Nutr. Bull.* 31: 234–41. https://doi.org/10.1177/156482651003100206.
DeRose, R. T., Ma, D-P., Kwon, I-S., Hasnain, S. E., Klassy, R. C. and Hall, T. C. 1989. Characterization of the kafirin gene family from sorghum reveals extensive homology with zein from maize. *Plant Mol. Biol.* 12:245–56. https://doi.org/10.1007/bf00043202.

Dicko, M. H., Gruppen, H., Traoré, A. S., Voragen, A. G. and Van Berkel, W. J., 2006. Sorghum grain as human food in Africa: Relevance of content of starch and amylase activities. *Afr. J. Biotechnol.* 5:384–95.

Duodu, K. G., Taylor, J. R. N., Belton, P. S. and Hamaker, B. R. 2003. Factors affecting sorghum protein digestibility. *J. Cereal Sci.* 38:117–31. https://doi.org/10.1016/s0733-5210(03)00016-x.

Dykes, L., Rooney, L. W., Waniksa, R. D. and Rooney, W. L. 2005. Phenolic compounds and antioxidant activity of sorghum grains of varying genotypes. *J. Agric. Food Chem.* 53:6813–18. https://doi.org/10.1021/jf050419e.

Dykes, L. and Rooney, L. W. 2006. Sorghum and millet phenols and antioxidants. *J. Cereal Sci.* 44:236–51. https://doi.org/10.1016/j.jcs.2006.06.007.

Earp, C. F., Doherty, C. A. and Rooney, L. W. 1983. Fluorescence microscopy of the pericarp, aleurone layer and endosperm cell walls of three sorghum cultivars. *Cereal Chem.* 60:408–10.

Earp, C. F., McDounough, C. M. and Rooney, L. W. 2004a. Microscopy of pericarp development in the caryopsis of *Sorghum bicolor* (L.) *Moench. J. Cereal Sci.* 39:21–7. https://doi.org/10.1016/s0733-5210(03)00060-2.

Earp, C. F., McDonough, C. M., Awika, J. and Rooney, L. W. 2004b. Testa development in the caryopsis of *Sorghum bicolor* (L.) *Moench. J. Cereal Sci.* 39:303–11. https://doi.org/10.1016/j.jcs.2003.11.005.

Elkhalil, E. A. I., El Tinay, A. H., Mohamed, B. E. and Elsheikh, E. A. E. 2001. Effect of malt pretreatment on phytic acid and in vitro protein digestibility of sorghum flour. *Food Chem.* 72: 29–32. https://doi.org/10.1016/s0308-8146(00)00195-3.

El-Nour, I. N. A., Peruffo, A. D. B. and Curioni, A. 1998. Characterization of sorghum kafirins in relation to their cross-linking behavior. *J. Cereal Sci.* 28:197–207.

Esen, A. 1987. A proposed nomenclature for the alcohol-soluble proteins (zeins) of maize (*Zea mays* L.). *J. Cereal Sci.* 5:117–28.

Evans, D. J., Schuessler, L. and Taylor, J. R. N. 1987. Isolation of reduced soluble protein from sorghum starchy endosperm. *J. Cereal Sci.* 5:61–5.

Evers, T. and Millar, S. 2002. Cereal grain structure and development: Some implications for quality. *J. Cereal Sci.* 36:261–84.

FAO. 1995. Chpt.4 Protein content and quality. Sorghum and millets in human nutrition (FAO Food and Nutrition Series, No. 27). http://www.fao.org/docrep/T0818E/T0818E00.htm.

Fernandez, M. G. S., Hamblin, M. T., Li, L., Rooney, W. L., Tuinstra, M. R. and Kresovich, S. 2008. Quantitative trait loci analysis of endosperm color and carotenoid content in sorghum grain. *Crop. Sci.* 48:1732–43.

Gaffa, T., Yoshimoto, Y., Hanashiro, I., Honda, O., Kawaski, S. and Takeda, Y. 2004. Physicochemical properties and molecular structures of starches from millet (*Pennisetum typhoides*) and sorghum (*Sorghum bicolor* L. Moench) cultivars in Nigeria. *Cereal Chem.* 81:255–60.

Gambin, B. L. and Borras, L. 2005. Sorghum kernel weight: Growth patterns from different positions within the panicle. *Crop Sci.* 45:553–61.

Garcia-Estepa, R. M., Guerra-Hernandez, E. and Garcia-Villanova, B. 1999. Phytic acid content in milled cereal products and breads. *Food Res. Int.* 32:217–21.

Gerrano, A. S., Labuschangne, M. T., van Biljon, A. and Shargie, N. G. 2016. Quantification of mineral composition and total protein content in sorghum [*Sorghum bicolor* (L.) Moench] genotypes. *Cereal Res. Commun.* 44:272–85.

Griess, J. K., Mason, S. C., Jackson, D. S., Galusha, T. D., Yaseen, M. and Pedersen, J. F. 2010. Environment and hybrid influences on food-grade sorghum grain yield and hardness. *Crop Sci.* 50:1480–9.

Gu, L., House, S. E, Rooney, L. W. and Prior, R. L. 2007. Sorghum bran in the diet dose dependently increased the excretion of catechins and microbrial-derived phenolic acids in female rats. *J. Agric. Food Chem.* 55: 5326–34.

Hagerman, A. E., Riedl, K. M., Jones, G. A., Sovik, K. N., Ritchard, N. T., Hartzfeld, P. W. and Riechel, T. L. 1998. High molecular weight plant polyphenolics (tannins) as biological antioxidants. *J. Agric. Food Chem.* 46:1887–92.

Hamaker, B. R., Kirleis, A. W., Mertz, E. D. and Axtell, J. D. 1986. Effect of cooking on the protein profiles and *in-vitro* digestibility of sorghum and maize. *J. Agric. Food Chem.* 34:647–9.
Hamaker, B. R., Mohamed, A. A., Habben, J. E., Huang, C. P. and Larkins, B. A. 1995. Efficient procedure for extracting maize and sorghum kernel proteins reveals higher prolamin contents than the conventional method. *Cereal Chem.* 72:583–8.
Hamaker, B. R. and Bugusu, B. A. 2000. Overview: Sorghum proteins and food quality. In AFRIPRO: Workshop on the proteins of sorghum and millets: Enhancing nutritional and functional properties for Africa; Belton, P. S., Taylor, J. R. N. (Eds). http://www.afripro.org.uk/papers/Paper08Hamaker.pdf (accessed 1 June 2017).
Hamblin, M. T., Salas Fernandez, M. G., Tuinstra, M. R., Rooney, W. L. and Kresovich, S. 2007. Sequence variation at candidate loci in the starch metabolism pathway in sorghum: Prospects for linkage disequilibrium mapping. *Crop Sci.* S125–34.
Heiniger, R. W., Vanderlip, R. L. and Kofoid, K. D. 1993. Caryposis weight patterns within the sorghum panicle. *Crop Sci.* 33:543–9.
Herald, T. J., Gadgil, P. and Tilley, M. 2012. High-throughput micro plate assays for screening flavonoid content and DPPH-scavenging activity in sorghum bran and flour. *J. Sci. Food Agric.* 92:2326–31.
Herald, T. J., Gadgil, P., Perumal, R., Bean, S. R. and Wilson, J. D. 2014. High-throughput micro-plate HCl-vanillin assay for screening tannin content in sorghum grain. *J. Sci. Food Agric.* 94:2133–6.
Hizukuri, S. 1985. Relationship between the distribution of the chain length of amylopectin and the crystalline structure of starch granules. *Carbohydr. Res.* 141: 295–06.
Holding, D. R. 2014. Recent advances in the study of prolamin storage protein organization and function. *Front. Plant Sci.* 5:276. doi:10.3389/fpls.2014.00276.
Hoseney, R. C., Davis, A. B. and Herbers, L. H. 1974. Pericarp and endosperm structure of sorghum grain shown by scanning electron microscopy. *Cereal Chem.* 51:552–8.
Hubbard, J. E., Hall, H. H. and Earle, F. R. 1950. Composition of the component parts of the sorghum kernel. *Cereal Chem.* 51:825–9.
Huber, K. C. and BeMiller, J. N. 2000. Channels of maize and sorghum starch granules. *Carbohydr. Polym.* 41:269–76.
Hwang, K. T., Weller, C. L., Cuppett, S. L. and Hanna, M. A. 2004. Policosanol contents and composition of grain sorghum kernels and dried distillers grains. *Cereal Chem.* 81:345–9.
Ioerger, B., Bean S. R., Tuinstra, M. R., Pedersen, J. F., Erpelding, J., Lee, K. and Herrman, T. 2007. Characterization of polymeric proteins from vitreous and floury sorghum endosperm. *J. Agric. Food Chem.* 55:10232–9.
Izquierdo, L. and Godwin, I. D. 2005. Molecular characterization of a novel methionine-rich δ-kafirin seed storage protein gene in sorghum (*Sorghum bicolor* L.). *Cereal Chem.* 82:706–10.
Jambunathan, R., Mertz, E. T. and Axtell, J. D. 1975. Fractionation of soluble proteins of high lysine and normal sorghum grain. *Cereal Chem.* 52:119–21.
Kaluza, W. Z., McGrath, R. M., Roberts, T. C. and Schroder, H. H. 1980. Separation of phenolics of *Sorghum bicolor* (L.) Moench grain. *J. Agric. Food Chem.* 28:1191–6.
Kaufman, R. C., Herald, T. J., Bean, S. R., Wilson, J. D. and Tuinstra, M. R. 2013. Variability in tannin content, chemistry, and activity in a diverse group of tannin containing sorghum cultivars. *J. Sci. Food Ag.* 93:1233–41.
Kayodé, A. P., Linnemann, A. R., Hounhouigan, J. D., Nout, M. J. R. and van Boekel, A. J. S. 2006. Genetic and environmental impact on iron, zinc, and phytate in food sorghum grown in Benin. *J. Agric. Food Chem.* 54:256–62. https://doi.org/10.1021/jf0521404.
Kean, E. G., Bordenave, N., Ejeta, G., Hamaker, B. R. and Ferruzzi, M. G. 2011. Carotenoid bioaccessibility from whole grain and decorticated yellow endosperm sorghum porridge. *J. Cereal Sci.* 54:450–9.
Khush, G. S., Lee, S., Cho, J.-I. and Jeon, J.-S. 2012. Biofortification of crops for reducing malnutrition. *Plant Biotechnol. Rep.* 6:195–202.
Kimber, C. T., Dahlberg, J. A. and Kresovich, S. 2013. The gene pool of Sorghum bicolor and its improvement. In Paterson, A. H. (Ed.) *Genomics of the Saccharinae, Plant Genetics and Genomics: Crops and Models.* New York: Springer, pp. 23–41.

Kirleis, A. W., Crosby, K. D. and Housley, T. L. 1984. A method for quantitatively measuring vitreous endosperm area in sectioned sorghum grain. *Cereal Chem.* 61:556–8.

Kumar, V., Sinha, A. K., Makkar, H. P. S. and Becker, K. 2010. Dietary roles of phytate and phytase in human nutrition: A review. *Food Chem.* 120: 945–59.

Kumar, A. A., Reddy, B. V. S., Ramaiah, B., Sahrawat, K. L. and Pfeiffer, W. H. 2012. Genetic variability and character association for grain iron and zinc contents in sorghum germplasm accessions and commercial cultivars. *Eur. J. Plant Sci. Biotechnol.* 6:1–5.

Kumar, A. A., Reddy, B. V. S., Ramaiah, B., Sahrawat, K. L. and Pfeiffer, W. H. 2013. Gene effects and heterosis for grain iron and zinc concentrations in sorghum [*Sorghum bicolor* (L.) Moench]. *Field Crops Res.* 146:86–95.

Laidlaw, H. K. C., Mace, E. S., Williams, S. B., Sakrewski, K., Mudge, A. M., Prentis, P. J., Jordan, D. R. and Godwin, I. D. 2010. Allelic variation of the β-, γ- and δ-kafirin genes in diverse *Sorghum* genotypes. *Theor. Appl. Genet.* 121:1227–37.

Landry, J. and Moureaux, T. 1970. Heterogeneity of corn seed glutelin: Selective extraction and amino acid composition of the 3 isolated fractions. *Bull. Soc. Chim. Biol.* 52:1021–37.

Landry, J. 1997. Comparison of extraction methods for evaluating zein content of maize grain. *Cereal Chem.* 74:188–9.

Lawton, J. W. and Faubion, J. M. 1989. Measuring kernel hardness using the tangential abrasive dehulling device. *Cereal Chem.* 66:519–24.

Lee, W. J., Pedersen, J. F. and Shelton, D. R. 2002. Relationship of sorghum kernel size to physiochemical, milling, pasting, and cooking properties. *Food Res. Int.* 35:643–9.

Li, E., Hasjim, J., Singh, V., Tizzotti, M., Godwin, I. D. and Gilbert, R. G. 2013. Insights into sorghum starch biosynthesis from structure changes induced by different growth temperatures. *Cereal Chem.* 90:223–30.

Lichtenwalner, R. E., Ellis, E. B. and Rooney, L. W. 1978. Effect of incremental dosages of the waxy gene of sorghum on digestibility. *J. Anim. Sci.* 46:1113–19.

Lipkie, T. E., De Moura, F. F., Zhao, Z.-Y., Alberten, M. C., Che, P., Glassman, K. and Ferruzzi, M. G. 2013. Bioaccessibility of carotenoids from transgenic provitamin A biofortified sorghum. *J. Agric. Food Chem.* 61:6754–5771.

Liu, L., Maier, A., Klocke, N., Yan, S., Rogers, D. Tesso, T. and Wang, D. 2013. Impact of deficit irrigation on sorghum physical and chemical properties and ethanol yield. *Trans. ASAE.* 56:1541–9. https://doi.org/10.13031/trans.56.10153.

Lu, Y., Zhao, G., Li, Y., Fan, J., Ding, G., Zhao, J., Ni, X., Xu, Y. and Wang, W. 2013. Identification of two novel *waxy* alleles and development of their molecular markers in sorghum. *Genome* 56:283–8.

Mahgoub, S. E. O. and Elhag, S. A. 1998. Effect of milling, soaking, malting, heat-treatment and fermentation on phytate levels of four Sudanese sorghum cultivars. *Food Chem.* 61: 77–80.

Makohka, A. O., Oniang'o, R. K., Njorge, S. M. and Kamar, O. K. 2002. Effect of traditional fermentation and malting on phytic acid and mineral availability from sorghum (*Sorghum bicolor*) and finger millet (*Eleusine coracana*) grain varieties grown in Kenya. *Food Nutr. Bull.* 23:241–5.

Matalanis, A. M., Campanella, O. H. and Hamaker, B. R. 2009. Storage retrogradation behavior of sorghum, maize and rice starch pastes related to amylopectin fine structure. *J. Cereal Sci.* 50:74–81.

Maxson, E. D. and Rooney, L. W. 1972. Evaluation of methods for tannin analysis in sorghum grain. *Cereal Chem.* 49:719–29.

Mazhar, H., Chandrashekar, A. and Shetty, H. S. 1993. Isolation and immunochemical characterization of the alcohol-extractable proteins (kafirins) of *Sorghum bicolor* (L.) Moench. *J. Cereal Sci.* 17:83–93.

McIntyre, C. L., Drenth, J., Gonzalez, N., Henzell, R. G. and Jordan, D. R. 2008. Molecular characterization of the waxy locus in sorghum. *Genome* 51:524–33.

Moreau, R. A., Harron, A. F., Powell, M. J. and Hoyt, J. L. 2016. A comparison of the levels of oil, carotenoids, and lipolytic enzyme activities in modern lines and hybrids of grain sorghum. *J. Am. Oil Chem. Soc.* 93:569–73.

Murray, S. C., Sharma, A., Rooney, W. L., Klein, P. C., Mullet, J. E., Mitchell, S. E. and Kresovich, S. 2008. Genetic improvement of sorghum as a biofuel feedstock: I. QTL for stem sorghum and grain nonstructural carbohydrates. *Crop Sci.* 48:2165–79.

Nucere, N. J. and Sumrell, G. 1979. Protein fractions from five varieties of grain sorghum: Amino acid composition and solubility properties. *J. Agric. Food Chem.* 27:809–12.

Ogbonna, A. C. 2011. Current developments in malting and brewing trials with sorghum in Nigeria: A review. *J. Inst. Brew.* 117:394–400.

Oria, M. P., Hamaker, B. R., Axtell, J. D. and Huang, C.-P. 2000. A highly digestible sorghum mutant cultivar exhibits a unique folded structure of endosperm protein bodies. *Proc. Natl. Acad. Sci.* 97:5065–70.

Osborne, T. B. 1907. *The Proteins of the Wheat Kernel.* Washington, D.C.: Carnegie Institute.

Osborne, T. B. 1924. *The Vegetable Proteins.* 2nd ed. London: Longmans Green.

Osman, M. A. 2004. Changes in sorghum enzyme inhibitors, phytic acid, tannins and in vitro protein digestibility occurring during Khamir (local bread) fermentation. *Food Chem.* 88:129–34.

Paiva, C. L., Queiroz, V. A. V., Simeone, M. L. F., Schaffert, R. E., de Oliveria, A. C. and da Silva, C. M. 2017. Mineral content of sorghum genotypes and the influence of water stress. *Food Chem.* 214:400–5.

Pedersen, J. F., Martin, C. R., Felker, F. C. and Steele, J. L. 1996. Application of the single kernel wheat characterization technology to sorghum grain. *Cereal Chem.* 73:421–3.

Pedersen, J. F. and Kofoid, K. D. 2003. Variability and relationships among 12-hour IVDMD, starch, oil, protein, and physical characteristics of 16 sorghum conversion lines. *Euphytica* 130:261–6.

Pederson, J. F., Bean, S. R., Graybosch, R. A., Park, S. H. and Tilley, M. 2005. Characterization of waxy grain sorghum lines in relation to granule-bound starch synthase. *Euphytica* 144:151–6.

Pomeranz, Y. 1986. Comparison of screening methods for indirect determination of sorghum hardness. *Cereal Chem.* 63:36–8.

Price, M. L., Van Scoyoc, S. and Butler, L. G. 1978. A critical evaluation of vanillin reaction as an assay for tannin in sorghum. *J. Agric. Food Chem.* 26:1214–18.

Rao, A. V. and Rao, L. G. 2007. Carotenoids and human health. *Pharmacol. Res.* 55: 207–16.

Reichert, R. D., Youngs, C. G. and Oomah, B. D. 1981. Measurement of grain hardness and dehulling quality with a multi-sample, tangential abrasive dehulling device (TADD). In Rooney, L. W. and Murty, D. S. (Eds) *Proc. Int. Symp. Sorghum Grain Quality.* Patancheru, India: International Crops Research Institute for Semi-Arid Tropics, pp. 186–93.

Reichert, R., Mwararu, M. and Mukuru, S. 1988. Characterization of colored grain sorghum lines and identification of high tannin lines with good dehulling characteristics. *Cereal Chem.* 65:165–70.

Rhodes, D. H., Hoffman, L., Rooney, W. L., Ramu, P., Morris, G. P. and Kresovich, S. 2014. Genome-wide association study of grain polyphenol concentrations in global sorghum [*Sorghum bicolor* (L.) Moench] germplasm. *J. Agric. Food Chem.* 62:10916–27.

Rhodes, D. H. and Kresovich, S. 2016. Sorghum [*Sorghum bicolor* (L.) Moench] genotypes with contrasting polyphenol compositions differentially modulate inflammatory cytokines in mouse macrophages. *J. Chem.* Article ID 9640869. doi:10.1155/2016/9640869.

Rhodes, D. H., Hoffman Jr., L. , Rooney, W. L., Herald, T. J., Bean, S., Boyles, R., Brenton, Z. W. and Kresovitch, S. 2017. Genetic architecture of kernel composition in global sorghum germplasm. *BMC Genomics* 18:15. doi:10.1186/s12864-016-3403-x.

Rodriguez-Amaya, D. B. and Kimura, M. 2004. *HarvestPlus handbook for carotenoid analysis* (Vol. 2). Washington: International Food Policy Research Institute (IFPRI).

Rooney, L. W. 1978. Sorghum and pearl millet lipids. *Cereal Chem.* 55:584–90.

Rooney, L. W. and Miller, F. R. 1982. Variation in the structure of the kernel characteristics of sorghum. In Mertin, J. V. (Ed.) *Proceedings of the International Symposium on Sorghum Grain Quality.* Patancheru, A.P., India: ICRISAT, pp. 143–62.

Rooney, W. L. 2000. Genetics and cytogenetics. In Smith, C. W. and Frederiksen, R. A. (Eds) *Sorghum: Origin, History, Technology, and Production.* New York: Wiley. pp. 261–308.

Rooney, L. W. and Serna-Saldivar, S. O. 2000. Sorghum. In Kulp, K. (Ed.), *Handbook of Cereal Science and Technology*, 2nd Ed. Boca Raton, FL: CRC Press, pp. 149–76.

Salas Fernandez, M. G., Hamblin, M. T., Li, L., Rooney, W. L., Tuinstra, M. R. and Kresovich, S. 2008. Quantitative trait loci analysis of endosperm color and carotenoid content in sorghum grain. *Crop Sci.* 48:1732–43.

Salas Fernandez, M. G., Kapran, I., Souley, S., Adbou, M., Maiga, I. H., Acharya, C. B., Hamblin, M. T. and Kresovich, S. 2009. Collection and characterization of yellow endosperm sorghums from West Africa for biofortification. *Genet. Resour. Crop Evol.* 56:991–1000.

Sang, Y., Bean, S. R., Seib, P. A., Pedersen, J. F. and Shi, Y.-C. 2008. Structure and functional properties of sorghum starches differing in amylose content. *J. Agric. Food Chem.* 56:6680–5.

Sastry, L. V. S., Paulis, J. W., Bietz, J. A. and Wall, J. S. 1986. Genetic variation of storage proteins in sorghum grain: Studies by isoelectric focusing and high-performance liquid chromatography. *Cereal Chem.* 63:420–7.

Sattler, S. E., Singh, J., Haas, E. J., Guo, L., Sarath, G. and Pedersen, J. F. 2009. Two distinct *waxy* alleles impact the granule-bound starch synthase in sorghum. *Mol. Breed.* 24:349–59.

Seckinger, H. L. and Wolf, M. J. 1973. Sorghum protein ultrastructure as it relates to composition. *Cereal Chem.* 50:455–65.

Serna-Saldivar, S. and Rooney, L. W. 1995. Structure and chemistry of sorghum and millets. In Dendy, D. V. (Ed.), *Structure and Chemistry of Sorghum and Millets*. St. Paul, MN: AACC, pp.69–124.

Shakoor, N., Ziegler, G., Dilkes, B. P., Brenton, Z., Boyles, R., Connolly, E. L., Kresovich, S. K. and Baxter, I. 2016. Integration of experiments across diverse environments identifies the genetic determinants of variation in Sorghum bicolor seed element composition. *Plant Physiol.* 170: 1989–98.

Shapter, F. M., Dawes, M. P., Lee, L. S. and Henry, R. J. 2010. Aleurone and subaleurone morphology in native Australian wild cereal relatives. *Austral. J. Bot.* 57:688–96.

Shull, J. M., Chandrashekar, A., Kirleis, A. W. and Ejeta, G. 1990. Development of sorghum (*Sorghum bicolor* (L.) Moench) endosperm in varieties of varying hardness. *Food Struct.* 9:253–67.

Shull, J. M., Watterson, J. J. and Kirleis, A. W. 1991. Proposed nomenclature for the alcohol-soluble proteins (kafirins) of *Sorghum bicolor* (L. Moench) based on molecular weight, solubility, and structure. *J. Agric. Food Chem.* 39:83–7.

Shull, J. M., Watterson, J. J. and Kirleis, A. W. 1992. Purification and immunocytochemical localization of kafirins in *Sorghum bicolor* (L. Moench) endosperm. *Protoplasma* 171:64–74.

Singh, V., Moreau, R. A. and Hicks, K. B. 2003. Yield and phytosterol composition of oil extracted from grain sorghum and its wet-milled fractions. *Cereal Chem.* 80:126–9.

Smith, J. S. C. 1994. RP-HPLC for varietal identification in cereals and legumes: Sorghum. In Kruger, J. E. and Bietz, J. A. (Eds), *High Performance Liquid Chromatography of Cereal and Legume Proteins*. St. Paul, MN: AACC, pp.201–5.

Subba Rao, M. and Muralikrishna, G. 2002. Evaluation of the antioxidant properties of free and bound phenolic acids from native and malted finger millet (Ragi, *Eleusine coracana* Indaf-15). *J. Agric. Food Chem.* 50:889–92.

Subramanian, V., Seetharama, N., Jambunathan, R. and Rao, P. V. 1990. Evaluation of protein quality of sorghum [*Sorghum bicolor* (L.) *Moench*]. *J. Agric. Food Chem.* 38:1344–7.

Sukumaran, S., Xiang, W., Bean, S. R., Pedersen, J. F., Tuinstra, M. R., Tesso, T. T., Hamblin, M. T. and Yu, J. 2012. Genetic structure of a diverse sorghum collection and association mapping for grain quality. *Plant Genome* 5:126–35.

Sullins, R. D. and Rooney, L. W. 1974. Microscopic evaluation of the digestibility of sorghum lines that differ in endosperm characteristics. *Cereal Chem.* 51:134–42.

Syahariza, Z. A., Li, E. and Hasjim, J. 2010. Extraction and dissolution of starch from rice and sorghum grains for accurate structural analysis. *Carbohyd. Polym.* 82: 14–20.

Taleon, V., Dykes L., Rooney W. L. and Rooney L. W. 2014. Environmental effect on flavonoid concentrations and profiles of red and lemon-yellow sorghum grains. *J. Food Comp. Anal.* 34:178–85.

Taylor, J. R. N. 1983. Effect of malting on the protein and free amino nitrogen composition of sorghum. *J. Sci. Food Agric.* 34:885–92.

Taylor, J. R. N. and Schüssler, L. and van der Walt, W. H. 1984a. Fractionation of proteins from low-tannin sorghum grain. *J. Agric. Food Chem.* 32:149–54. https://doi.org/10.1021/jf00121a036.

Taylor, J. R. N., Novellie, L. and Liebenberg, N. vd. W. 1984b. Protein body degradation in the starchy endosperm of germinating sorghum. *J. Exp. Bot.* 36:1287–95.

Taylor, J. R. N., Novellie L. and Liebenberg, N. vd. W. 1984c. Sorghum protein body composition and ultrastructure. *Cereal Chem.* 61:69–73.

Taylor, J. R. N. and Schüssler, L. 1986. The protein compositions of the different anatomical parts of sorghum grain. *J. Cereal Sci.* 4:361–9.

Taylor, J. R. N. and Belton, P. S. 2002. Sorghum. Pp. 25–92 In Belton, P. S. and Taylor, J. R. N. (Eds) *Pseudocereals and Less Common Cereals: Grain Properties and Utilization Potential.* Berlin: Springer-Verlag.

Taylor, J. R. N., Belton, P. S., Beta, T. and Duodu, K. G. 2014. Increasing the utilization of sorghum, millets and pseudocereals: Developments in the science of their phenolic phytochemicals, biofortification and protein functionality. *J. Cereal Sci.* 59:257–75.

Thaipong, K., Boonprakob, U., Crosby, K., Cisneros-Zevallos, L. and Byrne, D. H. 2006. Comparison of ABTS, DPPH, FRAP and ORAC assays for estimating antioxidant activity from guava fruit extracts. *J. Food Comp. Anal.* 19:669–75.

Thitisaksakul, M., Jimenez, R. C., Aria, M. C. and Beckles, D. M. 2012. Effects of environmental factors on cereal starch biosynthesis and composition. *J. Cereal Sci.* 56:67–80.

Towo, E., Matuschek, E. and Svanberg, U. 2006. Fermentation and enzyme treatment of tannin sorghum gruels: Effects on phenolic compounds, phytate and in vitro accessible iron. *Food Chem.* 94:369–76.

Urbano, G., Lopez-Jurado, M., Aranda, P., Vidal-Valverde, C., Tenorio, E. and Porres, J. 2000. The role of phytic acid in legumes: Antinutrient or beneficial factor? *J. Physiol. Biochem.* 56:283–94.

Wall, J. S. and Paulis, J. W. 1978. Corn and sorghum grain proteins. In Pomeranz, Y. (Ed.), *Advances in Cereal Science & Technology*, Vol.II. St. Paul, MN: AACC, pp.135–219.

Wang, Y. and Frei, M. 2011. Stressed food – The impact of abiotic environmental stresses on crop quality. *Agric. Ecosyst. Environ.* 141:271–86.

Waniksa, R. D., Hugo, L. F. and Rooney, L. W. 1992. Practical methods to determine the presence of tannins in sorghum. *J. Appl. Poult. Res.* 1:122–8.

Waniska, R. D. 2000. Structure, phenolic compounds, and antifungal proteins of sorghum caryopses. In Chandrashekar, A., Bandyopadhyay, R. and Hall, A. J. (Eds) Technical and institutional options for sorghum grain mold management: Proceedings of an international consultation, 18–19 May 2000, ICRISAT, Patancheru, India. Patancheru 502 324, Andhra Pradesh, India: International Crops Research Institute for the Semi-Arid Tropics, pp. 72–106.

Waniska, R. D. and Rooney, L. W. 2000. Structure and chemistry of the sorghum caryopsis. In Smith, C. W. and Frederiksen, R. A. (Eds), *Sorghum: Origin, History, Technology, and Production.* New York: Wiley, pp.649–88.

Watterson, J. J., Shull, J. M., and Kirleis, A. W. 1993. Quantitation of α-, β-, and γ-kafirins in vitreous and opaque endosperm of *Sorghum bicolor. Cereal Chem.* 70:452–7.

Weaver, C. A., Hamakar, B. R. and J. D. Axtell. 1998. Discovery of grain sorghum germplasm with high uncooked and cooked in vitro protein digestibilities. *Cereal Chem.* 75:665–70.

Wolf, K. L. and Liu, R. H. 2007. Cellular antioxidant activity (CAA) assay for assessing antioxidants, foods, and dietary supplements. *J. Agric. Food Chem.* 55:8896–907.

Woo, Y. M., Hu, D. W. N., Larkins, B. A. and Jung, R. 2001. Genomics analysis of genes expressed in maize endosperm identifies novel seed proteins and clarifies patterns of zein gene expression. *Plant Cell* 13:2297–317.

Wu, G., Johnson, S. K., Bornman, J. F., Bennett, S., Singh, V. and Fang, Z. 2016. Effect of genotype and growth temperature on sorghum grain physical characteristics, polyphenol content, and antioxidant activity. *Cereal Chem.* 93: 419–25.

Wu, G., Johnson, S. K., Bornman, J. F., Bennet, S. J. and Fang, Z. 2017. Changes in whole grain polyphenols and antioxidant activity of six sorghum genotypes under different irrigation treatments. *Food Chem.* 214: 199–207.

Xu, J.-H. and Messing, J. 2008. Organization of the prolamins gene family provides insight into the evolution of the maize genome and gene duplication in grass species. *Proc. Natl. Acad. Sci.* 105:14330–5.

Xu, J.-H. and Messing, J. 2009. Amplification of prolamin storage protein genes in different subfamilies of the Poaceae. *Theor. Appl. Genet.* 119:1397–412.

Yaseen, M., Griess, J. K., Mason, S. C. and Jackson, D. S. 2014. Potential use of GGE bi-plot analysis to evaluate white sorghum grain attributes for end uses. *Pak. J. Agri. Sci.* 51:263–71.

Zeleznak, K. and Varriano-Marston, E. 1982. Pearl millet (*Pennisetum americanum* (L.) Leeke) and grain sorghum (*Sorghum bicolor* (L.) Moench) ultrastructure. *Am. J. Bot.* 69:1306–13.

Zhan, X., Wang, D., Tuinstra, M., Bean, S., Seib, P. A. and Sun, X. S. 2003. Ethanol and lactic acid production as affected by sorghum genotype and location. *Ind. Crop. Prod.* 18:245–55.

Zhu, F. 2014. Structure, physicochemical properties, modifications, and uses of sorghum starch. *Compr. Rev. Food Sci. Food Saf.* 13:597–610.

Zimmermann, M. B. and Hurrell, R. F. 2007. Nutritional iron deficiency. *Lancet* 370:511–20.

Chapter 2

The domestication, spread and uses of sorghum as a crop

F. M. Shapter, A. Crowther, G. Fox, I. D. Godwin and L. Watson-Fox, University of Queensland, Australia; I. J. C. Hannah, AGR Industries, Australia; and S. L. Norton, Agriculture Victoria, Australia

1 Introduction

Sorghum's resilience under abiotic stress, particularly water and nutrient deficiencies, underpins its domestication and spread. Under suboptimal growing conditions, sorghums can produce greater biomass and grain yield, more economically than many other cereals. While the nutritional value of the grain for human and animal feed can be compromised by the protein–starch matrix, sorghum grain has a high energy and feed value (Dahlberg et al., 2011). It is a gluten-free grain high in micronutrients and antioxidants, making it a primary target for crop improvement programmes and novel application research over the past five decades.

Sorghum is cultivated worldwide, with production peaking in 1985 at just over 77.5 million tonnes. Production patterns have fluctuated since, and in 2014 the Food and Agriculture Organisation of the United Nations (FAO) reported production of just over 67.8 million tonnes worldwide. Since 1961 the percentage of world sorghum production attributed to the five main continental masses has only radically altered in Africa (26.1–42.7%) and Asia (38–41.1%), while production in the Americas, Europe and Oceania have remained reasonably consistent, with averages of 40.5%, 1.2% and 1.9%, respectively (FAO, 2016).

http://dx.doi.org/10.19103/AS.2017.0015.01

1.1 An overview of the *Sorghum* genus

1.1.1 Diversity and geographic provenance

Sorghum bicolor subsp. *bicolor* (L.) Moench, sorghum, is extensively cultivated in marginal rainfall areas of the tropics and subtropics, with selected varieties widely grown in temperate climates. The *Sorghum* genus has extremely varied geographic, morphological and genetic diversity. The genus has approximately 25 recognised species that are separated into the five taxonomic sections, namely *Eusorghum* (containing the domesticated species), *Chaetosorghum*, *Heterosorghum*, *Parasorghum* and *Stiposorghum*, which contain the undomesticated species (Garber, 1950). Of these species, 17 are native, 14 of which are endemic, to Australia (Lazarides et al., 1991).

The *Eusorghum* species include the cultivated *Sorghum bicolor* subsp. *bicolor*, *Sorghum* × *almum* Parodi, *Sorghum bicolor* subsp. *verticilliflorum* (Steud.) de Wet ex Wiersema and J. Dahlb (a progenitor of cultivated sorghum), *Sorghum bicolor* nothosubsp. *drummondii* (Steud.) de Wet ex Davidse, the widespread weedy species *Sorghum halepense* (L.) Pers. and *Sorghum propinquum* (Kunth) Hitchc. The *Eusorghums* originate from Africa and Asia and have $2n$ = 20 or 40 chromosomes (de Wet and Harlan, 1971; Doggett, 1988; Duvall and Doebley, 1990; Price et al., 2005a). The sections *Chaetosorghum* and *Parasorghum* are monotypic, containing *Sorghum macrospermum* E. D. Garber, endemic to a small area of the Northern Territory of Australia, and *Sorghum laxiflorum* F. M. Bailey, native to northern Australia and Papua New Guinea, with both species having $2n$ = 40 chromosomes (Garber, 1950; Lazarides et al., 1991; Price et al., 2005b).

The *Parasorghum* section contains seven species *Sorghum grande* Lazarides, *Sorghum leiocladum* (Hack.) C. E. Hubb., *Sorghum matarankense* E. D. Garber and Snyder, *Sorghum nitidum* (Vahl) Pers., *Sorghum purpureosericeum* (Hochst. ex. A. Rich.) Asch. and Schweinf., *Sorghum timorense* (Kunth) Buse and *Sorghum versicolor* Andersson. These species range in ploidy from $2n$ = 10 to 20, and are native to northern monsoonal Australia, Africa and Asia (Garber, 1950; Lazarides et al., 1991; Phillips, 1995; Price et al., 2005a). Section *Stiposorghum* is comprised of ten species ranging in ploidy from $2n$ = 10, 20, 30 to 40, with all endemic to northern Australia: *Sorghum amplum* Lazarides, *Sorghum angustum* S. T. Blake, *Sorghum brachypodum* Lazarides, *Sorghum bulbosum* Lazarides, *Sorghum ecarinatum* Lazarides, *Sorghum exstans* Lazarides, *Sorghum interjectum* Lazarides, *Sorghum intrans* F. Muell. Ex Benth., *Sorghum plumosum* (R. Br.) P. Beauv., and *Sorghum stipoideum* (Ewart and Jean White) C. A. Gardner and C. E. Hubb. (Garber, 1950; Lazarides et al., 1991; Price et al., 2005a).

The geographic distribution of *Sorghum* species is shown in Fig. 1. These distributions represent the natural geographic origin of species, and do not include the areas where sorghum is cultivated, nor the non-native distributions of the weedy species.

The range of environments and climatic conditions where *Sorghum* species naturally inhabit is diverse, resulting in a broad array of traits that could potentially be beneficial to cultivated sorghum. Undomesticated sorghums can colonise a wide range of soil and moisture conditions, including very hot, dry, nutrient-limited environments. Over a wide range of microenvironments the sorghums have demonstrated both their ability to survive and adapt. Many of the undomesticated *Sorghum* species have developed resistances to the suite of pests and diseases that affect sorghum grain production globally. Interestingly, many Australian undomesticated species contain resistances to the major pest/diseases of Africa and America, which are not yet present within Australia (Bapat and Mote, 1982; Franzmann and Hardy, 1996; Kamala et al., 2002; Komolong et al., 2002; Sharma and

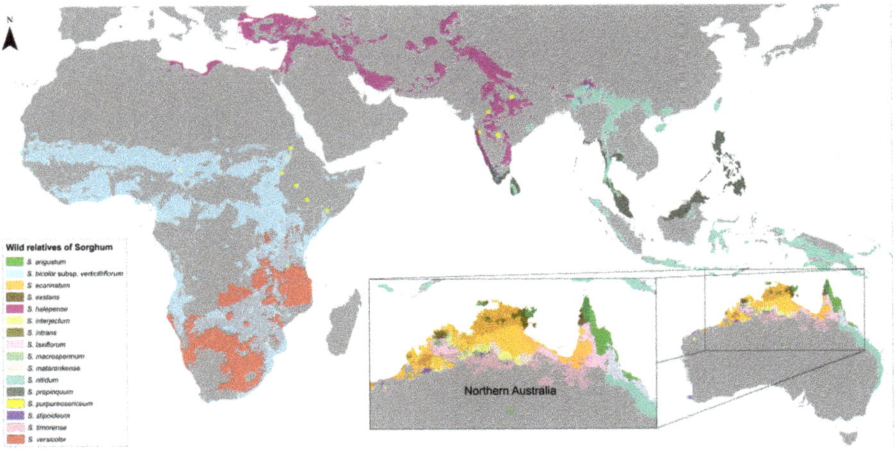

Figure 1 Distribution of wild sorghum species globally, showing Australia as the origin for a majority of tertiary gene pool species (Source: Castañeda-Álvarez et al., 2016).

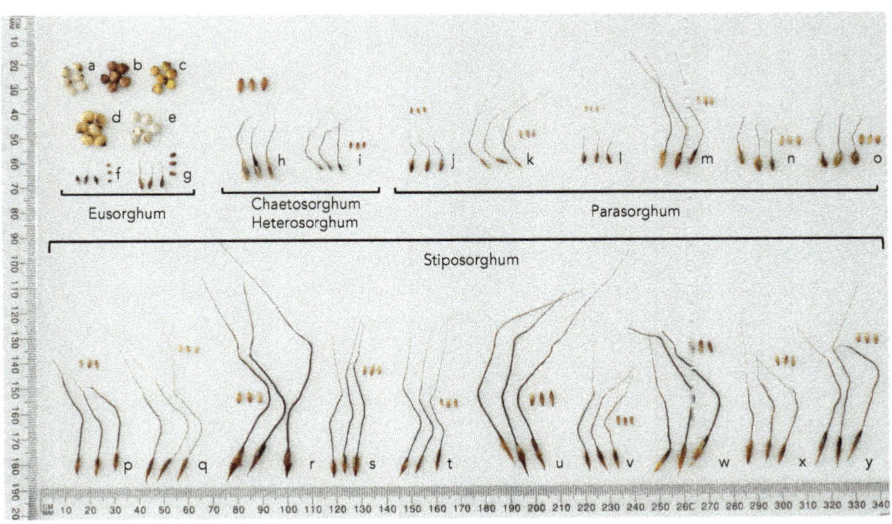

Figure 2 Morphological variation in *Sorghum* seed showing the wide range of diversity in the seed structures and caryopsis size. Letters on figure denote species, with *a–e*: *S. bicolor* AGG 322649, 322618, 322620, 322666 and 322611 caryopses, respectively. Then the seed and caryopsis of: *f*: *S. propinquum* 302546; *g*: *S. halepense* 300167; *h*: *S. macrospermum* 322277; *i*: *S. laxiflorum* 302503; *j*: *S. leiocladum* 300170; *k*: *S. matarankense* 302521; *l*: *S. nitidum* 302539; *m*: *S. timorense* 302660; *n*: *S. purpureosericeum* 321133; *o*: *S. versicolor* 321126; *p*: *S. amplum* 302623; *q*: *S. angustum* 302604; *r*: *S. brachypodum* 302480; *s*: *S. bulbosum* 302646; *t*: *S. ecarinatum*; *u*: *S. exstans* 302577; *v*: *S. interjectum* 302563; *w*: *S. intrans* 302390; *x*: *S. plumosum* 302489; *y*: *S. stipoideum* 302644 (Adapted from Dillon et al., 2007).

Table 1 Number of accessions for each recognised sorghum species held in the major *ex situ* plant genetic resources holdings globally: *AUS165* – AGG, Australia: Australian Grains Genebank; *BGR001* – IPGR, Bulgaria: Institute for Plant Genetic Resources 'K.Malkov'; *COL003* – CIAT, Colombia: Centro Internacional de Agricultura Tropical; *ETH013* – ILRI, Ethiopia: International Livestock Research Institute; *HUN003* – RCA, Hungary: Institute for Agrobotany; *IND002* – ICRISAT, India: International Crop Research Institute for the Semi-Arid Tropics; *ROM002* – ICCPT Fundul, Romania: Research Institute for Cereals and Technical Plants Fundulea; *ROM007* – BRGV Suceava, Romania: Suceava Genebank; *ROM023* – USAMVB Timis, Romania: University of Agricultural Sciences and Veterinary Medicine Timisoara; *UKR 005* – IK, Ukraine: Institute of Grain Growing; *UKR008* – UDS, Ukraine: Ustymivka Experimental Station of Plant Production; *USDA GRIN* – National Center for Genetic Resources Preservation

Accepted taxonomy	Taxonomic synonym	AUS165 AGG	BGR001 IPGR	COL003 CIAT	ETH013 ILRI	HUN003 RCA	IND002 ICRISAT	ROM002 ROM007 ROM023	UKR005 UKR008	USDA GRIN
Eusorghum										
Sorghum × almum		20	6	3	11	2	6	–	–	30
Sorghum bicolor		5 359	984	–	9 432	751	38 593	13	628	52 575
Sorghum bicolor subsp. *verticilliflorum*	*S. arundinaceum*	1	–	1	10	1	–	–	–	286
Sorghum bicolor nothosubsp. drummondii	*S. × drummondii*	5	35	–	5	41	174	49	99	113
Sorghum halepense		2	4	–	–	2	24	3	–	148
Sorghum propinquum		3	–	–	–	–	–	–	–	7
Chaetosorghum										
Sorghum macrospermum		3	–	–	–	–	1	–	–	–
Heterosorghum										
Sorghum laxiflorum		24	–	–	–	–	1	–	–	7

Parasorghum									
Sorghum grande	2	–	–	–	–	–	–	–	–
Sorghum leiocladum	27	–	–	–	–	–	–	–	3
Sorghum matarankense	9	–	–	–	–	2	–	–	–
Sorghum nitidum	10	–	1	–	–	–	–	–	12
Sorghum purpureosericeum	4	–	–	–	–	9	–	–	8
Sorghum timorense	43	–	–	–	–	3	–	–	3
Sorghum versicolor	7	–	3	–	–	3	–	–	10
Stiposorghum									
Sorghum amplum	2	–	–	–	–	–	–	–	2
Sorghum angustum	11	–	–	–	–	–	–	–	8
Sorghum brachypodum	3	–	–	–	–	–	–	–	2
Sorghum bulbosum	22	–	–	–	–	–	–	–	7
Sorghum ecarinatum	12	–	–	–	–	–	–	–	1
Sorghum exstans	4	–	–	–	–	–	–	–	3
Sorghum interjectum	20	–	–	–	–	–	–	–	2
Sorghum intrans	29	–	–	–	–	6	–	–	12
Sorghum plumosum	56	–	–	–	–	1	–	–	20
Sorghum stipoideum	42	–	–	–	–	5	–		9

Source: AGG, 2016; Genesys, 2016; USDA GRIN, 2016.

Franzmann, 2001). The native Australian sorghums exhibit a wide range of panicle and plant architecture (Bhattacharya et al., 2011), as well as highly varied seed morphologies (Fig. 2). Disease tolerances could be attributed to different morphological structures preventing infection or to underlying genetic traits.

Sorghum improvement programmes are dependent on access to genetic diversity and breeding tools, either traditional or modern techniques. Access to germplasm is generally through genebanks which conserve germplasm under long-term *ex situ* conditions. Genebanks aim to conserve the full diversity of crop species, including modern cultivated varieties, traditional landraces (village varieties), crop wild relatives (CWRs), specialised genetic stocks, and more recently, mutant populations. There are more than 1750 agricultural genebanks globally, with 11 of the world's most valuable collections held within the Consultative Group for International Agricultural Research genebanks (FAO, 2010). Many of the global genebanks operate within a global network under international conventions and treaties such as the Convention on Biological Diversity, and the International Treaty for Plant Genetic Resources for Food and Agriculture (the Treaty) that aim to promote the fair and equitable sharing of genetic resources and the benefits from their use. To date, more than 7.4 million accessions of seed are conserved in genebanks worldwide.

Sorghum traditional varieties, or landraces, and wild species have been collected from Africa, Asia, the Americas and India since the early 1900s. It is reported that initial efforts to build a global collection of sorghum started with the transfer of the Rockefeller Foundation collection from Mexico to India in 1957 (Rosenow and Dahlberg, 2000). While a centralised collection of Ethiopian sorghums began in the Jimma Agricultural Technical School in the late 1950s (Rosenow and Dahlberg, 2000), a major targeted collection of sorghum germplasm was made in 1973 by Gebrekidan and Ejeta from north eastern Ethiopia (Shewayrga et al., 2008). Since then, there have been targeted missions to capture *Sorghum bicolor* landraces and traditional village cultivars, and more recently, the wild species of sorghum. These valuable collections are now conserved in many genebanks globally, with collections ranging in size from a few accessions, or distinct samples, to more than 50 000 accessions (Table 1). The largest collections are comprised primarily of cultivated sorghum (*Sorghum bicolor*), both modern and traditional landrace varieties, and their close relatives from section *Eusorghum* primary and secondary gene pool species. These species have only minor genetic incompatibilities with *Sorghum bicolor*, with some of these species used extensively in crossing programmes, in genetic studies and mapping.

The more distantly related crop wild relative species from sections *Chaetosorghum*, *Heterosorghum*, *Parasorghum* and *Stiposorghum* form the tertiary gene pool, and are significantly underrepresented in *ex situ* genebank collections. Table 1 summarises the larger genebank collections of the recognised taxonomic species of cultivated and wild sorghums globally.

The tertiary gene pool *Sorghum* species are conserved primarily in the Australian Grains Genebank that has been undertaking targeted wild sorghum collecting activities in northern Australia since the 1980s. Some of these tertiary gene pool species are poorly represented as they are rare or located in extremely isolated areas, while other species are highly represented as they are widespread and easily accessible. Recent research by Dempewolf et al. (2014) and Castañeda-Álvarez et al. (2016) highlights the gaps in global genebank collections of CWR of agricultural crop species, which emphasises the underrepresentation of CWRs in *ex situ* genebank collections. Future wild sorghum collection activities will be targeted towards poorly represented species, and also aimed at representing the geographic spread of these species in line with the recommendations of recent gap analyses.

2 Early evidence of sorghum use

2.1 Archaeological evidence for wild sorghum use

The earliest documented evidence of the human use of sorghum is from Africa, where wild races of *Sorghum bicolor* ssp. *verticilliflorum* are widely distributed in the savannah regions. Research has identified starch granules, found on the edges of flaked stone tools from rock shelters in Mozambique, dating to the Middle Stone Age. These finds suggest wild sorghum grains were collected, processed and presumably consumed by hunter-gatherers in eastern Africa since at least 100 000 years before present (BP) (Mercader, 2009).

Aside from this very early finding, the majority of evidence for the pre-domestication exploitation of sorghum dates from the early Holocene and comes from a number of sites across the Sahara region. Together, these data sets not only demonstrate that sorghum was widely available at this time, but also attest to its general importance as a wild food source throughout the eastern savannah zone. Important findings include the remains of charred grains and spikelets from several domestic huts and pits at Nabta Playa in the Western Desert of South Egypt, as well as grains from Farafra and Abu Ballas, also in Egypt (Barakat and Fahmy, 1999), and Kadero, e-Damer, Abu Darbein and Aneibis in Sudan (Krzyżaniak, 1991; Magid, 1989; Stemler, 1990). The Nabta Playa remains, which provide some of the clearest examples of wild-type morphology, were identified as *Sorghum bicolor* ssp. *arundinaceum* (Desv.) based on spikelet form (broad, smooth articulation scars; none attached to a pedicel), and the small size and shape of the grains (dorsally flattened; narrow obovate to oval lateral outline; thick coriaceous glumes longer than the grain) (Dahlberg et al., 1995; Dahlberg and Wasylikowa, 1996b). Aside from carbonised remains, other archaeological evidence of morphologically wild sorghum use comes from impressions of temper in Neolithic ceramics from Central Sudan (Magid, 1989; Stemler, 1990).

Novel techniques, such as the analysis of microscopic silica phytoliths and starch granules in dental calculus, are also adding to the picture, with a recent study of remains from Neolithic cemeteries in North and Central Sudan demonstrating millet consumption at 7500–6500 BP (Madella et al., 2014). Many of these sites also record evidence of the use of other wild savannah grasses, such as *Panicum*, *Paspalum*, *Setaria* and *Pennisetum*.

Outside Africa, wild sorghum species were utilised by indigenous hunter–gatherer populations in Australia. The grains of *Sorghum intrans* (or *Sorghum stipoideum*), *Sorghum plumosum* and *Sorghum macrospermum* were an important source of starchy food for the Tagoman Aboriginal tribe in the Katherine district in the Northern territory (Arndt, 1961). Of these, *Sorghum macrospermum*, although limited in distribution, was likely to have been the most important as it produces abundant caryopses of a size similar to that of a small grain of polished rice, and is the only Australian species to have nitrogen and phosphorus concentrations in the grains similar to that of *Sorghum bicolor* (Cook and Andrew, 1991; Lazarides et al., 1991). *Sorghum leiocladum* was used by indigenous hunter–gatherer populations in Australia to prepare flours for baking bread (Flood, 1980), but the antiquity of such practices is yet to be demonstrated archaeologically. Starch granules from grass seeds have been recovered from grindstones dating from both Pleistocene and Holocene contexts (Balme et al., 2001; Fullagar et al., 2008, 2015), but none of these archaeobotanical remains have been identified as sorghum thus far.

2.2 Detecting domestication traits in the archaeological record

Sorghum was domesticated in sub-Saharan Africa, most likely in the grasslands of the eastern Sudanic region (Sudan, Chad or Ethiopia) where its wild relatives are common and diverse. Recent genetic studies indicate the different domesticated races may have undergone multiple independent domestications (Lin et al., 2012). This complication goes some way towards explaining the great challenge facing archaeobotanists as they have attempted to unravel the story of sorghum use and domestication across time and space. Very little is known of the early domestication history of sorghum in Africa, and archaeological studies of sorghum's transition from wild to domestic forms have been thwarted by poor preservation obscuring defining morphological characteristics (Deakin et al., 1998b). For example, sorghum found in archaeological sites is often charred from proximity to a heat source and can be difficult to classify according to race or level of domestication due to distortion, causing a 'popped' appearance (Dahlberg and Wasylikowa, 1996).

Biehl et al. (1999) recognised the need to find alternatives to pure morphological identification of archaeological sorghum using gas chromatography and mass spectrometry to identify charred sorghum grains from Nabta Playa, southern Egypt, but the results were inconsistent. Other studies have trialled infrared spectroscopic analysis (Dahlberg et al., 1995), ancient DNA (Rowley-Conwy et al., 1999) and image analysis (Dahlberg and Wasylikowa, 1996) with mixed success. Inadequate archaeobotanical sampling of sites in the Sudanic savannah belt has also led to a lack of critical information for determining where and when sorghum was taken into cultivation (Zohary et al., 2012). In fact, among African cereals, sorghum has a particularly poor archaeological assemblage and data set for tracking morphological changes during domestication (Fuller and Hildebrand, 2013).

These problems notwithstanding, the main physical change involved in the domestication of sorghum was the development of non-shattering seed panicles, meaning that instead of shedding naturally when mature, the grains are retained in the panicle, only to be removed by threshing after harvesting. This characteristic can be identified morphologically as the loss of abscission at the attachment between the rachilla and spikelet base – in other words, a tough rachis develops. Other morphological identifiers of domesticates that can be observed in the archaeological record include a decrease in the rachis internode length, an increase in the number of branches within the inflorescence and an increase in seed size, making the panicles more compact and higher yielding than their wild types (Harlan et al., 1973; Dillon et al., 2007). Conversely, wild sorghum seeds are smaller than any known domesticate and have smooth articulation scars at the base of the spikelets (Wasylikowa et al., 1995).

2.3 Archaeological evidence for domestication in Africa

As noted above, archaeologists attempting to unravel the domestication history of sorghum have faced a number of challenges, not the least of which is the lack of archaeobotanical evidence currently available. Fuller (2015) highlights 'major gaps in the evidence' for the sorghum story in Africa, citing archaeobotanical verification for early use of wild sorghums (early to mid-Holocene), then a dearth of finds spanning the gap between this time period and the appearance of advanced domestic taxa in the Nubian Nile (Meroitic period). Pelling (2005) documents the presence of a highly advanced free-threshing sorghum domesticate at the Fezzan in southwest Libya tied to the Garamantian civilisation (c. 900 BC–AD 500). As Fuller (2015) points out, the presence of such an advanced domesticate implies a long history of domestication for which there is no current archaeobotanical evidence. The presence of domesticated sorghums alongside other African crops (*Pennisetum glaucum*, pearl millet; *Eleusine coracana*, finger millet; *Lablab purpureus*, hyacinth bean; and *Vigna unguiculata*, cowpea) in India by 2000–1700 BC (Fuller, 2003) heightens the need for more archaeobotanical investigation in the savannah region of Africa between Lake Chad and Ethiopia.

Researchers at different stages over the last 50 years have constructed sorghum domestication and dispersal hypotheses using the best information at hand. Each has been set aside in turn as new evidence has come to light. One hypothesis based on non-archaeological evidence such as linguistic relationships and cultural group movements argued for early sorghum domestication between 6000 and 2000 BC (Doggett, 1988; Harlan, 1976), while an 'Indian Ocean' hypothesis suggested wild sorghum was exported, domesticated in India and then reintroduced into Africa in domestic form at a later date (Haaland, 1999). Other researchers used reliable archaeological evidence, which dated domestic sorghum to the BC/AD transition, to postulate a relatively late entrance of domesticated sorghums into African agriculture (Deakin et al., 1998a; Magid, 1989; Wigboldus, 1990). The latter theory has recently been undermined, however, by new archaeobotanical evidence for domesticated sorghums present in north Sudan by at least 1500 BC (Beldados and Costantini, 2011; Fuller, 2014).

The oldest domesticated sorghums, previously known from eastern North Africa, come from Qasr Ibrim, Egyptian Nubia (AD 100) (Rowley-Conwy, 1991) and Jebel et Tomat on the White Nile, Sudan (AD 245) (Clark and Stemler, 1975). The latter site contained spikelets attached to pedicles and panicle branchlets with grains still enclosed by the glumes, clearly identifiable as *Sorghum bicolor* ssp. *bicolor* (Clark and Stemler, 1975). Wild sorghum spikelet bases have also been studied in charred material at Nabta Playa and dated to c. 8000–7000 BC (Dahlberg et al., 1995; Wasylikowa and Dahlberg, 1999; Wasylikowa et al., 1995; Wendorf et al., 1992). Domestic sorghum is present in north Sudan (Nubia) by about 2500 BP (Fuller, 2004b), but southern Sudan has no published archaeobotanical remains (Hildebrand et al., 2010). The Nubian evidence, which derives from the site of Kawa, has been directly AMS radiocarbon dated to 2450 ± 40 BP (Beta-194234, 780–400 cal. BC), placing it in the Napatan period; however, the ^{13}C fractionation on this date is rather high and more typically associated with C3 grasses, so this sorghum identification needs to be regarded with some caution.

Directly dated primitive *bicolor* sorghums from both Umm Muri (Fuller, 2004a: 2180 ± 40 BP [Beta-194236], 350–150 cal. BC) and Qasr Ibrim (Clapham and Rowley-Conwy, 2007: various dates from ca. 100 BC onwards) have been useful temporal indicators of domestication. A number of other finds from Sudan not only indicate that primitive bicolor domesticates were widespread, but also suggest that replacement by the advanced, free-threshing durra race was underway by this time, and indeed this race may already have been present at Napatan Kawa (Fuller, 2004b). These finds include archaeobotanical remains from the Meroe civilisation dating back to the last centuries BC that indicate the cultivation of sorghum (Fuller, 2014, 2015), remains from Jebel Tomat (associated with a radiocarbon date of AD 100–400; Clark and Stemler, 1975) and the recently identified and directly dated sorghum from Abu Geili (1790 ± 40 BP [Beta-194245], cal. AD 100–350; Fuller, 2014). Evidence from Dangeil indicates that sorghum was used there as a substitute for filling traditional bread moulds, presumably designed for emmer/barley bread, as worship offerings (Anderson et al., 2007).

By far the most comprehensive record to date of sorghum use in Africa comes from Qasr Ibrim, Egyptian Nubia (near the Egyptian-Sudanese border) (Rowley-Conwy, 1991; Rowley-Conwy et al., 1999). The site consists of a rich organic assemblage, preserved by desiccation, and represents a continuous morphological series of plant remains spanning nearly 3000 years. As part of this assemblage, four main stages in the domestication process of sorghum have been identified (de Wet and Harlan, 1971; Deakin et al., 1998b):

1. 1000 BC–AD 100: From first settlement at c. 1000 BC until AD 100, only local wild forms such as *Sorghum bicolor* (L.) Moench subsp. *verticilliflorum* race arundinaceum (Desf.) have been reported to be present.
2. AD 100–1200: Domesticated examples (morphologically consistent with race *bicolor* cultivars) are reported to be present. These finds consist of the most primitive cultivated race (*Sorghum bicolor* subsp. *bicolor* race bicolor) and decrease in frequency after AD 1500.
3. AD 42–640: A bouquet of ripe panicles (radiocarbon dated to the fifth to seventh centuries AD) that are morphologically intermediate between *bicolor* and *durra* cultivars appeared.
4. AD 800–1200: By AD 1200, the more advanced free-threshing cultivated form (*Sorghum bicolor* subsp. *bicolor race* durra) appeared.

In a further study, Deakin et al. (1998a) attempted to determine the genetic relationship between these archaeobotanical remains by extracting DNA from seeds dating between 800 BC and AD 1800.

2.4 Archaeological evidence for spread throughout Africa

The spread of domesticated sorghum, along with other African millets and pulses, through sub-Saharan Africa has long links to the dispersal of iron-working, pottery-using cultures by Bantu language-speaking groups from around the late first millennium BC to first millennium CE (e.g. Oliver, 1966; Huffman, 1970, 2006, 2007; Phillipson, 2005; see de Maret (2013) for a recent review). While aspects of this model still have some currency, it is now widely seen as overly simplistic and masking the complexity underlying the dispersal process. For example, increasing recognition has been given to interactions between migrating agriculturalists and incumbent hunter–gatherers and pastoralists in these regions, as well as the more variable nature by which crops spread and were taken up in different regions (e.g. Crowther et al., in press; Lane, 2004, 2015; Kusimba and Kusimba, 2005; Wright, 2005; Shipton et al., 2013; Vansina, 1995). Linguistic data now suggest that Bantu agriculturalists obtained sorghum and possibly iron-working from Nilo-Saharan-speaking groups in the northern Great Lakes/southern Sudan region before dispersing southwards and eastwards towards the Indian Ocean coast and southern Africa (e.g. Ehret, 1998; Westengen et al. (2014)). Archaeological evidence of carbonised remains demonstrates that sorghum reached the interior of eastern Africa by at least AD 400 (Giblin and Fuller, 2011), the east coast and offshore islands of Kenya and Tanzania and the Comoros by AD 700 (Crowther et al., 2016; Walshaw, 2015), and southern Africa by about the same time (Sinclair et al., 1993; Sinclair, 1982; Pwiti, 1996; Denbow and Wilmsen, 1986) or possibly earlier (Robertson, 2000). Despite sorghum's importance in the more arid parts of Madagascar, at least in recent history, no archaeological finds of this crop have been recovered anywhere from the remote African island to date (Boivin et al., 2013).

2.5 Archaeological evidence from India, Asia and Europe

Somewhat paradoxically, the earliest evidence for cultivated and domesticated sorghum occurs not in Africa, but South Asia, where as mentioned above, sorghum was introduced along with four other major African crops (pearl millet, finger millet, hyacinth bean and cowpea) by the early second millennium BC (Fuller, 2003; Boivin and Fuller, 2009). How these crops reached India remains a puzzle, though Boivin and Fuller (2009) propose either Omani or Indian (Gujarati Harappan) seafaring traders as the most likely agents. Indeed, the fact that the earliest finds in India are found in the eastern Harappan periphery in Saurashtra and the Ganges-Yamuna regions (Fuller and Boivin, 2009), combined with the lack of archaeobotanical evidence for any African crops in the Arabian peninsula in the relevant time frame (noting that claims for domesticated sorghum in Yemen and Oman by the third millennium BC (Potts, 1993; Cleuziou and Costantini, 1980) are now largely dismissed on grounds of misidentification; e.g. Rowley-Conwy et al., 1999; Tengberg, 2003; Willcox, 1992; de Moulins et al., 2003; Fuller, 2002) seem to point to Indian merchants as the more likely candidates, though the question still remains open.

In southwest Asia, Europe and the Mediterranean, sorghum arrived relatively late, most likely in the medieval period. Its geographical spread was generally confined to the warmer, more southerly latitudes, so for example, in Europe, it was mainly grown in this period only in southern Spain. Central and East Asia appear to be a relatively late Old-World recipients of the crop. Although there are unsubstantiated claims of earlier archaeological finds (Kimber, 2000), textual sources suggest that sorghum only reached southwest China in early first millennium AD (Hagerty, 1941), possibly via trade through Central Asia (Sogdhiana) (Stevens et al., 2016). Domesticated sorghum completed its global spread in the last 300–400 years, with its introduction to North America in the 1700s through the slave trade from West Africa and to Australia following the widespread establishment of European colonies in the late 1800s (Dicko et al., 2006).

2.6 From savannah grass to grain crop

Cultivated sorghum (*Sorghum bicolor* subsp. *bicolor*) is a diploid ($2n = 20$), and is believed to have been derived from domestication of wild progenitor, *Sorghum bicolor* subsp. *verticilliflorum*, which grows widely across sub-Saharan Africa from Ethiopia west to Nigeria and as far south as Madagascar (Wiersema and Dahlberg, 2007). In the modern era, *Sorghum verticilliflorum* is reportedly still collected and eaten as a wild grain when there is local food scarcity (de Wet, 1978). The considerable conjecture regarding the domestication of sorghum has been discussed; however, it is generally accepted that there are five races of sorghum, namely Bicolor, Guinea, Caudatum, Kafir and Durra, based predominantly on floral morphology and centre of diversity (Harlan and de Wet, 1972; Dahlberg, 2000).

Recent molecular studies have largely confirmed the distinctness of accepted racial structures (Brown et al., 2011; Mace et al., 2013; Morris et al., 2013), with some key differences, notably among the Guinea sorghums. The open-headed Bicolor sorghums are the most divergent from modern sorghums, and it has been postulated that these represent the oldest domesticated sorghums, from which were derived the other races (de Wet, 1978). For example, selection for adaptation to the wet tropics led to the Guineas, and there is some evidence that the Durras were derived outside Africa in the Sind–Punjab after the introduction of Bicolor types over 3000 years ago.

2.7 Domestication to modern breeding

Modern hybrid sorghum cultivars have only been in cultivation for 60 years, with the first commercial hybrids made available by the DeKalb seed company in the United States in 1956 (Berenji et al., 2011). Given sorghum was domesticated approximately 8000 years ago (Wendorf et al., 1992) in the Horn of Africa (Ethiopia, Sudan and South Sudan), a significant proportion of genetic improvement and selection was performed before the modern era.

It is clear there were separate species of *Sorghum* outside Africa, most notably the Asian sorghum *Sorghum propinquum*, the Eurasian weed species *Sorghum halepense* (Johnson grass) and the 17 wild species native to Australia. Both *Sorghum propinquum* and *Sorghum halepense* are interfertile with *Sorghum bicolor*, although there are some barriers to fertility with *Sorghum propinquum*, and the tetraploid *Sorghum halepense* forms sterile triploids. Whole-genome sequencing (WGS) of *Sorghum propinquum* and genome assembly using *Sorghum bicolor* as a scaffold has demonstrated that 22% of the genome did not map to cultivated sorghum (Mace et al., 2013), and there was no evidence of interspecific introgression.

Having gone through the bottleneck of domestication (Doebley, 1989), the diversity in modern cultivated sorghum is expected to be significantly lower than that of the wild progenitor species. Numerous studies have shown this to be the case, with domesticated landraces demonstrating lower diversity at many loci and even allelic fixation in some cases. This reduction in allelic diversity, the 'domestication syndrome' can be the result of neutral selection, whereby allelic diversity is reduced effectively by chance during the domestication bottleneck. However, in other cases the reduction in allelic diversity may be the direct result of selection, either by natural or human (artificial) selection (Doebley et al., 2006). Both natural and artificial selection may be in the form of purifying selection (where one allele is favoured over all others) or balancing selection (where two or three alleles are favoured, possibly by differential adaptation). Artificial selection is commonly in performance traits, such as yield and yield components (seed size, seed number), nutrient use efficiency (Massel et al., 2016) and resistance to pests and diseases. However, human preferences in crop end use for feed and fibre qualities can also lead to a directed reduction in diversity, often balancing selection when more than one end use is favoured (Campbell et al., 2016).

2.8 Beyond landraces

Sorghum was introduced into the Caribbean with the slave trade and cultivation in the United States was recorded in the 1700s by Benjamin Franklin. It was primarily used as a broomcorn and minor animal feed. As a grain crop, sorghum was grown in Texas in the late 1700s as an animal feed, where it was known as milo. Numerous planned introductions were made from Africa in the latter part of the 1800s, and plant breeding began in the twentieth century to produce the 'combine' sorghums. Combine sorghums were of shorter stature and could be harvested in a combine harvester. The first releases were in Oklahoma, with cultivar Beaver in 1928, followed by the famous variety, Wheatland, released in 1931. This proved immensely popular and was soon grown in Australia and Latin America. A disease-resistant selection from Wheatland, known as Martin's Milo, was released in 1937, and soon became the number one variety in the United States until the first F1 hybrids were released. DeKalb Seeds released the first F1 hybrid in 1956, with high-yielding hybrids soon replacing the inbred varieties throughout the developed world. Within seven years of their introduction, hybrids more than doubled sorghum yield in the United States (Maunder, 2000).

3 The sorghum conversion (SC) programme

There was a need for more genetic diversity in modern hybrid breeding programmes worldwide, yet much of the diversity was in the form of tropically adapted, tall and photoperiod-sensitive (late flowering) landraces, which were difficult to breed with. The sorghum conversion (SC) programme was instigated in Texas in the 1960s to convert tropical landraces to temperately adapted, shorter stature lines. The original programme used the line BTx406 to 'convert' landraces with the backcrossing of some of the four dwarfing (dw) and four maturity (ma) alleles into the line, converting a landrace into an SC line which maintained as many of the original landrace qualities as possible. The outcome was to produce plants of a manageable height with photoperiod insensitivity, such that they would flower in the field in more temperate or subtropical environments such as the United States, Australia and Argentina. Hundreds of landraces were 'converted' through this programme, and many have been used widely in breeding programmes to introgress traits such as drought tolerance, acid soil tolerance and various pest and disease resistances (Rosenow, 1992).

4 Introgression of distant wild crop relatives

While the use of undomesticated *Sorghum* species for both food and crop improvement has been extensive, analyses of the genetic diversity in sorghum landrace collections suggest a high degree of polymorphism still remains underutilised (Abu Assar et al., 2005; Deu et al., 2006; Kayode et al., 2006). Additionally, modern breeding programmes have not focused on the use of the secondary and tertiary gene pool, beyond the Eusorghums. This is largely due to the strong sterility barriers between the more distantly related species (Harlan and de Wet, 1971). Phylogenetic analysis would suggest *Sorghum macrospermum* and *Sorghum laxiflorum* were most closely related to cultivated sorghum varieties and therefore should have the greatest potential for successful crosses, though morphologically they were quite distinct (Dillon et al., 2007).

In the case of *Sorghum macrospermum*, *Sorghum angustum* and *Sorghum nitidum*, identification of the exact nature of the sterility barrier, combined with the discovery of an *S. bicolor* accession containing a recessive gene (inhibition of alien pollen ¼ iap) which overrode the pollen–pistil incompatibilities, and subsequent application of reproductive technologies such as embryo rescue and tissue culture were able to overcome these sterility barriers and resulted in the growth of verified hybrid seedlings (Hodnett et al., 2005;

Laurie and Bennett, 1989; Price et al., 2005a,b, 2006). The ability to generate these secondary gene pool hybrids provides the potential for harnessing the suite of novel pest and disease resistances and environmental adaptability that is abundant in the undomesticated species but has been underutilised to date.

5 Population genetics and whole genome sequencing

The availability of next-generation sequencing has led to great advances in understanding the nature of genetic diversity, and the evolution of crop species and their progenitors. The sorghum genome was sequenced and assembled by a consortium funded by the US Department of Energy (Paterson et al., 2009), and made available on the Phytozome website (https://phytozome.jgi.doe.gov/pz/portal.html#!info?alias=Org_Sbicolor, last accessed 10 September 2016) as an international resource. Geneticists have used this resource to enhance our understanding of the genetic architecture of key traits, key pathways, and the evolution and domestication of sorghum and other grass species. Amongst cereals, it has a relatively small genome of 750 million base pairs with significant homology to both rice and maize. Due to its broad range of biotic and abiotic tolerances, increasing knowledge of the gene sequence and function amongst the diversity in the sorghum gene pool will have potential benefits to breeding programmes of other cereals as well (Mace et al., 2013; Mullet et al., 2002).

A re-sequencing study of diverse sorghums has further supported the proposal (Deu et al., 2006) that the Guinea margaritiferums represent a secondary and later domestication event in West Africa (Mace et al., 2013). The Guinea margaritiferums are quite distinct from the Guinea sorghums, with the grains usually wholly covered by glumes, and are mostly boiled or steamed and eaten as rice (de Wet, 1978). They are grown mostly in the high-rainfall areas of West Africa, commonly on acidic soils. This group of sorghums includes a great genetic resource for further sorghum improvement. Within the wild, weedy, landrace and improved sorghums, Mace et al. (2013) identified 91 fixed genes which were polymorphic in a small sample of only two Guinea margaritiferums. Another study has shown that genes for aluminium (Al) tolerance are quite rare in sorghum, and are found predominantly in the Guinea and Guinea margaritiferums from West Africa (Caniato et al., 2011). Additionally, genotype-by-sequencing data (Morris et al., 2013) concluded that there was considerable geographic and edaphic adaptation among the worldwide collection of sorghums. Lasky et al. (2015) also concluded that a substantial proportion of genetic diversity was explained by environment, with one of the strongest effects being the selection against one of the major maturity genes (Ma1) in sorghums from temperate environments.

During domestication and subsequent improvement, a number of gene families and pathways reveal signatures of selection. One of the major classes of disease resistance genes, the nucleotide-binding site genes, has been shown to be under strong selection pressure with both purifying and balancing selection in action (Mace et al., 2014). They also demonstrated there were numerous deleterious presence/absence variants which contribute to rapid evolution in NBS-encoding genes. Over half of the 114 genes involved in seed storage starch biosynthesis displayed signatures of selection (Campbell et al., 2016), which is not surprising given these are major determinants of seed size, yield and end use quality. The key enzyme in converting glucose to the form required for starch polymerisation, AGPglucose, consists of two sub-units, encoded by shrunken2 (sh2) and brittle2 (bt2), both of which were under selection, with sh2 under strong purifying selection. Interestingly, quite a number of genes were under balancing selection, possibly reflecting divergent end uses, and these included pullulanase, a starch debranching enzyme shown to alter starch digestibility (Gilding et al., 2013) and starch phosphorylase. Another study has also identified over 240 genes involved in nitrogen uptake and assimilation, many of which are under strong purifying or balancing selection (Massel et al., 2016).

The use of genotyping by sequencing information has also revealed a number of key findings regarding recent sorghum breeding and improvement. Most interesting, a survey of 580 SC lines from the SC programme revealed high levels of recurrent parent (BTx406) in regions associated with Ma and Dw genes. This was marked at distal ends of chromosomes 7 (Dw3) and 9 (Dw3), yet was overwhelmingly the case on chromosome 6. Chromosome 6 contains Dw2 and Ma1, and in 80% of the SC lines there was very little chromosome 6 from the original landrace, suggesting much of the landrace genetic diversity contained on this chromosome had not been captured in modern breeding programmes (Thurber et al., 2013).

6 Human use of sorghum

Sorghum is cultivated for many different end uses (Table 2). It has a long history of exploitation by humans, primarily as a food source but archaeobotanical research indicates it also is likely to have had a range of other uses including in making building and craft materials, traditional African medicines, dye production and alcoholic beverages (Mercader, 2009). Sorghum is grown for food uses, including porridges (West and South Africa, parts of East Asia), fermented flat breads (East Africa), unleavened flat breads (India) and is also used for a variety of fermented beverages, including beer and spirits. 'If we had a way to accurately tally the world's vast and complex drinking practices, sorghum is clearly one of the world's most imbibed plants' (Amy Stewart, 2013, *The Drunken Botanist*, Algonquin Books). Importantly, once the grain is harvested, the stover of the plant is then also used as animal feed, building materials and firewood.

In the developed world, the primary use of the grain is as feed in intensive animal industries. Other variants are grown specifically for biomass (animal feed and biofuels), sugar/syrup production (sweet sorghums), and there are also sorghums grown specifically for their open panicle, described as broomcorns (Berenji et al., 2011). These disparate end uses require plants with different architecture and grain qualities, and have a major effect on selection of landraces and modern cultivars.

6.1 Grain for human consumption

Sorghum grain comes in a spectrum of colours from white to black. Unblemished white and other light colours are considered more visually attractive and have less of the off-flavours which restrict its consumption by humans. However, light grains can take on a discoloured appearance after poor weather conditions prior to harvest (Smith and Frederiksen, 2000). Darker coloured grains, such as red, have become more commonly produced. Fortuitously, the coloured sorghum grains have higher levels of healthy compounds, including a broad

Table 2 Human utilization of sorghums (Anglani, 1998; Berenji et al., 2011; Cook and Andrew, 1991; Dahlberg et al., 2011; Regassa and Wortmann, 2014; Taylor et al., 2006; Smith and Frederiksen, 2000)

	Traditional and novel human foods	Beverages	Animal feed	Non-food uses
Grain	Composite breads Leavened breads (gluten free) Leavened/fermented flat breads (e.g. injera, kisra, dosa, dosai) Unleavened flatbreads (e.g. roti, chapatti, tortilla) Snacks; tortilla chips, muffins, cakes, biscuits, 'popcorn' Porridge Par-boiled or pearled as a rice substitute Noodles and pasta Weaning formulations Couscous, semolina	Malted beers (clear and cloudy lager/pale ale/stout) Malted non-alcoholic beverages Distilled (e.g.) Maotai/kaoliang Chinese baijiu	Primarily used for ruminants, but also monogastrics where alternatives are limited Raw grain Milled grain Steam flaked Alkali/enzyme treated Maltings (post-fermentation) Native pastures	Bioethanol Biodiesel Edible bioplastic films and coatings Sorghum wax Nutraceuticals Industrial starch applications
Stalk/stover/leaf	Syrups ('sorghum molasses') Chewing cane	Chinese baijiu	Silage Hay Standing fodder Green chop Bagasse (post extraction) Native pastures	Biomass for bioenergy/biogas/biofuel Brooms (includes panicle) Traditional woven goods Traditional housing and constructions Straw (fuel pellets, particle boards, paper) Biofertiliser

family of phytochemicals, which provide significant benefits in terms of human health. In broad terms, the main components of sorghum grain considered important for human food are starch, proteins, fibre, phytochemicals and anti-nutritional factors.

6.1.1 Starch

In all cereals, including sorghum, starch is the most abundant component and the main food source for the growing germ during germination; however, humans have taken advantage of this energy source. While starch is comprised of only two molecules, amylopectin and amylose, both are made up solely of glucose units. Starch is approximately 75% amylopectin and 25% amylose, and as such, amylopectin is the most abundant component in the grain accounting for around 40% of total grain composition.

While starch is made up solely of glucose units, the synthesis of amylose and amylopectin is complex with up to 114 genes involved. Through the advancement of genomics and transcriptomics, a number of studies have highlighted genetic variation due to the starch synthase families of genes (SSI, IIa, GBSSI & II) as well as the starch branching enzyme (SBE I, II & IIb) and debranching genes (DBE) (Boyer, 1984; Hill et al., 2012; Mutisya et al., 2003; Pedersen et al., 2005; Shapter et al., 2009). Several mutations in these genes, involved in chain length elongation, have been identified in landraces, wild types and domesticated sorghums. Several mutations in SBE have shown positive effects on the digestibility of starch (Gilding et al., 2013).

In both cultivated and wild-type sorghums the endosperm of the grain is filled with large, rounded to polygonal, densely packed starch granules which reduce in size, approaching the pericarp (Fig. 3). It is generally considered that endosperm morphology transitions between two regions in sorghums; the central/floury endosperm, made of more rounded, large, polygonal starch granules with little matrix, some air spaces and few if any protein bodies, to the outer vitreous or corneous endosperm which has smaller more rigidly polygonal granules, denser protein matrix and high density of protein bodies (Fig. 3). However, this variation in the presence/quantity of the inter-granular matrix, the size and shape of the starch granules and the density of protein bodies throughout the endosperm can differ greatly between sorghum varieties and species (Shapter et al., 2008; Smith and Frederiksen, 2000).

Additional notable differences have also been described in some of the wild-type Australian sorghums: the presence of small rice-like starch granules in the endosperm, a lack of differentiation into the two regions, the presence of intra-granular channels and the presence of a novel proteinaceous sub-aleurone layer in some species (Shapter et al., 2008, 2009). Since the cooking and nutritional properties of sorghum are impacted by the proportion of floury versus corneous region and by the amount and type of protein present in the endosperm (Duodu et al., 2003; Serna-Saldivar and Rooney, 1995; Lindeboom et al., 2004; Tesso et al., 2006) and the presence of channels in the starch granules (Fannon et al., 2003, 2004; Benmoussa et al., 2006), it is likely that the differentiation in the endosperm morphology will affect the digestibility of starch and protein, milling properties, texture and hence the end uses (Anglani, 1998; Shewry and Halford, 2002; Wong et al., 2009). Regardless of varietal variations, the protein/starch interactions ubiquitous amongst sorghums, result in lower digestibility than other cereals, once cooked (Hamaker et al., 1987; Duodu et al., 2002, 2003). The protein matrix appears to act as a barrier to starch gelatinisation and digestibility, and also to hydrolysis by amylolytic enzymes to varying degrees, proportionate to its composition (Wong et al., 2009).

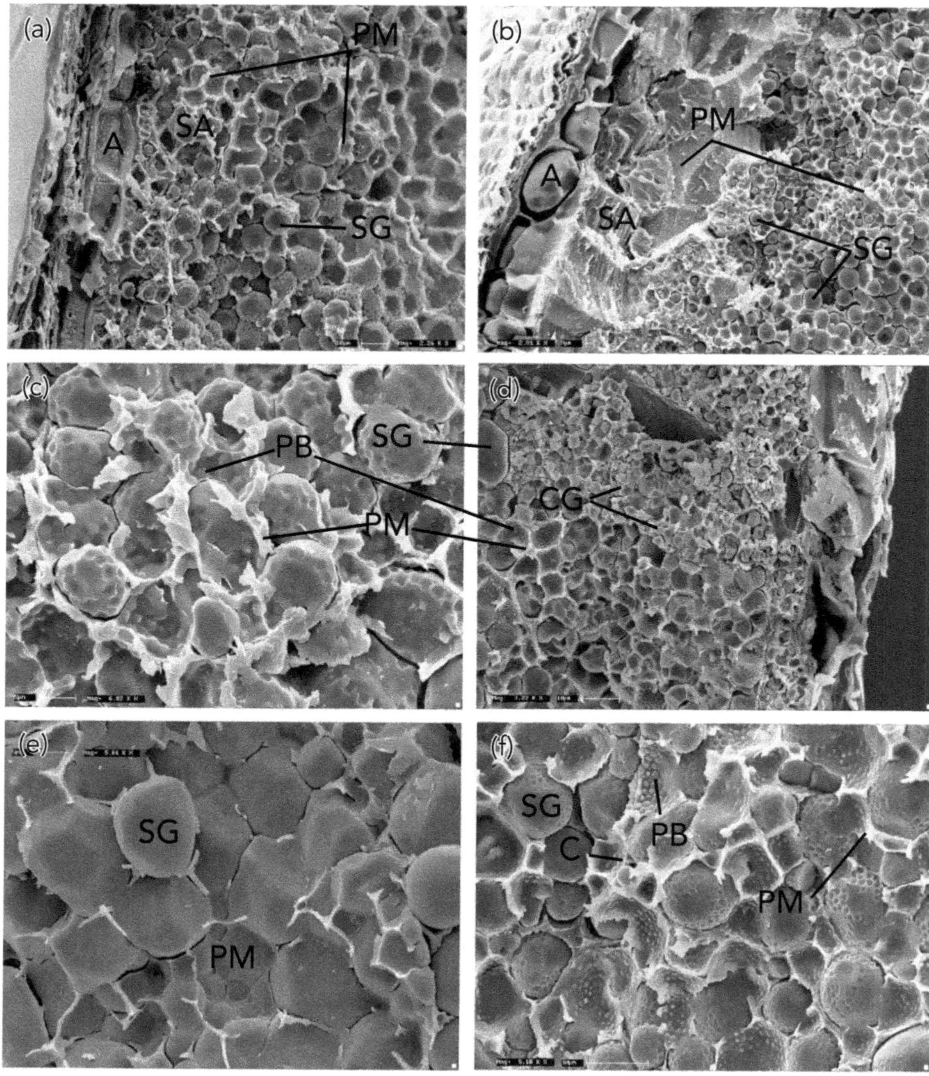

Figure 3 Endosperm variation amongst sorghum species. Plates (a) *Sorghum bicolor* and (b) *Sorghum stipoideum* showing the pericarp through to the vitreous endosperm (~2000× magnification). Note the differences in sub-aluerone morphology and starch granule shape. Plates (c) *Sorghum bicolor* (~6000×) and (d) *Sorghum leiocladum* (~3000×) show some of the difference observed in the vitreous endosperm. Of particular note is the presence of rice-like compound granules in the wild sorghum. Plates (e) *Sorghum bicolor* and (f) *Sorghum angustum* (~5000×) demonstrate the variability of the central floury endosperm morphology, where many of the natives retained high proportions of protein matrix and protein bodies. Plate (c) shows the novel sub-aleurone layer observed in some wild Australian sorghums. A – aleurone, SA – Sub-aleurone, SG – starch granules, PB – protein body, PM – protein matrix, CG – compound granules, C – channels/pores in the starch granules.

6.1.2 Protein

Protein content of the sorghum grain varies considerably between cultivars, races and species, but within each group the content is dependent on the abiotic and biotic conditions during grain fill. It is distributed between the pericarp (~3%), endosperm (~80%) and germ (~16%), and stored predominantly in the protein bodies and matrix (Smith and Frederiksen, 2000). Compared to other cereals, cooked sorghum has a reduced protein digestibility (Duodu et al., 2003). *In vitro* research has demonstrated up to 15% reduction in digestible protein in sorghum after cooking (Hamaker et al., 1987). Despite the issues with protein indigestibility, sorghum is considered a good source of dietary amino acids, with the exception of lysine, and to a lesser extent threonine and tryptophan. In developing countries, where cereals are a major source of dietary protein, this can be rectified with the inclusion of a legume component in the diet and/or the use of high-lysine cultivars, to overcome this deficit (Smith and Frederiksen, 2000).

Sorghum proteins are classified based on their solubility: water soluble (albumins), salt soluble (globulins), alcohol soluble (prolamins: kafirins in sorghum) and alcohol reduced (cross-linked glutelins and kafirins), with the latter two groups the most abundant and containing the storage proteins. In recent years, however, the literature reflects a simpler bi-grouping of kafirins versus non-kafirins (Wong et al., 2009). Protein distribution varies within the grain, the germ storing albumins and globulins, while the endosperm is predominantly kafirins and glutelins (Smith and Frederiksen, 2000). The prolamin, kafirin, is the most important of the storage proteins, due to its implications in reducing the digestibility of sorghum.

The kafirins form a spherical structure or protein body which is imbedded in the starch granule surface but also cross links to the glutenin matrix. While totally different structurally, kafirins possibly play a similar role to puroindolines in wheat or hordoindolines in barley, in connecting the storage protein matrix to the starch granules. The kafirins are comprised of four different proteins: alpha, beta, gamma and delta (Izquierdo and Godwin, 2005; Laidlaw et al., 2010). Together, these proteins form a protein body (Fig. 3). Other cereal species have similar protein bodies including maize (zeins) and wheat (gliadin) (Shewry and Halford, 2002). However, despite the structural and genomic similarities between zeins and kafirins, the kafirins have a higher degree of cross linkage and are more hydrophobic, which is the basis of their propensity to form disulphide bonds during cooking, thus reducing the digestibility of sorghum (Duodu et al., 2003; Wong et al., 2009).

The four kafirin proteins have slightly different amino acid sequences, with the alpha and gamma having low levels of cysteine, whereas beta and delta have high level of cysteine. These sulphur amino acids are responsible for internal protein cross linking, make up the surface layers of the protein body, and are responsible for the cross linking to the glutenin matrix. Beta kafirin is considered the main kafirin responsible for limiting the digestibility of starch due to its strong cross linking within the kafrin body as well as external cross linkage. A number of kafirin mutants have been identified where one of the kafirins is absent. These mutations are natural variants but have the potential to increase the nutritional value of sorghum for human and animals.

Lastly, in most of the western world, there is an ever-present concern about gluten and its serious effect in coeliacs and those with varying degrees of gluten intolerance. The good news is sorghum does not contain the amino acid sequences in the endosperm storage proteins, specific to form gluten, and therefore like rice, provides a 'gluten-free' cereal alternative (Ciacci et al., 2007).

6.1.3 Fibre

Sorghum grain, like most cereals can be a rich source of dietary fibre. Most of the fibre is stored in the pericarp, with some present in the vestigial endosperm walls. Average values for sorghum place the insoluble fibre at ~8% and soluble fibre closer to 1%, as unlike other cereals, sorghum is relatively low in non-starch polysaccharides, such as beta-glucan and arabinoxylan (Fox et al., 2012). Due to the concentration of fibre in the pericarp, the fibre content of sorghum-based food is proportional to the degree of polishing or milling undertaken. Similar to most sorghum traits, the thickness of the pericarp is quite variable between species. However, unlike most cereals, the percentage of non-soluble fibre increases during digestion due to the assimilation of protein complexes and enzyme-resistant starch fractions (Knudsen and Munk, 1985; Smith and Frederiksen, 2000).

6.1.4 Phytochemicals

The phytochemical levels in sorghum may well entitle it to be called a super cereal, though the concentrations present vary significantly with both genotype and environment. Phytochemicals are secondary metabolites which do not participate directly in basic biochemical processes to sustain growth of the plant. However, some phytochemicals play a role defending the plant against biotic stresses (Johnson, 2013). Phytochemicals can be grouped as polyphenols, terpenes, glucosinolates, polysaccharides, lectins and alkaloids (Campos-Vega and Oomah, 2013). While only pigmented varieties of sorghum contain tannins, most contain flavonoids and all contain some phenolic acids (PA) (Dykes and Rooney, 2006). The greater concentrations of phytochemicals are contained in the pericarp, aleurone and testa, although some are present in the endosperm. In a review by Awika and Rooney (2004), sorghum phytochemicals were broken down into three groups: phenolic compounds (PC), phytosterols and policosanols (Fig. 4). From the PC found in sorghum, the flavonoids and PA are the most important (Awika and Rooney, 2004). All these PC have diverse functions such as providing pigmentation to the plant, acting as phytoalexins against microbial infections, anti-feedants against insects and animals, attractants for pollination, playing an antioxidant role and providing protection against UV light radiation (Naczk and Shahidi, 2004).

There is mounting evidence to suggest a diet high in fruits, vegetables and cereals significantly reduces the probability of suffering cardiovascular disease, cancer, diabetes and stroke (Crozier et al., 2009). This is particularly the case when there is minimal processing, that is, eating raw or intact portions of these foods. Cereals, especially those with coloured seed coats and sub-layers, that is, sorghum, are rich in phytochemicals and these compounds may participate in preventing the onset of some diseases.

Antioxidant activity has been attributed to phytochemicals, where it helps to combat free radical proliferation in tissues which if left uncontrolled could lead to irreversible cell damage that could trigger threatening diseases such as cancer or cardiovascular disease (Stefoska-Needham et al., 2015). Sorghum contains several phytochemicals with antioxidant activity, mainly flavonoids and PA (Awika and Rooney, 2004). Awika reported that the antioxidant activity of some sorghum varieties was higher than the one of most common fruits (Awika, 2003). The antioxidant activity of brown and black sorghums is superior to blueberries and strawberries by a substantial margin (Table 3) and notably, those of brown sorghum are significantly higher than those of black. Additionally, brown sorghum bran fractions had a higher total phenolic content (TPC) (66.3 mg gallic acid

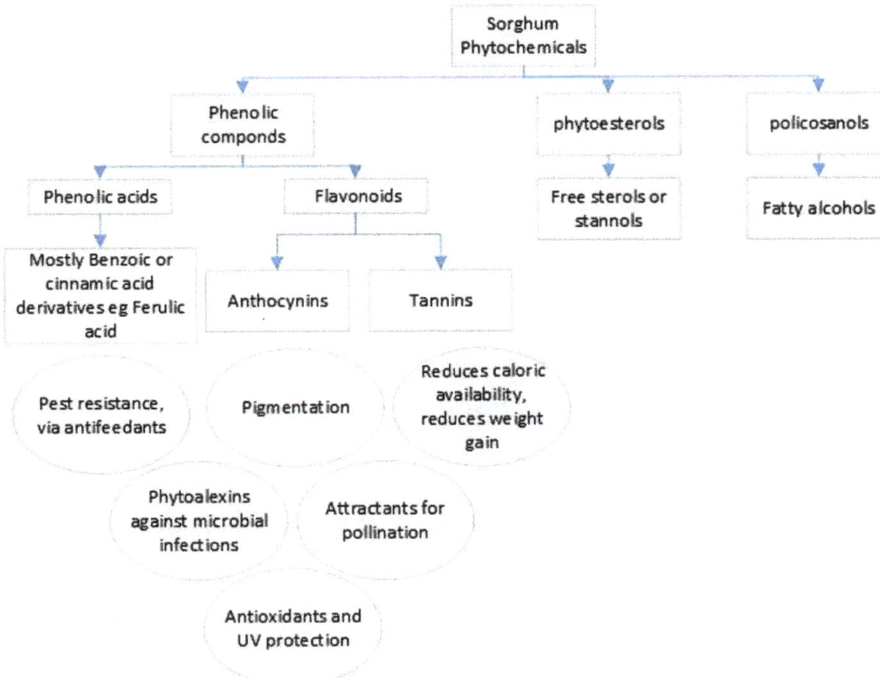

Figure 4 Breakdown of phytochemicals isolated in sorghum (Adapted from Awika and Rooney, 2004 and Dykes and Rooney, 2006).

equivalents (GAE)/g) compared to black sorghum (26.0 mg GAE/g); hence, it can be inferred that the higher antioxidant activity attributed to brown sorghum may come from its higher TPC (Awika, 2003).

Although the potential antioxidant activity has been demonstrated through *in vitro* studies, there is a lack of clinical evidence or *in vivo* studies that effectively demonstrate

Table 3 Antioxidant activity of sorghum compared to some common fruits

Sorghum types/fruits	ORAC[a] (μmol TE[b]/g)
Black sorghum	1008
Brown sorghum	3120
Blueberries	87–873
Strawberries	356–400
Grapes	100
Oranges	80–152

[a]Oxygen radical absorbance capacity.
[b]Trolox equivalents.
Source: Adapted from Awika (2003).

that PCs provide such benefits, due to the complexity of carrying out such studies (Crozier et al., 2009; Robbins, 2003).

Processing has a major impact on the amount of phytochemicals delivered in food. The commercial flour milling process removes the outer layers of grain and separates them from the germ and endosperm. The endosperm is reduced to a fine flour, then used as an ingredient, and the outer layers, the bran, are commonly used as animal feed. It is important to note most of the phytochemical compounds are in the outer sorghum layers which are removed during flour milling, and those retained in the sorghum flour are relatively low in health properties. Hence, a return to whole-meal foods or where the bran fractions are added as an ingredient would return the intrinsic health value of sorghum foods.

6.1.5 Anti-nutritional factors

The two main anti-nutritional factors associated with sorghum grain are tannins and phytic acids. Sorghum has received more attention than other cereals with regard to anti-nutritional factors due to its tannin content. Tannins have a biological role in protecting the grains from insect and bird predation, fungi and weathering. In high tannin varieties, the phenols can be extracted from the leaf sheath, glumes and pericarp, as a traditional source of dyes. However, these phenols can also result in carcass spotting in chickens, and subsequent rejection at abattoirs. In animal diets, the presence of sorghum tannins can reduce the digestibility and nutrient absorption by 3–15%. Similarly, in human diets, the presence of tannins has been linked to reductions in digestibility of starch and proteins (Shawrang et al., 2011; Smith and Frederiksen, 2000). Phytic acid, which is predominantly found in the aleurone, forms phytates by chelating cationic elements, for example, potassium, zinc, manganese, magnesium, iron and calcium as well as some B vitamins, rendering them unavailable during absorption, thus reducing the nutritional value of sorghum and other cereals (Liu et al., 2004).

6.2 Sorghum production for stockfeed

Both grain and forage sorghum varieties are used for animal production worldwide (Dahlberg et al., 2011). By dry matter (DM) yield, the majority of sorghum production (grain and forage) is utilised by ruminants, with a smaller yet significant proportion in the monogastric industries, and in pet foods and other boutique or traditional animal industries. The choice of cultivar is determined by the environment, the production system in which it is required and the production goal; maintenance or productivity gains. In some parts of the world, specific regions of Africa and Australia for example, sorghum species like *Sorghum leiocladum* and *Sorghum nitidum* form a significant proportion of the native sward and are an integral part of extensive grazing operations (Cook and Andrew, 1991). As a cultivated pasture plant sorghum is grown for green forage, silage, hay and grain, for use in both intensive industries and as a supplementary feed in some extensive production units.

Destruction of the seed from grinding or milling ruptures the starchy endosperm, improving access to the starch by endogenous or exogenous starch-degrading enzymes, facilitating the reduction of starch to its simplest form of small glucose chains or glucose. The reduction of the grain into small particles improves its digestibility in animals (Liu et al., 2013) where rapid liveweight gains are required.

6.2.1 Green forages

Sorghum is used as a green forage in tropical and subtropical agriculture because of its water use efficiency, high DM yield and hardiness. It is able to withstand periods of water stress and then quickly recover with subsequent rainfall events. In general, the Sudan varieties are used for grazing systems because of their rapid growth under ideal conditions. While sorghum's reputation has been built on its ability to be able to tolerate dry seasons and low-input systems, conversely under irrigated and fertilised management systems its DM production can exceed other cereals and grasses. In its early vegetative state forage varieties are of moderate to high nutrient value, depending on environmental inputs; however, once the crop approaches its reproductive state, its nutritive value reduces rapidly. The key to effective utilization of sorghum is to maintain the crop at its peak nutritive value-to-DM production ratio. The metabolisable energy (ME) of the Sudan, Sudan crosses and hybrid varieties rapidly declines as the plant approaches physical maturity until an elevation at grain set, whereas the sweet sorghums increase in sugar content, and hence ME, towards maturity which has implications for grazing, ensiling and hay production management.

Properly managed forage sorghum can support multiple grazing events per season. Under ideal pasture management, and when meeting nutrient input demands for nitrogen, potassium and to some degree sulphur, intensive management can result in very high DM production (16–22 tonnes DM) of moderate nutrient value (~12–16% protein and a ME of 9 MJ/kg DM). Intensive applications of sorghum under high input, high utilization are generally restricted to intensive dairy grazing systems and some beef fattening/backgrounding. Alternatively, forage sorghum is also utilised in lower input, lower utilization feeding systems with resultant lower production outputs per land area or per animal.

One of the challenges of using some sorghum varieties for animal production is its formation of prussic acid complexes which when ingested by ruminants form a cyanide-like compound, which can result in rapid death of the grazing animals. Prussic acid content is elevated in rapidly growing young material and drought-affected plants, though the actual content varies between species, with the Sudan types having the highest risk (Goff et al., 2011). Due to selective breeding programmes there are now low prussic acid cultivars available. A recent report showed how near infrared could quickly identify hydrogen cyanide (prussic acid) in sorghum plant material which would assist in identifying sorghum material at high risk (Fox et al., 2012). Grazing management plans which include minimum plant height requirements (cultivar specific), monitoring stock for 48 h, introducing them to the sorghum pasture when not hungry and provision of a sulphur lick have all been used to mediate the risks. In addition to prussic acid, stock are also susceptible to nitrate poisoning where the soil nitrogen concentration is high and conditions (water stress, overcast weather or herbicide application) have seen DM production restricted. Similar grazing management strategies apply, with cattle not being fed the pasture under these conditions, not being introduced to pasture when hungry and monitoring stock for signs of toxicity.

6.2.2 Sorghum hay and silage

Sorghum is often cultivated for its capacity to make large amounts of hay from limited areas, in short time frames. The same DM yields versus nutritional value relationships apply as per green forage, therefore crop management and end use production system requirements are the key drivers for determining the optimal time for mowing a forage crop for hay production. Equally, the caveats regarding nitrate and prussic acid poisoning are not mitigated simply by using a crop for hay rather than green forage and should be considered. Optimum yield and quality balance occurs under irrigation and high fertiliser inputs over high planting rates, with the subsequent forage being cut under 1.5 m high. This can result in 5–6 cutting cycles per crop, with a final production of up to 16 tonnes/hectare/season with good nutritive value maintained.

Depending on the variety, sorghum silage provides a good-quality forage fibre source to subtropical dairy and beef finishing systems. Both sweet sorghums and grain cultivars are used for silage forage systems, typically being direct cut and pitted. High yields, high DM and high nutritive value can be achieved with good management. Sweet sorghum and grain varieties are direct cut at early maturation, or soft- to mid-dough grain stage respectively, to produce the optimum yield/nutrient axis. Sorghum can be utilised to a lesser extent, to produce higher quality silage by cutting at earlier vegetative stages; however, the material needs to be wilted prior to ensiling.

Mutation-induced brown midrib (*bmr*) sorghums, developed in the late 1970s, were shown to have significant reduction in lignin present in their stems and leaves and associated increases in *in vitro* DM digestibility (Porter et al., 1978). Despite the increases in digestibility, associated issues with lodging in the *bmr* varieties led to the inclusion of the brachytic height genes, which decreased the internode elongation and reduced lodging (Morgan and Finlayson, 2000). Research and breeding have continued to develop new *bmr*-based varieties of sorghum for forage and silage production, with many recent studies investigating their conversion efficiencies for milk production in high-producing dairy herds. A 2008 study showed brown midrib sorghum–sudangrass silage was an effective alternative to corn silage, when used as 35–45% of their DM ration (Dann et al., 2008). This was further supported by work done in 2015 which identified that the first cut or ratoon crop of *bmr* varieties had comparable intake and performance as corn silage-based diets (Bernard and Tao, 2015).

6.3 Sorghum grain

Sorghum grain is a valuable energy source for ruminants in subtropical and tropical production systems. It is a hard grain, with a very high starch content, relatively low protein, but one of the highest gross energy values amongst the cereals. However, the availability of its starch reduces the ME content significantly. Unlike other grains used for animal feed, the starch molecules in sorghum are encapsulated by a kafirin/lignin matrix which reduces digestion in monogastrics and ruminants alike. A number of physical and chemical processes are used to increase its digestibility prior to feeding the grain. This includes mechanical means of increasing the surface area (e.g. steam flaking, grinding, hammer milling, roller milling) for enzymes and microbial digestion, and chemical processes such as hydration or alkali and enzymes treatments. Novel technologies including microwave and other direct heat processes have also been reported to cause an increase in starch availability (Theurer et al., 1999).

6.4 Syrup production

Sweet sorghums are used to produce syrup, often known under the misnomer of molasses, which actually applies only to the extract produced from sugarcane. Production occurs mainly in the United States and peaked in 1946, at approximately 136 million litres/

year. Due to poor cost efficiencies, limited advances in post-harvest technologies and a requirement to process the stalk immediately after harvest to maximise quality and quantity, production occurs in small on-farm operations. Leaves are stripped and the stalks crushed to extract the syrup which is then clarified through a process of heating and/or filtering. While the syrup can be refined to a crystalline sugar, this is rarely done, as cane sugar processing is more cost-effective (Dahlberg et al., 2011; Regassa and Wortmann, 2014; Rooney and Waniska, 2000).

6.5 Non-food uses of sorghum

Cultivation of broomcorn sorghum for the production of corn brooms, remains a significant industry in South Eastern Europe. What is referred to as 'straw brooms' in Western society is actually made from the panicle fibres of an Old-World variety of sorghum, which has since been bred to optimise fibre length, fineness and appearance (Berenji et al., 2011). In addition to its use in broom production, broomcorn has also been hybridised with sweet and grain sorghums for application in biomass production.

Due to sorghum's high yields, growth season and stress tolerances, there has been a large amount of research committed to the use of sorghum as a renewable energy source via the production of sustainable biomass for biofuels and bioenergy (Dahlberg et al., 2011; Regassa and Wortmann, 2014). Bioethanol can be produced from fermenting sorghums, starch-rich grains, similar to maize-based ethanol production, or from the stover of the sweet sorghums from extracted stem juices or directly fermenting the chopped stems. There are many modifications to the cultivation and processing methodologies which can be applied depending on the starting product, industrial plant and scale of the operation (Dykes and Rooney, 2006; Regassa and Wortmann, 2014). One of the benefits of using sweet sorghum for bioethanol applications is the functionality of the by-products. The grain can be used as a stockfeed, as can the bagasse with additional supplements. Bagasse can also be sold as a biofertiliser to improve soil organic matter or in paper manufacture, providing supplemental income streams.

With the wide genetic pool available in sorghum and detailed production data available, selections and cross-breeding programmes for biomass production tailored to the needs of specific bioenergy/biofuels industries are already possible. Sweet sorghum production has been predicted to reach a potential of 6000 L/ha, with an energy return on investment of 3:1; however, a gap in large-scale processing technology and the associated transport and storage requirements has limited this potential (Regassa and Wortmann, 2014).

7 Conclusion

Fifth amongst the cereal crops in terms of production, sorghum has arguably the greatest adaptability, genetic, phenotypic and functional diversity of all the cereal species. While it was domesticated in tropical regions, modern breeding, and particularly the SC programme, have broadened the adaptation of its cultivars. Sorghum has been shown to perform well in temperate crop-growing areas including in central Europe, the US Midwest and in southern parts of Argentina. Current predictions strongly suggest that tropical regions are expanding to higher latitudes, and the climate uncertainty will lead to higher incidence of drought and heat. Modern sorghum cultivars are ideally adapted to hotter and drier conditions, and it is predicted that sorghum cultivation will increase over the next few decades. Its high biomass under these conditions will also increase sorghum's demand as a source of biomaterials and bioenergy as we reduce our reliance on non-renewable sources. The confluence of technology and knowledge enabling our ability to accurately identify and harness this diversity for specific production requirements will undoubtedly see sorghum's utility and production increasing.

8 References

Abu Assar, A. H., Uptmoor, R., Abdelmula, A. A., Salih, M., Ordon, F. and Friedt, W. 2005. Genetic variation in sorghum germplasm from Sudan, ICRISAT, and USA assessed by simple sequence repeats (SSRs). *Crop Science*, 45, 1636–44.

Anderson, J., D'Andrea, A., Logan, A. and Salah, M. 2007. Bread moulds from the Amun Temple at Dangeil–An addendum. *Sudan & Nubia*, 11, 89–93.

Anglani, C. 1998. Sorghum for human food – A review. *Plant Foods for Human Nutrition*, 52, 85–95.

Arndt, W. 1961. Indigenous sorghum as food and in myth: The Tagoman Tribe. *Oceania*, 32, 109–12.

Awika, J. M. 2003. *Antioxidant Properties of Sorghum*. Ann Arbor: Texas A&M University.

Awika, J. M. and Rooney, L. W. 2004. Sorghum phytochemicals and their potential impact on human health. *Phytochemistry*, 65, 1199–221.

Balme, J., Garbin, G. and Gould, R. A. 2001. Residue analysis and palaeodiet in arid Australia. *Australian Archaeology*, 53, 1–6.

Bapat, D. R. and Mote, U. N. 1982. Sources of shootfly resistance in Sorghum. *Journal of the Maharashtra Agricultural University*, 7, 238–40.

Barakat, H. and Fahmy, A. G. E. D. 1999. Wild grasses as 'Neolithic' food resources in the Eastern Sahara. In: van der Veen, M. (Ed.), *The Exploitation of Plant Resources in Ancient Africa*. Boston, MA: Springer US.

Beldados, A. and Costantini, L. 2011. Sorghum exploitation at Kasala and its environs, north eastern Sudan in the second and first millennium BC. *Nyame Akuma*, 75, 33–9.

Benmoussa, M., Suhendra, B., Aboubacar, A. and Hamaker, B. R. 2006. Distinctive sorghum starch granule morphologies appear to improve raw starch digestibility. *Starch – Starke*, 58, 92–9.

Berenji, J., Dahlberg, J., Siroka, V. and Latkovic, D. 2011. Origin, history, morphology, production, improvement, and utilization of broomcorn [*Sorghum bicolor* (L.) Moench] in Serbia. *Economic Botany*, 65, 190–208.

Bernard, J. K. and Tao, S. 2015. Short communication: Production response of lactating dairy cows to brachytic forage sorghum silage compared with corn silage from first or second harvest. *Journal of Dairy Science*, 98, 8994–9000.

Bhattacharya, A., Rice, N., Shapter, F. M., Norton, S. L. and Henry, R. J. 2011. Sorghum. In: Kole, C. (Ed.), *Wild Crop Relatives: Genomic and Breeding Resources*. Heidelberg, Dordrecht, London and New York: Springer.

Biehl, E., Wendorf, F., Landry, W., Desta, A. and Watrous, L. 1999. The use of gas chromatography and mass spectrometry in the identification of ancient Sorghum seeds. In: van der Veen, M. (Ed.), *The Exploitation of Plant Resources in Ancient Africa*. New York: Springer, pp. 47–53.

Boivin, N. and Fuller, D. Q. 2009. Shell middens, ships and seeds: Exploring coastal subsistence, maritime trade and the dispersal of domesticates in and around the ancient Arabian peninsula. *Journal of World Prehistory*, 22, 113–18.

Boivin, N., Crowther, A., Helm, R. and Fuller, D. Q. 2013. East Africa and Madagascar in the Indian Ocean world. *Journal of World Prehistory*, 26, 213–81.

Boyer, C. D. 1984. Soluble starch synthases and starch branching enzymes from developing seeds of sorghum. *Phytochemistry*, 24, 15–18.

Brown, P. J., Myles, S. and Kresovich, S. 2011. Genetic support for phenotype-based racial classification in sorghum. *Crop Science*, 51, 224–30.

Campbell, B. C., Gilding, E. K., Mace, E. S., Tai, S., Tao, Y., Prentis, P. J., Thomelin, P., Jordan, D. R. and Godwin, I. D. 2016. Domestication and the storage starch biosynthesis pathway: Signatures of selection from a whole sorghum genome sequencing strategy. *Plant Biotechnology Journal*, 14(12), 2240–53.

Campos-Vega, R. and Oomah, B. D. 2013. Chemistry and classification of phytochemicals. In: Tiwari, B. K., Brunton, N. P. and Brennan, C. S. (Eds), *Handbook of Plant Food Phytochemicals: Sources, Stability and Extraction*, 1st Edition. Oxford: John Wiley & Sons Ltd., pp. 5–48.

Caniato, F. F., Guimarães, C. T., Hamblin, M., Billot, C., Rami, J., Hufnagel, B., Kochian, L. V., Liu, J., Garcia, A. A. F., Hash, C. T., Ramu, P., Mitchell, S., Kresovich, S., Oliveira, A. C., de Avellar, G., Borém, A., Glaszmann, J., Schaffert, R. E. and Magalhaes, J. V. 2011. The relationship between population structure and aluminum tolerance in cultivated sorghum. *FLoS ONE*, 6, e20830.

Castañeda-Álvarez, N. P., Khoury, C. K., Achicanoy, H. A., Bernau, V., Dempewolf, H., Eastwood, R. J., Guarino, L., Harker, R. H., Jarvis, A., Maxted, N., Müller, J. V., Ramirez-Villegas, J., Sosa, C. C., Struik, P. C., Vincent, H. and Toll, J. 2016. Global conservation priorities for crop wild relatives. *Nature Plants*, 2, 16022.

Ciacci, C., Maiuri, L., Caporaso, N., Bucci, C., Del Giudice, L., Rita Massardo, D., Pontieri, P., Di Fonzo, N., Bean, S. R., Ioerger, B. and Londei, M. 2007. Celiac disease: *In vitro* and *in vivo* safety and palatability of wheat-free sorghum food products. *Clinical Nutrition*, 26, 799–805.

Clapham, A. J. and Rowley-Conwy, P. A. 2007. New discoveries at Qasr Ibrim, Lower Nubia. In: Cappers, R. (Ed.), *Fields of Change. Proceedings of the fourth International Workshop on African Archeobotany*. Groningan: Backhuis and Groningan University Library, pp. 157–64.

Clark, J. D. and Stemler, A. 1975. Early domesticated Sorghum from Central Sudan. *Nature*, 254, 588–91.

Cleuziou, S. and Costantini, L. 1980. Premiers elements sur l'agriculture protohistorique de l'Arabie orientale. *Paleorient*, 6, 245–51.

Cook, G. D. and Andrew, M. H. 1991. The nutrient capital of indigenous Sorghum species and other understory components of Savannas in north-western Australia. *Australian Journal of Ecology* 16, 375–84.

Crowther, A., Lucas, L., Helm, R., Horton, M. C., Shipton, C., Wright, H. T., Walshaw, S., Pawlowicz, M. C., Radimilahy, C., Douka, K., Picornell-Gelabert, L., Fuller, D. Q. and Boivin, N. 2016. Ancient crops provide first archaeological signature of the westward Austronesian expansion. *Proceedings of the National Academy of Sciences of the United States of America*, 113, 6635–40.

Crowther, A., Prendergast, M., Fuller, D. Q. and Boivin, N. in press. Subsistence mosaics, forager-farmer interactions and the transition to farming in eastern Africa. Quaternary International.

Crozier, A., Jaganath, I. B. and Clifford, M. N. 2009. Dietary phenolics: Chemistry, bioavailability and effects on health. *National Product Reports*, 26, 1001–43.

Dahlberg, J. A. 2000. Classification and characterization of sorghum. In: Smith, C. W. and Frederiksen, R. A. (Eds), *Sorghum: Origin, History, Technology, and Production*. New York: John Wiley and Sons.

Dahlberg, J. A. and Wasylikowa, K. 1996. Image and statistical analyses of early sorghum remains (8000 BP) from the Nabta Playa archaeological site in the Western Desert, southern Egypt. *Vegetation History and Archaeobotany*, 5, 293–9.

Dahlberg, J. A., Evans, J., Johnson, E., Hyman, M., Biehl, E. and Wendorf, F. 1995. Attempts to identify 8000-bp sorghum using image-analysis, infrared spectroscopy, and biotechnological procedures. *Acta Palaeobotanica*, 35, 167–73.

Dahlberg, J., Berenji, J., Sikora, V. and Latkovic, D. 2011. Assessing sorghum [*Sorghum bicolor* (L.) Moench] germplasm for new traits: Food, fuels and unique uses. *Maydica*, 56(1750), 85–92. http://eprints.icrisat.ac.in/7455/1/Maydica_56_1750_85-92_2011.pdf.

Dann, H. M., Grant, R. J., Contanch, K. W., Thomas, E. D., Ballard, C. S. and Rice, R. 2008. Comparison of brown midrib sorghum-sudangrass with corn silage on lactational performance and nutrient digestibility in Holstein dairy cows. *Journal of Dairy Science*, 91, 663–72.

de Maret, P. 2013. Archaeologies of the Bantu expansion. In: Mitchell, P. and Lane, P. (Eds), *The Oxford Handbook of African Archaeology*. Oxford: Oxford University Press.

de Moulins, D., Phillips, C. S. and Durrani, N. 2003. The archaeobotanical record of Yemen and the question of Afro-Asian contacts. In: Neumann, K., Butler, A. and Kahlheber, S. (Eds), *Food, Fuel and Fields: Progress in African Archaeobotany, Africa Praehistorica* 15. Köln: Heinrich-Barth Institute, pp. 213–28.

de Wet, J. M. J. 1978. Systematics and evolution of sorghum sect. sorghum (*Gramineae*). *American Journal of Botany*, 65, 477–84.

de Wet, J. M. and Harlan, J. 1971. The origin and domestication of *Sorghum bicolor*. *Economic Botany*, 25, 128–35.

Deakin, W., Rowley-Conwy, P. and Shaw, C. 1998a. Amplification and sequencing of DNA from preserved sorghum of up to 2800 years antiquity found at Qasr Ibrim. *Ancient Biomolecules*, 1, 27–41.

Deakin, W., Rowley-Conwy, P. and Shaw, C. 1998b. The sorghum of Qasr Ibrim: Reconstructing DNA templates from ancient seeds. *Ancient Biomolecules*, 2, 117–24.

Dempewolf, H., Eastwood, R. J., Guarino, L., Khoury, C. K., Müller, J. V. and Toll, J. 2014. Adapting agriculture to climate change: A global initiative to collect, conserve, and use crop wild relatives. *Agroecology and Sustainable Food Systems*, 38, 369–77.

Denbow, J. R. and Wilmsen, E. W. 1986. Advent and course of pastoralism in the Kalahari. *Science*, 234, 1509–15.

Deu, M., Rattunde, F. and Chantereau, J. 2006. A global view of genetic diversity in cultivated sorghums using a core collection. *Genome*, 49, 168–80.

Dicko, M. H., Gruppen, H., Traoré, A. S., Voragen, A. G. J. and van Berkel, W. J. H. 2006. Sorghum grain as human food in Africa: Relevance of content of starch and amylase activities *African Journal of Biotechnology*, 5, 384–95.

Dillon, S. L., Shapter, F. M., Henry, R. J., Cordeiro, G., Izquierdo, L. and Lee, L. S. 2007. Domestication to crop improvement: Genetic resources for sorghum and saccharum (*Andropogoneae*). *Annals of Botany*, 100, 975–89.

Doebley, J. 1989. Isozymic evidence and the evolution of crop plants. In: Soltis, D. E., Soltis, P. S. and Dudley, T. R. (Eds), *Isozymes in Plant Biology*. Dordrecht: Springer Netherlands.

Doebley, J., Gaut, B. S. and Smith, B. D. 2006 The molecular genetics of crop domestication. *Cell*, 127, 1309–21.

Doggett, H. 1988. *Sorghum*. UK: Longman Scientific & Technical.

Duodu, K. G., Taylor, J., Belton, P. and Hamaker, B. 2003. Factors affecting sorghum protein digestibility. *Journal of Cereal Science*, 38, 117–31.

Duodu, K. G., Nunes, A., Delgadillo, I., Parker, M. L., Mills, E. N. C., Belton, P. S. and Taylor, J. R. N. 2002. Effect of grain structure and cooking on sorghum and maize *in vitro* protein digestibility. *Journal of Cereal Science*, 35, 161–74.

Duvall, M. R. and Doebley, J. F. 1990. Restriction site variation in the chloroplast genome of sorghum (*Poaceae*). *Systematic Botany*, 15, 472–80.

Dykes, L. and Rooney, L. W. 2006. Sorghum and millet phenols and antioxidants. *Journal of Cereal Science*, 44, 236–51.

Ehret, C. 1998. *An African Classical Age: Eastern and southern Africa in World History, 1000 BC to AD 400*. Charlottesville: University of Virginia Press.

Fannon, J., Gray, J.A., Gunawan, N., Huber, K.C. and BeMiller, J.N. 2003. The channels of starch granules. *Food Science and Biotechnology*, 12, 700–4.

Fannon, J., Gray, J. A., Gunawan, N., Huber, K. C. and BeMiller, J. N. 2004. Heterogeneity of starch granules and the effect of granule channelization on starch modification. *Cellulose*, 11, 247–54.

FAO, F. A. A. O. O. T. U. N. 2010. The Second Report on the State of the World's Plant Genetic Resources for Food and Agriculture. Rome: FAO.

FAO, F. A. A. O. O. T. U. N. 2016. FAO 2014 Statistics on Sorghum [Online]. Rome: FAO. Available: http://faostat3.fao.org/browse/Q/QC/E.

Flood, J. 1980. *The Moth Hunters: Aboriginal Prehistory of the Australian Alps*. Canberra: Australian Institute of Aboriginal Studies.

Fox, G. P., O'Donnell, N. H., Stewart, P. N. and Gleadow, R. M. 2012. Estimating hydrogen cyanide in forage sorghum (*Sorghum bicolor*) by nearinfrared spectroscopy. *Journal of Agricultural and Food Chemistry*, 60, 6183–7.

Franzmann, B. A. and Hardy, A. T. 1996. Testing the host status of Australian indigenous sorghums for the sorghum midge. In: Foale, M. A., Henzell, R. G. and Kneip, J F. (Eds), *Proceedings of the Third Australian Sorghum Conference*. Tamworth, NSW, Australia, 20–22 February 1996. Melbourne, Vic: Australian Institute of Agricultural Science.

Fullagar, R. L. K., Field, J. H. and Kealhofer, L. 2008. Grinding stones and seeds of change: Starch and phytoliths as evidence of plant food processing. In: Rowan, Y. and Ebeling, J. (Eds), *New Approaches to Old Stones*. London: Equinox Publishing Ltd.

Fullagar, R. L. K., Hayes, E., Stephenson, B., Field, J., Matheson, C., Stern, N. and Fitzsimmons, K. 2015. Evidence for Pleistocene seed grinding at Lake Mungo, south-eastern Australia. *Archaeology in Oceania*, 50(Suppl. 1), 3–19.

Fuller D. Q. 2002. Fifty years of archaeobotanical studies in India: Laying a solid foundation. In: Settar, S. and Korisettar, R. (Eds), *Indian Archaeology in Retrospect, vol. III. Archaeology and Interactive Disciplines*. Delhi: Manohar, pp. 247–363.

Fuller, D. Q. 2003. African crops in prehistoric South Asia: A critical review. In: Neumann, K. and Butler, A. (Eds), *Food, Fuel and Fields: Progress in African Archaeobotany*. Köln: Heinrich-Barth Institut. 15, 239–71.

Fuller, D. Q. 2004a. The Central Amri to Kirbekan survey: A preliminary report on excavations and survey 2003–04. *Sudan & Nubia*, 8, 4–16.

Fuller, D. Q. 2004b. Early Kushitic agriculture: Archaeobotanical evidence from Kawa. *Sudan and Nubia*, 8, 70–4.

Fuller, D. Q. 2014. Agricultural innovation and state collapse in Meroitic Nubia: The impact of the Savannah package. In: Stevens, C. J., Nixon, S., Murray, M. A. and Fuller, D. Q. (Eds), *Archaeology of African Plant Use*. Walnut Creek: Left Coast Press.

Fuller, D. Q. 2015. The economic basis of the Qustul splinter state: Cash crops, subsistence shifts, and labour demands in the Post-Meroitic transition. In: Zach, M. (Ed.), *The Kushite World. Proceedings of the 11th International Conference for Meroitic Studies*. Vienna: Verein der Forderer der Sudanforschung, 1st Edition, 33–60.

Fuller, D. Q. and Boivin, N. 2009. Crops, cattle and commensals across the Indian Ocean: Current and potential archaeobiological evidence. *Etudes Océan Indien*, 42–43, 13–46.

Fuller, D. and Hildebrand, L. 2013. Domesticating plants in Africa. In: Mitchell, P. and Lane, P. (Eds), *Oxford Handbook of African Archaeology*. Oxford: Oxford University Press.

Garber, E. D. 1950. Cytotaxonomic studies in the genus Sorghum. *University of California Publications in Botany*, 23, 283–361.

Giblin, J. and Fuller, D. Q. 2011. First and second millennium AD agriculture in Rwanda: Archaeobotanical finds and radiocarbon dates from seven sites. *Vegetation History and Archaeobotany*, 20, 253–65.

Gilding, E. K., Frere, C. H., Cruickshank, A., Rada, A. K., Prentis, P. J., Mudge, A. M., Mace, E. S., Jordan, D. R. and Godwin, I. D. 2013a. Allelic variation at a single gene increases food value in a drought-tolerant staple cereal. *Nature Communications*, 4, 1483.

Goff, B., Moore, K. J., Fales, S. L. and Pedersen, J. F. 2011. Comparison of gas chromatography, spectrophotometry and near infrared spectroscopy to quantify prussic acid potential in forages. *Journal of the Science of Food and Agriculture*, 91, 1523–6.

Haaland, R. 1999. The puzzle of the late emergence of domesticated sorghum in the Nile Valley. In: Gosden, C. and Hather, J. G. (Eds), *The Prehistory of Food*. London: Routledge, pp. 397–418.

Hagerty, M. H. 1941. Comments on writings concerning Chinese sorghums. *Harvard Journal of Asiatic Studies*, 5, 234–60.

Hamaker, B. R., Kirleis, A. W., Butler, L. G., Axtell, J. D. and Mertz, E. T. 1987. Improving the *in vitro* protein digestibility of sorghum with reducing agents. *Proceedings of the National Academy of Sciences*, 84, 626–8.

Harlan, J. R. 1976. *Origins of African Plant Domestication*. The Hague: Walter de Gruyter.

Harlan, J. R. and de Wet, J. M. J. 1971. Toward a rational classification of cultivated sorghums. *Crop Science*, 12, 172–6.

Harlan, J. R. and de Wet, J. M. J. 1972. A simplified classification of cultivated sorghum. *Crop Science*, 12, 172–6.

Harlan, J. R., de Wet, J. and Price, E. G. 1973. Comparative evolution of cereals. *Evolution*, 27, 311–25.

Hildebrand, E. A., Brandt, S. A. and Lesur-Gebremariam, J. 2010. The Holocene archaeology of southwest Ethiopia: New insights fom the Kafa Archaeological Project. *African Archaeological Review*, 27, 255–89.

Hill, H., Lee, L. S. and Henry, R. J. 2012. Variation in sorghum starch synthesis genes associated with differences in starch phenotype. *Food Chemistry*, 131, 175–83.

Hodnett, G. L., Burson, B. L., Rooney, W. L., Dillon, S. L. and Price, H. J. 2005. Pollen–pistil interactions result in reproductive isolation between *Sorghum bicolor* and divergent *Sorghum* species. *Crop Science*, 45, 1403–9.

Huffman, T. N. 1970. The Early Iron Age and the spread of the Bantu. *South African Archaeological Bulletin*, 25, 3–21.

Huffman, T. N. 2006. Bantu migrations in southern Africa. In: Soodyall, H. (Ed.), *The Prehistory of Africa: Tracing the Lineage of Modern Man*. First Edition, Johannesburg: Witwatersrand University, 97–108.

Huffman, T. N. 2007. *Handbook to the Southern African Iron Age*. Pietermaritzburg: University of KwaZulu-Natal Press.

Izquierdo, L. and Godwin, I. D. 2005. Molecular characterization of a novel methionine-rich δ-kafirin seed storage protein gene in sorghum (*Sorghum bicolor* L.). *Cereal Chemistry*, 82, 706–10.

Johnson, I. T. 2013. Phytochemicals and health. In: Tiwari, B. K., Brunton, N. P. and Brennan, C. S. (Eds), *Handbook of Plant Food Phytochemicals: Sources, Stability and Extraction*, 1st Edition. Oxford: John Wiley & Sons Ltd, pp. 49–67.

Kamala, V., Singh, S. D., Bramel, P. J. and Rao, D. M. 2002. Sources of resistance to downy mildew in wild and weedy sorghums. *Crop Science*, 42, 1357–60.

Kayode, A. P. P., Linnemann, A. R., Nout, M. J. R., Hounhouigan, J. D., Stomph, T. J. and Smulders, M. J. M. 2006. Diversity and food quality properties of farmers' varieties of sorghum from Benin. *Journal of the Science of Food and Agriculture*, 86, 1032–9.

Kimber, C. T. 2000. Origins of domesticated sorghum and its early diffusion to India and China. In: Smith, C. W. and Frederiksen, R. A. (Eds), *Sorghum: Origin, History, Technology and Production*. New York: John Wiley & Sons.

Knudsen, K. E. B. and Munk, L. 1985. Dietary fibre contents and compositions of sorghum and sorghum-based foods. *Journal of Cereal Science*, 3, 153–64.

Komolong, B., Chakraborty, S., Ryley, M. and Yates, D. 2002. Identity and genetic diversity of the sorghum ergot pathogen in Australia. *Australian Journal of Agricultural Research*, 53, 621–8.

Krzyżaniak, L. 1991. Early farming in the middle Nile Basin: Recent discoveries at Kadero (central Sudan). *Antiquity*, 65, 515–32.

Kusimba, C. and Kusimba, S. 2005. Mosaics and interactions: East Africa, 2000 b.p. to the present. In: Stahl, A. B. (Ed.), *African Archaeology: A Critical Introduction*, 1st Edition, Oxford: Blackwell, 319–419.

Laidlaw, H. K. C., Mace, E. S., Williams, S. B., Sakrewski, K., Mudge, A. M., Prentis, P. J., Jordan, D. R. and Godwin, I. D. 2010. Allelic variation of the β-, γ- and δ-kafirin genes in diverse Sorghum genotypes. *Theoretical and Applied Genetics*, 121, 1227–37.

Lane, P. 2004. The 'moving frontier' and the transition to food production in Kenya. *Azania*, 39, 243–64.

Lane, P. 2015. Early agriculture in sub-Saharan Africa to c. 500 CE. In: Barker, G. and Goucher, C. (Eds), *The Cambridge World History*. Cambridge: Cambridge University Press, pp. 472–98.

Lasky, J. R., Upadhyaya, H. D., Ramu, P., Deshpande, S., Hash, C. T., Bonnette, J., Juenger, T. E., Hyma, K., Acharya, C., Mitchell, S. E., Buckler, E. S., Brenton, Z., Kresovich, S. and Morris, G. P. 2015. Genome-environment associations in sorghum landraces predict adaptive traits. *Science Advances*, 1(6), e1400218.

Laurie, D. and Bennett, M. D. 1989. Genetic variation in Sorghum for the inhibition of maize pollen tube growth. *Annals of Botany*, 64, 675–81.

Lazarides, M., Hacker, J. B. and Andrew, M. H. 1991. Taxonomy, cytology and ecology of indigenous Australian sorghums (*Sorghum* Moench: Andropogoneae: Poaceae). Australian Systematic Botany, 4, 591–635. http://dx.doi.org/10.1071/SB9910591.

Lin, Z., LI, X., Shannon, L. M., Yeh, C. T., Wang, M. L., Bai, G., Peng, Z., LI, J., Trick, H. N., Clemente, T. E., Doebley, J. F., Schnable, P. S., Tuinstra, M. R., Tesso, T. T., White, F. and Yu, J. 2012. Parallel domestication of the Shattering1 genes in cereals. *Nature Genetics*, 44, 720–4.

Lindeboom, N., Chang, P. R. and Tyler, R. T. 2004. Analytical, biochemical and physicochemical aspects of starch granule size, with emphasis on small granule starches: A review. *Starch – Starke*, 56, 89–99.

Liu, J. C., Irene, O., Truax, M. and Lott, J. N. A. 2004. Phytic acid-phosphorus and other nutritionally important mineral nutrient elements in grains of wild-type and low phytic acid (lpa1–1) rice. *Seed Science Research*, 14, 109–16.

Liu, S. Y., Selle, P. H. and Cowieson, A. J. 2013. Strategies to enhance the performance of pigs and poultry on sorghum-based diets. *Animal Feed Science and Technology*, 181, 1–14.

Mace, E. S., Tai, S., Gilding, E. K., LI, Y., Prentis, P. J., Bian, L., Campbell, B. C., HU, W., Innes, D. J., Han, X., Cruickshank, A., Dai, C., Frère, C., Zhang, H., Hunt, C. H., Wang, X., Shatte, T., Wang, M., SU, Z., LI, J., Lin, X., Godwin, I. D., Jordan, D. R. and Wang, J. 2013. Whole-genome sequencing reveals untapped genetic potential in Africa's indigenous cereal crop sorghum. *Nature Communications*, 4, 2320.

Mace, E., Tai, S., Innes, D., Godwin, I., HU, W., Campbell, B., Gilding, E., Cruickshank, A., Prentis, P., Wang, J. and Jordan, D. R. 2014. The plasticity of Nbs resistance genes in sorghum is driven by multiple evolutionary processes. *Bmc Plant Biology*, 14, 253.

Madella, M., García-Granero, J. J., Out, W., Ryan, P. and Usai, D. 2014. Microbotanical evidence of domestic cereals in Africa 7000 years ago. *PLoS ONE*, 9, e110177.

Magid, A. A. 1989. *Plant Domestication in the Middle Nile Basin: An Archeoethnobotanical Case Study*. Oxford: Oxford University Press.

Massel, K., Campbell, B. C., Mace, E. S., Tai, S., Tao, Y., Worland, B. G., Jordan, D. R., Botella, J. R. and Godwin, I. D. 2016. Whole genome sequencing reveals potential new targets for improving nitrogen uptake and utilization in sorghum bicolor. *Frontiers in Plant Science*, 7, 1544.

Mercader, J. 2009. Mozambican grass seed consumption during the Middle Stone Age. *Science*, 326, 1680–3.

Morgan, P. W. and Finlayson, S. A. 2000. Physiology and genetics of maturity and height. In: C. W. Smith and R. A. Frederiksen (Eds), *Sorghum: Origin, History, Technology, and Production*. New York: John Wiley & Sons, Inc, pp.227–59.

Morris, G. P., Ramu, P., Deshpande, S. P., Hash, C. T., Shah, T., Upadhyaya, H. D., Riera-Lizarazu, O., Brown, P. J., Acharya, C. B., Mitchell, S. E., Harriman, J., Glaubitz, J. C., Buckler, E. S. and Kresovich, S. 2013. Population genomic and genome-wide association studies of agroclimatic traits in sorghum. *Proceedings of the National Academy of Sciences*, 110, 453–8.

Mullet, J. E., Klein, R. R. and Klein, P. E. 2002. *Sorghum bicolor* – an important species for comparative grass genomics and a source of beneficial genes for agriculture. *Current Opinion in Plant Biology*, 5, 118–21.
Mutisya, J., Sathish, P., Sun, C., Andersson, L., Ahlandsberg, S., Baguma, Y., Palmqvist, S., Odhiambo, B., Aman, P. and Jansson, C. 2003. Starch branching enzymes in sorghum (*Sorghum bicolor*) and barley (*Hordeum vulgare*): Comparative analyses of enzyme structure and gene expression. *Journal of Plant Physiology*, 160, 921–30.
Naczk, M. and Shahidi, F. 2004. Extraction and analysis of phenolics in food. *Journal of Chromatography*, 1054, 95–111.
Oliver, R. 1966. The problem of the Bantu expansion. *Journal of African History*, 7, 361–76.
Paterson, A. H., Bowers, J. E., Bruggmann, R., Dubchak, I., Grimwood, J., Gundlach, H., Haberer, G., Hellsten, U., Mitros, T., Poliakov, A., Schmutz, J., Spannagl, M., Tang, H., Wang, X., Wicker, T., Bharti, A. K., Chapman, J., Feltus, F. A., Gowik, U., Grigoriev, I. V., Lyons, E., Maher, C. A., Martis, M., Narechania, A., Otillar, R. P., Penning, B. W., Salamov, A. A., Wang, Y., Zhang, L., Carpita, N. C., Freeling, M., Gingle, A. R., Hash, C. T., Keller, B., Klein, P., Kresovich, S., McCann, M. C., Ming, R., Peterson, D. G., Mehboob-ur-Rahman, Ware, D., Westhoff, P., Mayer, K. F. X., Messing, J. and Rokhsar, D. S. 2009. The *Sorghum bicolor* genome and the diversification of grasses. *Nature*, 457, 551–6.
Pedersen, J. F., Bean, S. R., Graybosch, R. A., Park, S. H. and Tilley, M. 2005. Characterization of waxy grain sorghum lines in relation to granule-bound starch synthase. *Euphytica*, 144, 151–6.
Pelling, R. 2005. Garamantian agriculture and its significance in a wider North African context: The evidence of the plant remains from the Fazzan project. *The Journal of North African Studies*, 10, 397–412.
Phillips, S. 1995. *Poaceae: Sorghum purpureosericeum* description. In: Hedberg, I. and Edwards, S. (Eds), *Flora of Ethiopia and Eritrea Volume 7*. Sweden: Swedish Science Press.
Phillipson, D. W. 2005. *African Archaeology*. Cambridge: Cambridge University Press.
Porter, K. S., Axtell, J. D., Lechtenberg, V. L. and Colenbrander, V. F. 1978. Phenotype, fiber composition, and *in vitro* dry matter disappearance of chemically induced brown midrib (bmr) mutants of sorghum. *Crop Science*, 18, 205–8.
Potts, D. T. 1993. The late prehistoric, protohistoric, and early historic periods in eastern Arabia (c. 5000–1200 BC). *Journal of World Prehistory*, 7, 163–212.
Price, H. J., Dillon, S. L., Hodnett, G., Rooney, W. L., Ross, L. and Johnston, J. S. 2005a. Genome Evolution in the Genus Sorghum (*Poaceae*). *Annals of Botany*, 95, 219–27.
Price, H. J., Hodnett, G. L., Burson, B. L., Dillon, S. L. and Rooney, W. L. 2005b. *A Sorghum bicolor × S. macrospermum* hybrid recovered by embryo rescue and culture. *Australian Journal of Botany*, 53, 579–82.
Price, H. J., Hodnett, G. L., Burson, B. L., Dillon, S. L., Stelly, D. M. and Rooney, W. L. 2006. Genome dependent interspecific hybridisation of *Sorghum bicolor* (*Poaceae*). *Crop Science*, 46, 2617–22.
Pwiti, G. 1996. Settlement and subsistence of prehistoric farming communities in the mid-Zambezi Valley, northern Zimbabwe. *South African Archaeological Bulletin*, 51, 3–6.
Regassa, T. H. and Wortmann, C. S. 2014. Sweet sorghum as a bioenergy crop: Literature review. *Biomass and Bioenergy*, 64, 348–55.
Robbins, R. J. 2003. Phenolic acids in foods: An overview of analytical methodology. *Journal of Agricultural and Food Chemistry*, 51, 2866–87.
Robertson, J. H. 2000. Iron Age archaeology in Central Zambia. *Azania*, 35, 1.
Rooney, L. W. and Waniska, R. D. 2000. Sorghum food and industrial utilization. In: Smith, C. W. and Frederiksen, R. A. (Eds), *Sorghum: Origin, History, Technology and Production*. New York: John Wiley & Sons, Inc.
Rosenow, D. T. 1992. Using germplasm from the world collection in breeding for disease resistance. *Sorghum and Millers Diseases – A second world review. Icrisat, Patancheru, Andhra Pradesh*, 502324, 319–24.

Rosenow, D. T. and Dahlberg, J. A. 2000. Collection, conversion and utilization of sorghum. In: Smith, C. W. and Frederiksen, R. A. (Eds), *Sorghum: Origin, History, Technology, and Production*. New York: John Wiley & Sons.

Rowley-Conwy, P. 1991. Sorghum from Qasr Ibrim, Egyptian Nubia, c. 800 BC-AD 1811: A preliminary study. In: Renfrew, J. (Ed.), *New Light on Early Farming: Recent Developments in Paleoethnobotany*. Edinburgh: Edinburgh University Press, pp. 191–212.

Rowley-Conwy, P., Deakin, W. and Shaw, C. H. 1999. Ancient DNA from sorghum. In: van der Veen, M. (Ed.), *The Exploitation of Plant Resources in Ancient Africa*. New York: Kluwer academic, pp. 55–61.

Serna-Saldivar, S. and Rooney, L. 1995. Structure and chemistry of sorghum and millets. In: Dendy, D. A. V. (Ed.), *Sorghum and Millets, Chemistry and Technology*. St Paul, MN: American Association of Cereal Chemists, pp. 69–124.

Shapter, F. M., Henry, R. J. and Lee, L. S. 2008. Endosperm and starch granule morphology in wild cereal relatives. *Plant Genetic Resources*, 6, 85–97.

Shapter, F. M., Dawes, M. P., Lee, L. S. and Henry, R. J. 2009. Aleurone and subaleurone morphology in native Australian wild cereal relatives. *Australian Journal of Botany*, 57, 688–96.

Sharma, H. C. and Franzmann, B. A. 2001. Host-plant preference and oviposition responses of the sorghum midge, *Stenodiplosis sorghicola* (Coquillett) (Dipt., Cecidomyiidae) towards wild relatives of sorghum. *Journal of Applied Entomology*, 125, 109–14.

Shawrang, P., Sadeghi, A. A., Behgar, M., Zareshahi, H. and Shahhoseini, G. 2011. Study of chemical compositions, anti-nutritional contents and digestibility of electron beam irradiated sorghum grains. *Food Chemistry*, 125, 376–9.

Shewayrga, H., Jordan, D. R. and Godwin, I. D. 2008. Genetic erosion and changes in distribution of sorghum (*Sorghum bicolor* L. (Moench)) landraces in north-eastern Ethiopia. *Plant Genetic Resource*, 6, 1–10.

Shewry, P. R. and Halford, N. G. 2002. Cereal seed storage proteins: Structures, properties and role in grain utilization. *Journal of Experimental Botany*, 53, 947–58.

Shipton, C., Helm, R., Boivin, N., Crowther, A., Austin, P. and Fuller, D. Q. 2013. Intersections, networks and the genesis of social complexity on the Nyali Coast of East Africa. *African Archaeological Review*, 30, 427–53.

Sinclair, J. 1982. Chibuene: An early trading city in southern Mozambique. *Paideuma*, 28, 149–64.

Sinclair, P., Morais, J., Adamowicz, L. and Duarte, R. T. 1993. A perspective on archaeological research in Mozambique. In: Shaw, T., Sinclair, P., Andah, B. and Okpoko, A. (Eds), *The Archaeology of Africa: Food, Metals and Towns*. London: Routledge.

Smith, W. S. and Frederikson, R. A. 2000. History of cultivar development in the United States: from memoirs of A.B. Maunder – sorghum breeder. In: Smith, C. W. and Frederiksen, R. A. (Eds), *Sorghum: Origin, History, Technology, and Production*. New York: John Wiley & Sons, pp. 191–223.

Stefoska-Needham, A., Beck, E. J., Johnson, S. K. and Tapsell, L. C. 2015. Sorghum: An underutilized cereal whole grain with the potential to assist in the prevention of chronic disease. *Food Reviews International*, 31, 401–37.

Stemler, A. 1990. A scanning Electron Microscopic analysis of plant impressions in pottery from Kadero, El Zakiab, Um Direiwa and Al Kadada. *Archéologie du Nil Moyen*, 4, 87–105.

Stevens, C. J., Murphy, C., Roberts, R., Lucas, L., Silva, F. and Fuller, D. Q. 2016. Between China and South Asia: A Middle Asian corridor of crop dispersal and agricultural innovation in the Bronze Age. *The Holocene*, 26(10), 1541–55.

Stewart, A. 2013. *The Drunken Botanist*. New York: Algonquin Books.

Taylor, T. R. N., Tilman, J. S. and Bean, S. R. 2006. Novel food and non-food uses for sorghum and millets. *Journal of Cereal Science*, 44, 252–71.

Tengberg, M. 2003. Archaeobotany in the Oman peninsula and the role of Eastern Arabia in the spread of African crops. In: Neumann, K., Butler, A. and Kahlheber, S. (Eds), *Food, Fuel and*

Fields: Progress in African Archaeobotany, Africa Praehistorica 15. Koln: Heinrich-Barth Institute, pp. 229–38.

Tesso, T., Ejeta, G., Chandrashekar, A., Huang. C.-P., Tandjung, A., Lewamy, M., Axtell, J. D. and Hamaker, B. R. 2006. A novel modified endosperm texture in a mutant highprotein digestibility/high-lysine grain sorghum (*Sorghum bicolor* (L.) Moench). *Cereal Chemistry*, 83, 194–201.

Theurer, C. B., Huber, J. T., Delgado-Elorduy, A. and Wanderley, R. 1999. Summary of steam-flaking corn or sorghum grain for lactating dairy cows. *Journal of Dairy Science*, 82, 1950–9.

Thurber, C. S., Ma, J. M., Higgins, R. H. and Brown, P. J. 2013. Retrospective genomic analysis of sorghum adaptation to temperate-zone grain production. *Genome Biology*, 14(6), 1–13.

Vansina, J. 1995. New linguistic evidence and the Bantu expansion. *Journal of African History*, 36, 173–95.

Walshaw, S. C. 2015. *Swahili Trade, Urbanization, and Food Production: Botanical Perspectives from Pemba Island, Tanzania*, 600–1500. Oxford: Archaeopress.

Wasylikowa, K. and Dahlberg, J. A. 1999. Sorghum in the economy of the Early Neolithic nomadic tribes at Nabta Playa, Southern Egypt. In: van der Veen, M. (Ed.), *The Exploitation of Plant Resources in Ancient Africa*. New York: Kluwer academic, pp. 511–31.

Wasylikowa, K., Schild, R., Wendorf, F., Królik, H., Kubiak-Martens, L. and Harlan, J. R. 1995. Archaeobotany of the early Neolithic site E-75-6 at Nabta Playa, western desert, south Egypt. *Acta Palaeobotanica*, 35, 133–55.

Wendorf, F., Close, A. E., Schild, R., Wasylikowa, K., Housley, R. A., Harlan, J. R. and Krolik, H. 1992. Saharan exploitation of plants 8,000 years BP. *Nature*, 359, 721–4.

Westengen, O. T., Okongo, M. A., Onek, L., Berg, T., Upadhyaya, H., Birkeland, S., Khalsa, S. D. K., Rings, K. H., Stenseth, N. C. and Brysting, A. K. 2014. Ethnolinguistic structuring of sorghum genetic diversity in Africa and the role of local seed systems. *Proceedings of the National Academy of Sciences of the United States of America*, 111, 14100–5.

Wiersema, J. H. and Dahlberg, J. 2007. The nomenclature of *Sorghum bicolor* (L.) Moench (*Gramineae*). *Taxon*, 56, 941–6.

Wigboldus, J. S. 1990. Disputable datings of early sorghum cultivation in the southern Old World: A case for tracing crop evolution and diffusion with help from history. In: Posey, D. A. and Overal, W. L. (Eds), *Ethnobiology, Implications and Applications*. Belem, Brazil: Museu Paraense Emílio Goeldi, pp. 291–300.

Willcox, G. 1992. Some differences between crops of Near Eastern origin and those from the tropics. In: Jarrige, C. (Ed.), *South Asian Archaeology 1989*. Madison: Prehistory Press, pp. 291–9.

Wong, J. H., Lau, T., Cai, N., Singh, J., Pedersen, J. F., Vensel, W. H., Hurkman, W., Wilson, J. D., Lemaux, P. G. and Buchanan, B. B. 2009. Digestibility of protein and starch from sorghum (*Sorghum bicolor*) is linked to biochemical and structural features of grain endosperm. *Journal of Cereal Science*, 49, 73–82.

Wright, D. K. 2005. New perspectives on early regional interaction networks of East African trade: A view from Tsavo National Park, Kenya. *African Archaeological Review*, 22, 111–40.

Zohary, D., Hopf, M. and Weiss, E. 2012. *Domestication of Plants in the Old World: The Origin and Spread of Domesticated Plants in Southwest Asia, Europe and the Mediterranean Basin*. Oxford: Oxford University Press.

Chapter 3

Growth, development and physiology of grain sorghum

M. Djanaguiraman, P. V. V. Prasad and I. A. Ciampitti, Kansas State University, USA

1 Introduction

Sorghum [*Sorghum bicolor* (L.) Moench] is the fifth most important coarse cereal. It is mainly grown in the semi-arid tropics for food and nutritional security. Sorghum is a staple cereal food crop for more than 500 million people throughout the world. The production and utilization of sorghum fall within two broad regions. The first region comprises countries in Africa and Asia, where sorghum is produced using traditional practices, at a small-scale, subsistence level and the grain is mainly used for human consumption. The second region includes some developed and a few developing countries, where the sorghum production system is mechanized and on a large scale, more inputs are utilized and yields are higher, with grains used for animal feed. The United States is the largest sorghum producer, followed by Mexico, Nigeria and India. The top exporters are the United States, Australia and Argentina. The average productivity of sorghum varies from 0.28 Mg ha^{-1} in Nigeria to 4.4 Mg ha^{-1} in Argentina (FAOSTAT, 2014). Across the globe, the average sorghum yield remains below 1 Mg ha^{-1}, because of the low level of inorganic fertilizer use, the low or non-existent use of pesticides and the use of traditional varieties or landraces. However, where intensive agriculture is practised, the yields are much higher and are comparable with those of other major coarse cereal crops. Over the past 25 years, sorghum production has increased steadily in the semi-arid tropics due to a combination of an increase in the area under cultivation and better techniques to improve yields.

http://dx.doi org/10.19103/AS.2017.0015.03

Sorghum is a genetic diploid with a base chromosome number of n = 10 and 2n = 2 × 10 = 20. Although there is evidence that the crop is an ancient tetraploid, it is a functional diploid. The genome of sorghum is smaller than maize, but larger than rice. Sorghum wild relatives and landraces, consisting of 235711 accessions, are housed in national and international gene banks. Sorghum has four wild races and five cultivated races (Harlan and Stemler, 1976). The four wild races of *Sorghum bicolor* are arundinaceum, virgatum, aethiopicum and verticilliflorum, and these are placed in *Sorghum bicolor* subspecies *verticilliflorum*, formerly subspecies *arundinaceum*. The five cultivated races are bicolor, guinea, caudatum, kafir and durra. This classification is based on the panicle type. Natural hybridization between these cultivated races has resulted in the evolution of ten hybrid races, namely guinea–bicolor, guinea–caudatum, guinea–kafir, guinea–durra, caudatum–bicolor, kafir–bicolor, kafir–caudatum, kafir–durra, durra–bicolor and durra–caudatum. Bicolor is widely distributed across the African Savannah and Asia; caudatum can be found in Central Sudan and surrounding areas; guinea in eastern and western Africa; durra in Arabia and Asia Minor, India, Myanmar, Ethiopia and along the Nile Valley; and kafir in south eastern Africa (Harlan and de Wet, 1972).

In the United States, sorghum is classified into four classes based on tannin content and grain colour, namely sorghum, tannin sorghum, white sorghum and mix sorghum. The sorghum and white sorghum groups are characterized by low tannin content, whereas tannin sorghum has a high tannin content. The mix sorghum group does not meet the requirements for any of the other classes. Alternatively, based on its use, sorghum is classified into four types, namely grain sorghum, sweet sorghum, broom sorghum and grass sorghum. Grain sorghum is mainly used as a principal food, broom sorghum is used for making brooms, sweet sorghum is used for making sweetener syrup and grass sorghum is grown for green feed and forage use.

Knowing how a sorghum crop develops is important in understanding how to manage the crop best. In this chapter, the various growth stages, from emergence to physiological maturity, are presented and fully described. In general, the sorghum growth stage can be classified into three major phases, namely: (i) vegetative; (ii) reproductive and (iii) grain filling. The duration of each growth stage presented in this chapter is an overview, because growth depends on the genotype and the growing environment. Other factors, such as soil fertility, insect or disease damage, moisture stress and weed competition, can also affect the duration of each growth stage. Sorghum growth and yield are significantly influenced by abiotic stresses like drought, high temperature, an elevated carbon dioxide (CO_2) level and the interaction of these stresses. The effects of abiotic stresses on sorghum growth and yield depend on the stage of development and the intensity and duration of the stress. Sorghum is most sensitive to drought stress during the booting stage, booting to anthesis and milk dough to the soft dough stages, compared to other growth stages (Lewis et al., 1974; Assefa et al., 2010). Sorghum is most sensitive to high temperature stress during gametogenesis and at flowering than at other growth stages (Prasad et al., 2008). In this chapter, sorghum origin, botany, growth stages and climatic requirements (photoperiod, rainfall and temperature) are presented. Regarding climatic requirements, the effects and physiological basis of drought, high temperature and elevated carbon dioxide tolerance are discussed.

2 Origin, spread and adaptation

Northeast Africa is believed to be the place of origin for sorghum. Sorghum was first domesticated in Africa, and cultivation then spread to India and China. Major sorghum-producing areas include the Great Plains of North America, the north central, north western and eastern mid-altitude areas of Ethiopia, Rwanda, northern and eastern Uganda and Central Tanzania, and the areas in Kenya and Tanzania that are east of Lake Victoria, Northeast China and Central India. For the United States, important sorghum-producing states are Kansas, Texas, Oklahoma, Nebraska, New Mexico and Missouri. Kansas is the largest producer, contributing to about 40–50% of total US sorghum production. Globally, sorghum is grown over an area of 45 million ha. Sudan accounts for 21% of Africa's sorghum production, followed by Nigeria (34%). Ethiopia, Tanzania, Uganda, Rwanda and Kenya account for 7%, 4%, 2%, 0.8% and 0.6%, respectively (Mitaru et al., 2012). India contributes to about 16% of global sorghum production.

The adaptive nature of sorghum includes its extensive root system, ability to tiller, different head size, diverse grain colour, perennial nature and photoperiod sensitivity. Sorghum is a C_4 plant, with better photosynthetic capacity and greater nutrient-use and water-use efficiency. Sorghum's high water-use efficiency has been associated with its characteristic of drought tolerance. Sorghum plants require up to 70–100 mm of water every 10 days in their early stages of growth. As growth progresses, the roots penetrate more deeply into the soil to tap into hidden water reserves and the plant progressively needs less water. Sorghum requires a total of 332 kg of water per kg of dry matter, whereas maize, barley and wheat require 368, 434 and 514 kg of water, respectively, for production of 1 kg of dry matter (Reddy et al., 2011). Sorghum's ability to survive in conditions where there is less moisture may be attributed to its ability to withstand the drought through differential root and leaf adaptations. Sorghum has a relatively greater tolerance to harsh environments, such as dry weather and high temperatures, in comparison to other cereal crops, such as maize, rice or wheat.

3 Botany

Grain sorghum is an annual, monoecious, erect plant with a solid stem, which can grow up to 5 m high, depending mainly on its photoperiod sensitivity. The first visible leaf at emergence is the coleoptile leaf. It differs from other leaves in that it has rounded leaf tip. The leaf collar marks the junction of the leaf blade and sheath and appears when the leaf is fully developed. The leaf sheath originates at a node on the plant stalk. The leaf sheath surrounds the stalk and provides mechanical strength. The flag leaf is the last and smallest leaf to emerge on the plant. The sorghum leaf has a prominent midrib and a typical leaf blade. The leaf sheath and stem are often covered with a waxy bloom. The sorghum plant produces 7–28 leaves with a parallel venation. The sorghum phyllotaxis (the arrangement of leaves on a plant) is alternate. The number of leaves varies among the genotypes. The leaves are 50–100 mm wide and 0.5–0.8 m long, depending on the genotype. Leaves and stems are often covered with a wax layer, which is an adaptation mechanism to tolerate drought and insect attacks (Kumar, 2016). The stem consists of nodes and internodes. The number of nodes and the number of leaves are the same. The stem diameter ranges from 0.5 to 5 cm near the base. A typical sorghum plant has 14–17 nodes and internodes.

Sorghum has an extensive fibrous root system which can grow as deep as 3 m. The sorghum root system consists of three types of root, that is, primary (seminal), secondary (adventitious) and nodal (crown) roots. Primary roots develop directly from an elongation of the radicle and subsequently die after about 4–5 days. They are easily identified by their continuity with the base of the culm. Secondary roots develop from the first and second internodes on the mesocotyl. They are branched and are mainly used for the absorption and uptake of water and nutrients. Nodal roots initiate from the base of the stem about 30–40 days after sowing. Nodal roots are thick above the ground level and thin in the soil, and produce many lateral roots. Seminal roots subsequently die after about 4–5 days. Nodal roots initiate from the base of the stem about 30–40 days after sowing.

The primary roots of sorghum grow more vertically. The post-embryonic roots are called crown and brace roots. Brace roots are formed at the nodes above the soil surface and are usually up to a maximum of three nodes above the sixth node. In addition to the subterranean root system, sorghum also forms a strong aerial root (nodal roots), permeating through the soil and ensuring better stability. The root system of sorghum was studied in detail by Miller (1916), who showed that the primary roots of sorghum are finer and more fibrous. Root distribution and activity depend on the soil type, the level of moisture, tillage and fertilizer application (Blum and Ritchie, 1984). Nakayama and van Bavel (1963) found that 90% of the root activity of irrigated sorghum is concentrated within an area measuring 91 cm deep and 38 cm wide from the plant.

The floral initiation (the primordial differentiation on the floral apex) occurs in the stem at 15–30 cm above the ground when the plants are about 50–75 cm tall. This phase occurs at the end of the vegetative stage. The time required for transformation from the vegetative primordial to the reproductive primordial is greatly influenced by the genotype and the environment. The main growth period follows the formation of a floral bud. Blum (2013) reported that sorghum hybrids take less time to reach the panicle initiation stage and more time to expand the panicle compared with their corresponding parents, leading to higher heterosis. Sorghum is characterized by an inflorescence (raceme) and grain (caryopsis) in the form of a panicle. The spikelets presented in the panicle are borne in pairs. The sorghum plant has an incomplete but perfect flower, with both its male and female parts located in the same head. Because stamen and pistil are present in the same position or flower, it is considered to be a self-pollinated crop. However, it shows 0–50% cross-pollination, depending on the genotype.

The sorghum panicle emerges from the flag leaf sheath and is supported by the portion of the stalk called the peduncle. A sorghum panicle is classified as loose, compact, calendric, conic, ellipsoid, oblong, lanceolate, ovoid, truncate and so forth, based on the different possible combinations of the length of rachis, the mode of variation of the primary panicle branches, the extent of the ramification of panicle branching, the angle which the whorls make with the rachis, the secondary branch zone of the primary branch and the density of the clustering of fertile spikelets (Aruna and Audilakshmi, 2008). The seed (grain) of sorghum is partially enclosed in the glumes, which can be removed during threshing. The shape of the seed is oval to round, and the colour varies from red, white, yellow, brown or shades thereof. If the pericarp alone is coloured, the seed is usually yellow or red. Pigment in both the pericarp and testa can result in a dark brown or reddish-brown seed colour.

The high photosynthetic capacity of C_4 sorghum plants is due to sorghum's unique mode of carbon assimilation, which concentrates CO_2 at the site of the enzyme RuBisCO. The CO_2 pump provides high rates of photosynthesis even at lower concentrations of CO_2. The functioning of C_4 photosynthesis is dependent upon the strict compartmentation of the CO_2 assimilatory enzymes into two distinct cell types, namely mesophyll and bundle sheath cells. The primary CO_2 assimilatory enzyme, namely phosphoenolpyruvate carboxylase, accumulates in the mesophyll cells, although the second carboxylase enzyme (RuBisCO) and the decarboxylating enzyme (NADP-dependent malic enzyme) are restricted in the bundle sheath cells. The atmospheric CO_2 is fixed in the mesophyll cells by phosphoenolpyruvate carboxylase to form a four-carbon compound, and then exported to the bundle sheath cells, where a decarboxylation occurs to yield CO_2 that is refixed, via the Calvin cycle, by the enzyme RuBisCO. The elevated CO_2 concentrations in the bundle sheath cells decrease the oxygenase activity of RuBisCO, and consequently, the process of photorespiration is decreased.

The function of the C_4 pathway is to provide CO_2 to the Calvin pathway and to refix any CO_2 originating from photorespiration. The C_4 pathway is not an alternative route for carbon fixation, but an additional one. Thus, it reduces the loss of energy that takes place in C_3 plants as a consequence of the oxygenase activity of the enzyme RuBisCO. As a consequence of this CO_2 pumping, the competitive inhibition of RuBisCO by oxygen is largely excluded, leading to a reduced rate of photorespiration. However, in C_3 plants, 30% of the net CO_2 fixed in photosynthesis is lost during photorespiration. Since the C_4 plants have a mechanism for concentrating CO_2 in the bundle sheath cells, their photosynthetic rates will not respond to rising CO_2 levels to the same extent as C_3 plants. Leaf photosynthetic rates are known to be sensitive to temperature. C_4 plants like sorghum have a higher temperature maximum for photosynthesis than C_3 plants. However, the relative differences are small at lower temperatures. Maximum photosynthesis occurs at a temperature of at least 30°C. The higher photosynthetic efficiency of sorghum is due to the presence of a specialized leaf anatomy called 'Kranz anatomy'.

Growing degree-days (GDDs) are a way of assigning a heat value to each day, following a calculation using the air temperature. The values are added together to give an estimate of the amount of seasonal growth that a plant has achieved. The GDD is the difference between the daily mean temperature and the base temperature. For sorghum, the base and upper temperatures are 10°C and 30°C, respectively. The duration from emergence to panicle initiation comprises a juvenile or pre-inductive phase followed by an inductive photoperiod-sensitive phase (Major and Kiniry, 1991). The rate of progress can be quantified by linear responses to a mean temperature and photoperiod (Craufurd et al., 1999). For short-day plants, under optimum day-length conditions, the duration of the photoperiod inductive phase is assumed to be 0°-days, and no delay in flowering occurs. For panicle initiation, a short- and long-season hybrid requires 495 and 740 cumulative GDD (°C), respectively. Similarly, for physiological maturity, it requires 1467 and 1848 cumulative GDD (°C) for the short- and long-season hybrids, respectively.

4 Growth and development

The growth and development of sorghum is broadly divided into vegetative and reproductive growth stages. The growth and development of the sorghum plant can be divided into three phases, namely: (i) vegetative, (ii) reproductive and (iii) grain filling. The vegetative phase consists of seed germination, emergence, three-leaf, five-leaf and the floral initiation stages. The reproductive phase includes flag leaf emergence, booting and the flowering stages. The grain filling phase encompasses soft dough, hard dough and physiological maturity (Murdy et al., 1994). Eastin (1972) reported three simplified sorghum growth stages, namely: (i) planting to panicle initiation, (ii) panicle initiation to flowering and (iii) flowering to physiological maturity. However, Vanderlip and Reeves (1972) described the temperate sorghum growth stages on a scale of 0–9 (see Fig. 1 and Table 1). These are common descriptors of growth and the development of sorghum, and are widely used in the literature. The stages and images of the description were recently revised by Ciampitti (2015) and are described below.

4.1 Stage 0 (Emergence)

Seed germination begins with the absorption of moisture. Sorghum is a warm-season crop that requires soil temperatures of at least 16°C to initiate the germination process. Emergence is considered to be the stage when the seedling breaks through the soil surface. This can be identified when the coleoptile is visible at the soil surface, but the total number of days depends on the soil temperature. Under optimum soil moisture and temperature conditions, germination occurs in 3–5 days and seedlings emerge in about 5–10 days. The radicle grows downwards into the soil and forms the first primary seminal roots. Emergence will vary depending on planting depth, soil moisture and temperature, soil compaction, residue cover and distribution and seed vigour. Planting should be done to synchronize with adequate moisture in the top 15 cm of soil profile. Optimum planting depth varies with soil moisture, but usually 2.5–5 cm is recommended. Planting within 10 days after the onset of the monsoon will result in good seedling emergence.

4.2 Stage 1 (Three-leaf stage)

Three leaves are fully expanded with a visible collar. At this stage, the plant is very sensitive to weed competition. The first manual weeding should be done to promote better seedling

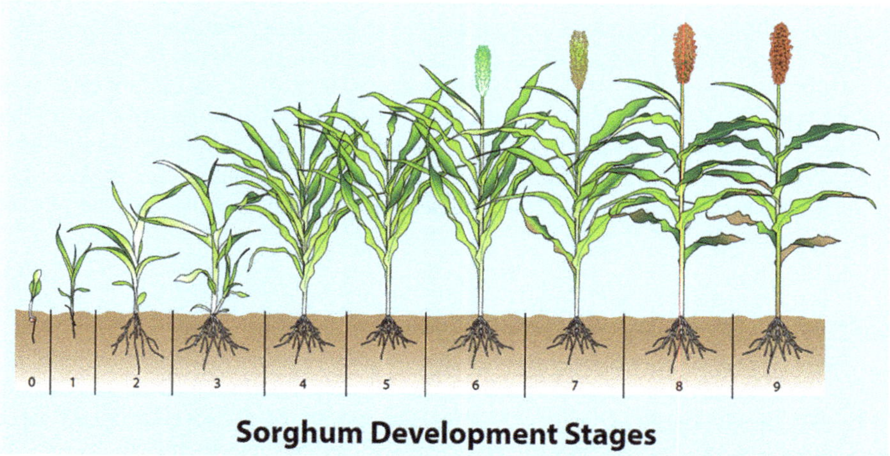

Figure 1 Sorghum growth and development stages, adapted from Ciampitti (2015). Stage 0 = Emergence; 1 = Three-leaf stage; 2 = Five-leaf stage; 3 = Growing point differentiation; 4 = Flag leaf visible; 5 = Boot stage; 6 = Half-bloom; 7 = Soft dough; 8 = Hard dough; 9 = Physiological maturity. Source: https://www.bookstore.ksre.ksu.edu/pubs/MF3234.pdf.

Table 1 Growth and development stages of grain sorghum

Growth stage	Name	Description
Stage 0	Emergence	Emergence occurs when the coleoptile is visible at the soil surface.
Stage 1	Three-leaf stage	The three-leaf stage occurs when the collars of three leaves can be seen without dissecting the plant (collar of third leaf is visible).
Stage 2	Five-leaf stage	The five-leaf stage occurs when the collars of five leaves can be seen without dissecting the plant, occurring approximately three weeks after emergence.
Stage 3	Growing point differentiation	The growing point of the sorghum plant changes from vegetative to reproductive.
Stage 4	Flag leaf visible in the whorl	All leaves, except the final three to four leaves, are fully expanded, representing about 80% of the total leaf area potential.
Stage 5	Boot stage	All leaves are fully expanded, maximizing leaf area and light interception. The head reaches maximum size and is encompassed by the flag leaf sheath.
Stage 6	Half-bloom	This stage is defined as the stage when half of the plants in a field have started to bloom.
Stage 7	Soft dough	The grain has a dough-like consistency and grain fill occurs rapidly.
Stage 8	Hard dough	By this stage, approximately three-fourths of the grain dry weight has been attained.
Stage 9	Physiological maturity	The maximum total dry weight of the plant has been reached.

growth and moisture conservation. The growing point is under the soil surface. This stage occurs 10–20 days after emergence, depending on soil moisture and temperature.

4.3 Stage 2 (Five-leaf stage)

Five leaves are fully expanded with a visible collar. The plant begins a rapid growth and nutrient uptake phase. The root system also expands rapidly. The growing point is still under the soil surface. The crop should be thinned to one seedling per hill in order to reduce competition for soil moisture. Plant densities should be less than three plants per 60 cm of row to promote tillering. A primary factor in establishing a sorghum stand is temperature. Sunny days with a temperature below 16°C promote tillering when the plants are in the four-to-six-leaf stage. This stage occurs 20–25 days after emergence.

4.4 Stage 3 (Panicle initiation or growing point differentiation stage)

The growing point changes from the vegetative to the reproductive phase, from producing leaves to forming heads. At this stage, the plant will have nine to ten leaves, depending on the maturity group, and the basal two to three leaves may become senesced. Culm growth increases rapidly following this stage. Panicle initiation can be identified by splitting the stalk with a sharp knife and can be seen under the microscope as an appearance of protrusions (raised bumps) on the surface of the plant's growing point. These are the primordial structures that form the panicle branches, and later differentiate into the spikelet branches and floral structures that are responsible for seed-set. Maximum plant growth and nutrient uptake rates are achieved. A top-dressing of nitrogen (N) fertilizer is recommended to hasten panicle growth. Inter-cultivation practices and second weeding during this stage enable root growth and conserve soil moisture. During this period, the plants are sensitive to stresses such as drought, high temperature and nutrient deficiency. A significant decrease in the number of seeds per panicle can occur if stress is impacting at this stage. It is considered to be the most critical period for grain production since seed numbers per plant account for 70% of the grain yield. This stage occurs 30–40 days after emergence.

4.5 Stage 4 (Flag leaf visible)

The flag leaf is the last leaf to emerge from the whorl and it will be smaller than the other leaves of the plant. This stage can be identified by observing the appearance of the tip of flag leaf in the whorl. During this stage, the plants will show a rapid stem elongation and all the leaves in the plants are fully expanded. The top three to four leaves will be photosynthetically active and the bottom three to four leaves will show symptoms of senescence and sometimes will be lost completely. When the flag leaf collar appears, the plant will be in the boot stage. Potassium uptake is around 40%, nitrogen 30%, phosphorus 20% and total growth is about 20% complete relative to the final nutrient content. This stage occurs about 50 days after emergence.

4.6 Stage 5 (Boot stage)

Boot stage can be identified when a swollen flag leaf sheath encloses the developing panicle, resembling a boot shape. All the leaves are fully expanded, and the panicle is fully developed (with a maximum potential head size) and enclosed in the flag leaf sheath. The upper stalk, the peduncle, begins to elongate, resulting in the exertion of the head from the flag leaf sheath. The final size of the peduncle varies with the genotype. This stage is very sensitive to abiotic stress, such as drought and high temperature stress. Previous research has indicated that drought and/or high temperature stress during this stage causes the incomplete exertion of the panicle, resulting in a reduced seed-set percentage. This stage occurs 50–60 days after emergence.

4.7 Stage 6 (Half-bloom)

Full exertion of the head occurs at this stage and 50% of the plants in the field are in some stage of bloom. For an individual plant, this stage is when the flowering reaches 50% of the head, with an approximate 50% of the total growth complete. In general, flowering will start 5–7 days after panicle exertion. Anthesis (flower opening) starts about two days after the complete emergence of the inflorescence (panicle) from the boot leaf. At the start of anthesis, the tip of the lemma and palea slightly opens, filaments elongate and the anther starts to emerge, followed by the stigma. Flowering starts from the top of the panicle and progresses to the bottom of the panicle. Maximum flowering occurs in the middle of this period, at about 3–6 days after the start of anthesis. The duration of flowering in sorghum is between 4 and 9 days. Drought or heat stress at this stage causes a poor seed-set percentage. Anthesis starts around midnight and proceeds up to 10 am, depending on the cultivar and the environmental conditions. Maximum anthesis occurs between 6 am and 8 am. The floret opens as a result of a swelling of the two lodicules at the base of the florets, which exerts pressure on the glumes, forcing them to open in about 10 min. Anthers protrude first, followed by the stigma, a condition termed 'protandry'. Anthers shed pollen (dehiscence) just before or shortly after anthesis, between 6 am and 7 am, depending on the cultivar and the weather. On a dry and clear day, dehiscence occurs soon after sunrise, but on a cool and damp day, it may extend until 10 am. Pollen in the anthers remains alive for 3–6 h. After dehiscence, pollen grains are alive for only 1 h, or even shorter, depending on the cultivar, temperature and the level of humidity. The stigma is receptive for about 10 days. Sorghum is a self-pollinating plant with cleistogamy. Pollination occurs with the shedding of pollen grains on the stigma. Pollen grains start to germinate immediately after contact with the stigma and develop pollen tubes through the style. Fertilization will be completed in 2 h. Sorghum pollen grains require light to germinate. Flowers remain open for about 30–90 min. After pollination, the glumes close, but the empty anthers and stigma still protrude. During this process, the most important physiological events leading to the formation of seeds, such as fertilization, embryo formation and seed-set, will happen. Compared to the final nutrient content, nutrient accumulation is 60% for phosphorus, 70% for nitrogen and above 80% for potassium. This stage occurs about 70 days after emergence.

4.8 Stage 7 (Soft dough)

Seed development progresses from milk through to the soft dough stage. This stage can be identified by squeezing the developing seed between the fingers to look for the presence of a little or no milky juice. Drought or heat stress can significantly decrease grain growth, leading to lighter and chaffy grains, and a small grain size. Kernels reach their maximum volume approximately 10 days after flowering, during the milk stage. The stem loses weight due to a remobilization process (from stem to grain). The soft dough stage occurs approximately 15–25 days after flowering.

4.9 Stage 8 (Hard dough)

At this stage, the grain cannot be compressed between the fingers. This stage is reached about 96 days after emergence. During this stage, approximately 75% of the grain weight has been achieved and nutrient uptake is almost complete. Drought stress during the hard dough stage can result in a shrivelled grain with a low seed size. The lower leaves lose functionality due to the remobilization of nutrients to the grains. A severe stress can still reduce grain weight, but not to the extent possible in the soft dough stage.

4.10 Stage 9 (Physiological maturity)

Physiological maturity occurs when a black layer appears immediately above the point of kernel attachment in the floret near the base of the kernel (see Fig. 2). During this stage, the grain has attained its full potential weight and has a moisture content of 30–35%. The plants can be harvested when the moisture content reaches approximately 20%.

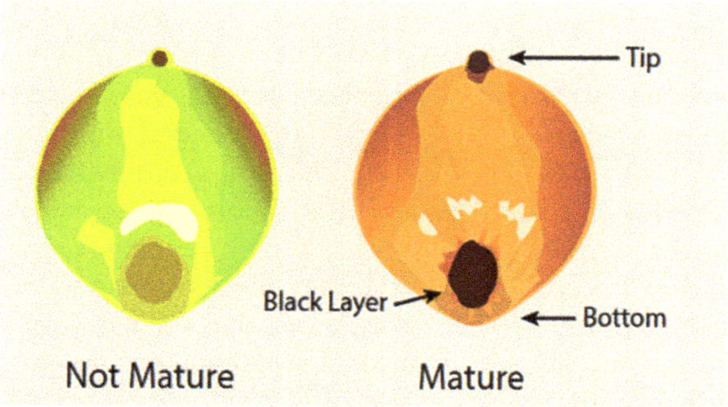

Figure 2 An immature and physiologically mature grain of the sorghum crop, adapted from Ciampitti (2015). Source: https://www.bookstore.ksre.ksu.edu/pubs/MF3234.pdf.

5 Climatic requirements

5.1 Photoperiod

Sorghum, photoperiodically, is a short-day plant that requires day lengths of less than 12 h 20 min to induce flowering. Hence, most of the global collection (~36000 accessions) flowers too late or is too tall in temperate environments. The duration from emergence to panicle initiation comprises a juvenile or pre-inductive phase and is followed by a photoperiod-sensitive inductive phase (Major and Kiniry, 1991). The rate of progress can be quantified by linear responses to the mean temperature and photoperiod (Craufurd et al., 1999). For short-day plants, under optimum day-length conditions, the duration of the photoperiod-inductive phase is assumed to be 0°-days, and no delay in flowering occurs.

In 1963, the Sorghum Conversion Program was established at the Texas Agricultural Experiment Station-USDA to convert these tropical lines to suit temperate conditions. This was achieved by introgressing recessive day-neutral flowering alleles and dwarf height genes into exotic backgrounds via a backcross scheme to recover the exotic genome in an early-flowering, combined-height inbred (Stephens et al., 1967). In 2009, efforts were made to convert selected sorghum accessions from the USDA-ARS germplasm collections. Accessions from the Sudan and Ethiopian collections were targeted for conversion because of the performance of the previously converted Ethiopian and Sudanese germplasm (grain yield) under temperate conditions. The selection criterion for the tropical accessions was grain yield (seed size and the number of seeds per panicle). As a result of the programme, numerous converted and partially converted lines have been used by sorghum breeding programmes to enhance a wide array of traits in modern sorghum cultivars and hybrids. As the photoperiod changes with latitude (both north and south), as well as with the time of the year in temperate regions, care should be taken when selecting cultivars or hybrids based on their photoperiod sensitivity and the duration of the critical photoperiod. Most of the new sorghum hybrids are photoperiod-insensitive.

Three major genetic components are involved in the varietal response to the photoperiod: (i) the basic vegetative phase (the minimum thermal time required for panicle initiation under optimum day length); (ii) the minimum/maximum optimal photoperiod (the critical photoperiod beyond which the vegetative period is influenced by changes in day length); and (iii) the photoperiod sensitivity slope (Major, 1980). Reciprocal transfer experiments between environments differing in photoperiod and thermoperiod revealed three distinct developmental phases leading to floral initiation in sorghum (Ellis et al., 1997). The duration of the initial juvenile or photoperiod-insensitive phase was shown to be regulated by temperature. A relatively low temperature of 24°C was optimal for a rapid end of the juvenile phase. Following the juvenile phase, sorghum became sensitive to the photoperiod but insensitive to temperature, with shorter photoperiods being more inductive to flowering. The photoperiod-sensitive phase ended some time before floral differentiation could be observed at the shoot meristem. This phase of latent floral initiation was not sensitive to either temperature or photoperiod (Ellis et al., 1997).

5.2 Temperature

Sorghum is grown in a wide range of climates and is well adapted to hot, semi-arid tropical environments. When sown under spring conditions, sorghum can experience cold temperature episodes during germination and seedling growth, leading to reduced growth and a lower yield (Maulana and Tesso, 2013). Temperatures below 20°C cause chilling stress, which greatly affects the agronomic performance of the sorghum crop. RuBisCO activity is a major limitation to carbon assimilation at this temperature (20°C) due to a reduced RuBisCO capacity and content (Kubien et al., 2003). Apart from this, a reduction in the photosynthetic rate is also due to the instability of pyruvate phosphate dikinase under cold conditions, which impairs the capacity to regenerate phosphoenolpyruvate in the mesophyll cells (Wang et al., 2008). Cold conditions coupled with excessive light causes photoinhibition (Pimentel et al., 2005). During this process, excess energy is transferred to the oxygen molecule from the chlorophyll, leading to the production of harmful, reactive oxygen species. To cope with this situation, plants have developed several mechanisms to protect themselves from excessive light. Photoprotection is achieved through a reduction of light absorption due to chloroplast movement (Kagawa and Wada, 2002) and a reduced antenna size (Bailey et al., 2001). In addition, the carotenoid pigments present in the photosystem (PS) can dissipate the excess energy through the xanthophyll cycle. During this process, violaxanthin is converted into zeaxanthin, causing conformational change in the antenna of photosystem PS II, which allows the dissipation of excess energy as heat (Demmig-Adams et al., 1989). Other important photoprotection processes include the cyclic electron flow around PS I, which prevents an over-reduction in the electron transport pathway (Munekage et al., 2004), and the repair of the damaged D1 protein of the PS II by the ATP-dependent metalloprotease (Lindahl et al., 2000). Studies have reported quantitative trait loci (QTLs) for germination and early growth parameters under cold conditions (Burow et al., 2010; Fiedler et al., 2016; Bekele et al., 2014), but photosynthesis and photoprotection were not evaluated.

Plants exposed to chilling temperatures suffer injury, exhibiting a reduction of or the cessation of growth. Initially, this phenomenon is reversible. However, if it persists for a long time, it is irreversible, resulting in cell death. Chilling adversely affects a wide range of processes, including the germination and establishment of the crop, photosynthesis, flowering, plant growth and grain yield (Ortiz et al., 2017). Chilling stress during germination and the establishment of seedlings causes reduced plant density and yield. Similarly, a reduction of the soil temperature during pollination and the grain development period decreases the yield. Sorghum landraces from the temperate regions of China were found to exhibit higher seedling emergence and greater seedling vigour under cold conditions than most sorghum cultivars (Franks et al., 2006). However, these landraces lack other desirable agronomic characteristics. A population developed from a cross between the Chinese landrace Shan Qui Red (cold tolerant) and SRN39 (cold sensitive) was employed for a QTL analysis of early-season cold performance in sorghum. Two QTLs for germination under cold and optimal temperatures – one on the linkage group SBI-03 and the second on the group SBI-07 – were identified. Another QTL located on the linkage group SBI-01 showed a strong association with seedling emergence and seedling vigour scores under early and late field planting (Knoll et al., 2008; Knoll and Ejeta, 2008). These QTLs can be useful for marker-assisted breeding to improve cold tolerance in sorghum. Adams and Thompson (1973) reported that yield increased by 10% when the soil was covered with clear plastic, because the soil temperature increased by about 2°C during the growth period. Paul (1990) showed that, for sorghum, a minimum temperature of 16°C is necessary for all physiological processes to occur. Low temperatures (<10°C) have resulted in decreased plant height and leaf area, dry matter accumulation and delayed flowering. If the average daily temperatures are lesser than 20°C, there will be an extension of the growth period from 10 to 20 days. A low temperature during anthesis affects panicle development, resulting in spike sterility. Low temperature stress at flowering was detrimental to yield due to lower grain numbers and smaller grain weight (Maulana and Tesso, 2013).

Average soil temperature for seed germination ranges from 4.6°C to 16.5°C. Optimum soil temperature for germination ranges from 21°C to 35°C, with a maximum temperature up to 40–48°C, beyond which it is considered lethal. Sorghum cannot tolerate cool weather and grows slowly below 20°C (Rao et al., 2015). For leaf appearance, the minimum temperature ranges from 7°C to 12°C, with an optimum between 26°C and 34°C. Temperatures above the optimum decrease the rate of leaf appearance. The leaf elongation rate linearly increases from 13°C to 32°C and declines above 34°C. The minimum and maximum temperatures for leaf elongation are 15.5°C and 43°C, respectively. The base temperature for panicle initiation is from 6.8°C to 10.4°C, and the optimum is from 25.9°C to 27.3°C. The response to temperatures above or below the optimum is linear and delays panicle initiation. Eastin et al. (1976) observed that night temperatures of 5°C above the optimum temperature reduced grain yield from 25% to 33%, and 10°C above the optimum reduced the yield by 50%. High or low temperatures during panicle initiation and the flowering stage significantly decrease the seed-set percentage. High temperatures above 32/22°C (daytime maximum/night-time minimum) cause the abortion of flowers, resulting in a lower seed-set and grain number (Prasad et al., 2006; Singh et al., 2015). The highest temperature-sensitive stages in grain sorghum are at the flowering stage and 10 days before flowering (Prasad et al., 2008; Djanaguiraman et al., 2014). High temperature stress during these periods causes a maximum reduction in seed-set, seed numbers and seed yields. Similarly, early seed-filling periods are also more sensitive to high temperature stress compared with later periods (Prasad et al., 2015; Djanaguiraman et al., 2014). Seed yield losses during the post-flowering stages are mainly due to decreases in seed size. The maximum potential grain number is set by the end of the floral initiation stage and the final grain yield is dependent on the number of filled grains and individual grain weight. The cardinal temperatures – minimum (base) temperature (below which growth ceases); optimum temperature (at which growth is maximum); and maximum (ceiling) temperature (above which growth ceases) – for vegetative growth were identified as 8°C, 34°C and 44°C (Alagarswamy and Ritchie, 1991). Similarly, the optimum and maximum temperatures for reproductive growth were identified as 32°C and 40°C (Prasad et al., 2006). The influence of environmental factors such as soil and air temperature, rainfall and day length controlling the growth stages of sorghum is presented in Table 2.

Grain weight (grain size) is a product of two factors: the rate and duration of the grain filling process. Both of these processes are highly sensitive to temperature. High temperatures (>36/26°C) decrease both the rate of grain filling and the grain filling duration (Prasad et al., 2006). Similarly, cool temperatures (<27/22°C) during early booting to maturity significantly decrease seed numbers and grain yield. A minimum temperature below 10°C during the pre-flowering stages (23–27 days before flowering) significantly decreases pollen viability, leading to lower seed-set and yields.

Concentrations of atmospheric carbon dioxide are rising at an unprecedented rate as a result of fossil fuel burning and land use change. The increasing concentration of atmospheric CO_2 has led to concerns about its potential effects on agricultural production. Elevated CO_2 levels have the potential to enhance crop system processes such as photosynthesis and the water-use efficiency of plants, leading to increased biomass production. Crops with a C_3 photosynthetic pathway (soybean) often exhibit a greater response compared to crops with a C_4 pathway (sorghum). A higher grain yield was observed in the C_3 plant (soybean) compared to the C_4 plant (sorghum) when CO_2 levels were elevated (Prior et al., 2005). Elevated CO_2 substantially decreases stomatal conductance and the transpiration rate. The partial closure of stomata often leads to an increase in leaf or tissue temperatures by about 1–2°C (Prasad et al., 2006). A higher tissue (leaf) temperature increases the vapour pressure of the water inside the leaves and causes a greater difference in leaf-to-air vapour pressure, which is the driving force for transpiration. However, the effects of

Table 2 Key environmental factors influencing components of yield at different growth stages of grain sorghum

Growth stages	Key environmental factors	Influencing yield component
Stage 0 – Emergence	Soil temperature and moisture	Plant density
Stage 1 – Three-leaf	Soil and air temperature and moisture	Plant density
Stage 2 – Five-leaf	Soil and air temperature and moisture	Plant biomass
Stage 3 – Panicle initiation	Day length, soil and air temperature and moisture	Biomass, number of leaves $plant^{-1}$ and number of spikelets per panicle
Stage 4 – Flag leaf	Day length, air temperature and soil moisture	Biomass and number of spikelets per panicle
Stage 5 – Boot leaf	Day length, air temperature and soil moisture	Panicle emergence, number of spikelets per panicle and seed numbers
Stage 6 – Half-bloom	Day length, air temperature and soil moisture	Seed numbers and partitioning
Stage 7 – Soft dough	Air temperature and soil moisture	Seed size and partitioning
Stage 8 – Hard dough	Air temperature and soil moisture	Seed size and partitioning
Stage 9 – Maturity	Air temperature and soil moisture	Seed size

elevated CO_2 in the single leaf cannot be extrapolated directly to the transpiration of the whole canopy under field conditions (Kadam et al., 2014).

5.3 Rainfall

Sorghum requires a rainfall level of 600–700 mm during the crop cycle, although a satisfactory grain yield can be obtained at a rainfall level of 400–500 mm. The yields of grain sorghum will be high if the annual rainfall is >800 mm and distributed during the reproductive stages. Seed germination and seedling establishment requires adequate soil moisture if a crop is to be properly established. The water requirement for sorghum plants gradually increases from the emergence of seedlings to flowering and then decreases gradually. Daily water use is at its highest during the booting to seed-set stages (7–10 mm day^{-1}). About 90% of the total water used by sorghum is extracted from a soil depth of 0–1.5 m. The rooting depth of mature sorghum plants can extend to 2.5 m. Moisture below a depth of 1 m is an important source of stored water at the end of the growing season. Soil management practices, namely zero or no tillage, can significantly decrease soil evapotranspiration by decreasing the level of incoming heat energy to the soil. The grain yield response to the water supply was found to be greater with no tillage than with conventional tillage. Stone and Schlegel (2006) indicated that every millimetre of water above 100 mm resulted in an additional 16.6 kg of grain. However, the relationship between grain yield and water is complex because yield is more sensitive to drought at certain growth stages.

Sorghum is sensitive to drought stress during the reproductive stage rather than the vegetative stage. Sorghum can tolerate drought stress for a short period, although prolonged drought can result in a decreased grain yield. Inuyama et al. (1976) reported a yield reduction of 16% and 36% due to 16 days and 28 days of drought stress, respectively, during the vegetative stage of sorghum. Eck and Musick (1979) reported that a short period (13–15 days) of drought stress during the early booting, heading and early grain filling stages did not have a significant effect on yield. However, drought stress of 27–28 days during the early booting, heading and early grain filling stages reduced the grain yield by 27%, 27% and 12%, respectively. Prolonged stress of 35 and 42 days at the beginning of the booting stage reduced the grain yield by 43% and 54%, respectively. For maximum yield, an adequate soil moisture should be available during the two critical stages of crop development. First, prior to booting and panicle exertion, during which the potential seed number per panicle was decided, and secondly, at flowering, to ensure seed-set and early grain filling. Drought stress during the reproductive stages can have an adverse effect on pollen and ovule development, fertilization and the premature abortion of fertilized ovules (Saini, 1997).

Sorghum plants exhibit a pre-flowering and post-flowering drought response. These two responses are controlled by different genetic mechanisms. Pre-flowering stress affects biomass, panicle size, grain number and grain yield (Sanchez et al., 2002), while post-flowering drought leads to premature leaf and stem senescence, lodging and a reduced seed size (Borrell et al., 2000). Genomic analysis in sorghum recombinant inbred lines derived from the cross between Tx7078 (pre-flowering-tolerant and post-flowering-susceptible) and B35 (pre-flowering-susceptible and post-flowering-tolerant) genotypes have shown six distinct genomic regions (Tuinstra et al., 1996), and these loci accounted for approximately 40% of the total phenotypic variation in yield under pre-flowering drought conditions and were stable in different environments. Kebede et al. (2001) identified four

QTLs associated with pre-flowering drought tolerance from RILs derived from the cross between SC56 and Tx7000.

Post-flowering drought tolerance in sorghum is associated with the stay green trait. Maintaining a functional green leaf area under drought stress during the grain filling period plays an important role in drought tolerance of sorghum, and is directly related to grain yield under drought conditions (Borrell et al., 2000, 2014). In sorghum different stay green sources are B35, SC56, E36-1 and KS19 (Kebede et al., 2001; Hash et al., 2003). B35-derived lines have a greater leaf area at flowering and a normal rate of leaf senescence, whereas KS19-derived lines have a smaller leaf area at flowering and a slower rate of leaf senescence (Borrell et al., 2000). Studies have shown four major QTLs that contribute to stay green traits in sorghum. QTLs *Stg1* and *Stg2* are located on LG-03, *Stg3* on LG-02 and *Stg4* on LG-05, and account for 20%, 30%, 16% and 10% of the phenotypic variance, respectively (Xu et al., 2000; Sanchez et al., 2002). Among these, *Stg2* was found to be the most important QTL and was consistent across the environment.

Drought stress during the post-flowering stage can cause a decreased photosynthetic rate, so that, consequently, grain filling is dependent on the stored carbohydrate reserves. One way to achieve a reasonable yield is to assess the remobilization rate of the carbohydrates that are produced under stressful conditions. Drought stress causes an imbalance in assimilate translocation from source to sink. During drought stress, most of the assimilates are translocated to the roots in order to improve water uptake (Leport et al., 2006). The export of assimilates from source to sink generally depends upon the rate of photosynthesis and the sucrose concentration in the leaves. Drought stress affects photosynthesis and decreases the leaf sucrose content, leading to a reduction in the export of sucrose from source to sink. Moreover, drought stress affects the ability of the sink to utilize the incoming assimilates efficiently. Beheshti and Fard (2010) have shown that, in sorghum, significant genetic variation exists for reserve accumulation and utilization, as well as the partitioning of assimilates in both drought stress and well-watered conditions. The genotype M5 has produced the highest grain yield under drought stress through the remobilization of stored reserves at pre- and post-anthesis.

Next to stomata, water loss from plants occurs through its cuticle. High cuticular wax minimizes nonstomatal water loss from the plants. Sorghum produces the highest amount of epicuticular waxes among various cereals. On weight basis, sorghum produces approximately 52 mg g^{-1} wax, which is three times higher than that produced by maize (17 mg g^{-1}) and two times higher than that produced by durum wheat (25 mg g^{-1}) (Araus et al., 1991). Epicuticular wax accumulation varies greatly with species, organ, stage of development and environmental conditions. Epicuticular wax is hydrophobic in nature, and it is considered to be an important trait, imparting drought tolerance (Islam et al., 2009). In sorghum, epicuticular wax is deposited on the abaxial leaf blade and sheath and culms during pre-flowering and at the maturity stages. The wax composition of the sorghum leaf sheath shows the highest (96%) level of free fatty acids, with a chain length of 16–33 carbons, of which C_{28} and C_{30} represent 78% and 20% of the constituents, respectively (Jenks et al., 2000). A significant negative correlation was observed between the leaf epicuticular wax load and night-time conductance, suggesting that epicuticular wax may enhance the water-use efficiency of sorghum by regulating night-time water loss (Burow et al., 2008). Identification of the genes involved in epicuticular wax deposition is likely to be significant in improving drought tolerance in sorghum.

Osmotic adjustment is another important trait that has been associated with drought tolerance. Osmotic adjustment refers to the lowering of the osmotic potential of the plant by the accumulation of compatible solutes in response to a deficit of water. The various compatible solutes are proline, sucrose, fructose, mannitol, pinitol, glycine betaine, malate and citrate. Osmotic adjustment is an inherited trait that has been associated with sustained yields under drought stress in many crops (Blum, 2005). In sorghum, the genes *OA1* and *OA2* are reported to be involved in osmotic adjustment (Basnayake et al., 1995). Maize and sorghum synthesize glycine betaine, whereas rice does not (Rathinasabapathi et al., 1993). Moreover, within maize and sorghum, there are certain genotypes that do not accumulate glycine betaine (Mickelbart et al., 2003). Besides glycine betaine, other solutes, such as proline, K^+ and sugars, were also reported to contribute to osmotic adjustment in sorghum (Premachandra et al., 1995).

Sorghum at elevated carbon dioxide levels has shown a differential response, namely, there is no effect on growth at either a high or a low water supply (Mo et al., 1992), or it has stimulated growth under drought conditions (Owensby et al., 1996). Ottman et al. (2001) have shown that when water was limited and CO_2 level was elevated, the harvest index of sorghum was not decreased due to delayed leaf senescence. Elevated levels of CO_2 can ameliorate the negative effects of drought in sorghum by maintaining a higher photosynthetic rate due to an increase in water-use efficiency, which is achieved by a reduction in stomatal conductance (De-Souza et al., 2015). Studies on the growth of sorghum under elevated CO_2 levels and drought stress showed an increase of the net assimilation rate of 23%, due to a decrease of 32% in stomatal conductance (Wall et al., 2001). This increased sorghum's water-use efficiency by 17% (Conley et al., 2001). Sorghum yield is a function of the number of harvested panicles, seeds per panicle and individual seed weight, and all of these were affected by the duration and severity of drought stress. Eck and Musick (1979) reported that yield decreases as a result of drought stress at the early boot stage were due to reduced seed size and seed number, and that a yield reduction due to drought stress at the heading or later stages was due to reduced seed size.

6 Conclusion and future trends

Grain sorghum follows a predictable pattern of growth from planting through to the reproductive stage. The timing and duration of each growth stage are closely related to air temperature and the genetic background. Sorghum development has been separated into three phases, namely the vegetative stage, the reproductive stage and the grain fill stage. In temperate sorghum, the growth and development phases have been classified on a growth scale of 0–9 stages. While sorghum is well adapted to tropical and subtropical climates, in the future, as a result of climate change, sorghum-producing areas will face increased exposure to water deficits and high and cold temperatures. Understanding sorghum growth and development is essential for comprehending the plant's response to environmental stresses and for providing knowledge that can inform management decisions. Sorghum has been reported to have greater developmental plasticity, and if we are to exploit that plant trait, then knowledge on sorghum growth and development and its climatic requirements will be useful. To increase the current productivity of sorghum, the use of high-yielding varieties and proper management practices must be adopted.

Proper identification of growth stages and understanding physiological processes at each stage and its response to inputs and environmental will help develop best crop management practices. Future research on matching genotypes with appropriate environment and management practices for improving yields and resource use efficiency will be critical. Incorporating novel and improved algorithms of yield formation and impact of agronomy, biotic and abiotic pests in crop simulation models will allow large-scale evaluations of innovations. Several national institutes (Universities and National Agricultural Research System) and international organizations (e.g. International Crops Research Institute for the Semi-Arid Tropics) and private industries are involved in sorghum research. Products and outcomes of innovative research and capacity building activities will help in improving sorghum yields. Sustained funding from donor agencies such as United States Agency for International Development, Australian Centre for International Agricultural Research, French Agricultural Research Centre for International Development, Department for International Development and Bill and Melinda Gates Foundation and other donors is critical to address the needs of farming communities and industry.

7 Acknowledgements

We thank the following entities for sustained support to sorghum research: Kansas Grain Sorghum Commission, United Sorghum Checkoff Program, K-State Center for Sorghum Improvement, and United States Agency for International Development through Feed the Future Innovation Lab for Collaborative Research on Sustainable Intensification; and on Sorghum and Millet. Contribution no. 18-141-B from Kansas Agricultural Experiment Station.

8 References

Adams, J. E. and Thompson, D. O. 1973. Soil temperature reduction during pollination and grain formation of corn and grain sorghum. *Agron. J.*, 65:60–3.

Alagarswamy, G. and Ritchie, J. T. 1991. Phasic development in CERES sorghum model. In: T. Hodges (Ed.), *Predicting Crop Phenology*. CRC Press, Boca Raton, FL, pp. 143–52.

Araus, J. L., Febrero, A. and Vendrell, P. 1991. Epidermal conductance in different parts of durum wheat grown under Mediterranean conditions: The role of epicuticular waxes and stomata. *Plant Cell Environ.*, 14:545–58.

Aruna, C. and Audilakshmi, S. 2008. Reproductive biology and breeding behavior of sorghum. In: Reddy, B. V. S., Ramesh, S., Ashok Kumar, A. and Gowda, C. L. L. (Eds), *Sorghum Improvement in the New Millennium*. International Crops Research Institute for the Semi-Arid Tropics, Patancheru, Andhra Pradesh, India, pp. 28–30.

Assefa, Y., Staggenborg, S. A. and Prasad, P. V. V. 2010. Grain sorghum water requirement and responses to drought stress: A review. *Crop Manage.* Online. doi:10.1094/CM-2010-1109-01-RV.

Bailey, S., Walters, R. G., Jansson, S. and Horton, P. 2001. Acclimation of *Arabidopsis thaliana* to the light environment: The existence of separate low light and high light responses. *Planta*, 213:794–801.

Basnayake, J., Cooper, M., Ludlow, M. M., Henzell, R. G. and Snell, P. J. 1995. Inheritance of osmotic adjustment to water stress in three grain sorghum crosses. *Theor. Appl. Genet.*, 90:675–82.

Beheshti, A. R. and Fard, B. B. 2010. Dry matter accumulation and remobilization in grain sorghum genotypes (*Sorghum bicolor* L. Moench) under drought stress. *AJCS*, 4:185–9.

Bekele, W. A., Fiedler, K., Shiringani, A., Schnaubelt, D., Windpassinger, S., Uptmoor, R., Friedt, W. and Snowdon, R. J. 2014. Unravelling the genetic complexity of sorghum seedling development under low-temperature conditions. *Plant Cell Environ.*, 37:707–23.

Blum, A. 2005. Drought resistance, water-use efficiency, and yield potential: Are they compatible, dissonant, or mutually exclusive? *Aust. J. Agric. Res.*, 56:1159–68.

Blum, A. 2013. Heterosis, stress, and the environment: A possible road map towards the general improvement of crop yield. *J. Exp. Bot.*, 64:4829–37.

Blum, A. and Ritchie, J. T. 1984. Effect of soil surface water content on sorghum root distribution in the soil. *Field Crops Res.*, 8:169–76.

Borrell, A. K., Hammer, G. L. and Henzell, R. G. 2000. Does maintaining green leaf area in sorghum improve yield under drought? II. Dry matter production and yield. *Crop Sci.*, 40:1037–48.

Borrell, A. K., Mullet, J. E., George-Jaeggli, B., van Oosterom, E. J., Hammer, G. L., Klein, P. E. and Jordan, D. R. 2014. Drought adaptation of stay-green sorghum is associated with canopy development, leaf anatomy, root growth, and water uptake. *J. Exp. Bot.*, 65:6251–63.

Burow, G. B., Franks, C. D. and Xin, Z. 2008. Genetic and physiological analysis of an irradiated bloomless mutant (epicuticular wax mutant) of sorghum. *Crop Sci.*, 48:41–8.

Burow, G. B, Burke, J. J., Xin, Z. and Franks, C. D. 2010. Genetic dissection of early-season cold tolerance in sorghum (*Sorghum bicolor* (L.) Moench). *Mol. Breed.*, 28:391–402.

Ciampitti, I. A. 2015. *Sorghum Growth and Development*. Kansas State University, Manhattan, KS. https://www.bookstore.ksre.ksu.edu/pubs/MF3234.pdf

Conley, M. M., Kimball, B. A., Brooks, T. J., Pinter Jr., P. J., Hunsaker, H. J., Wall, G. W., Adam, N. R., La Morte, R. L., Matthias, A. D., Thompson, T. L., Leavitt, S. W., Ottman, M. J., Cousins, A. B. and Triggs, J. M. 2001. CO_2 enrichment increases water-use efficiency in sorghum. *New Phytol.*, 151:407–12.

Craufurd, P. Q., Mahalakshmi, V., Bidinger, F. R., Mukuru, S. Z., Chantereau, J., Omanga, P. A., Qi, A., Roberts, E. H., Ellis, R. H., Summerfield, R. J. and Hammer, G. L. 1999. Adaptation of sorghum: Characterization of genotypic flowering responses to temperature and photoperiod. *Theor. Appl. Genet.*, 99:900–11.

Demmig-Adams, B., Winter, K., Kruger, A. and Czygan, F. C. 1989. Zeaxanthin synthesis, energy dissipation, and photoprotection of photosystem II at chilling temperatures. *Plant Physiol.*, 90:894–8.

De-Souza, A. P., Cocuron, J.-P., Garcia, A. C., Alonso, A. P. and Buckeridge, M. S. 2015. Changes in whole-plant metabolism during the grain-filling stage in sorghum grown under elevated CO_2 and drought. *Plant Physiol.*, 169:1755–65.

Djanaguiraman, M., Prasad, P. V. V., Murugan, M., Perumal, M. and Reddy, U. K. 2014. Physiological differences among sorghum (*Sorghum bicolor* L. Moench) genotypes under high temperature stress. *Environ. Exp. Bot.*, 100:43–54.

Eastin, J. D. 1972. Photosynthesis and translocation in relation to plant development. In: N. G. P. Roa and L. R. House (Eds), *Sorghum in the Seventies*. Oxford and IBH Publishing Co., New Delhi, India, pp. 214–16.

Eastin, J. D., Brooking, I. and Taylor, S. O. 1976. Influence of temperature on sorghum respiration and yield. In: *Agronomy Abstracts*. ASA, Madison, WI.

Eck, H. V. and Musick, J. C. 1979. Plant water stress effect on irrigated sorghum. I. Effect on yield. *Crop Sci.*, 19:586–92.

Ellis, R. H., Craufurd, P. Q., Summerfield, R. J. and Robert, E. H. 1997. Effect of photoperiod, temperature and asynchrony between thermoperiod and photoperiod on development to panicle initiation in sorghum. *Ann. Bot.*, 79:169–78.

FAOSTAT. 2014. Food and Agriculture Organization of the United Nations. www.faostat.org.

Fiedler, K., Bekele, W. A., Matschegewski, C., Snowdon, R., Wieckhorst, S., Zacharias, A. and Uptmoor, R. 2016. Cold tolerance during juvenile development in sorghum: A comparative analysis by genome wide association and linkage mapping. *Plant Breed.*, 135:598–606.

Franks, C. D., Burow, G. B. and Burke, J. J. 2006. A comparison of U.S. and Chinese sorghum germplasm for early season cold tolerance. *Crop Sci.*, 46:1371–6.

Harlan, J. R. and de Wet, J. M. J. 1972. A simplified classification of cultivated sorghum. *Crop Sci.*, 12:172.

Harlan, J. R. and Stemler, A. 1976. The races of sorghum in Africa. In: J. R. Harlan, J. M. J. de Wet and A. Stemler (Eds), *Origins of African Plant Domestication*. Mouton, The Hague, the Netherlands, pp. 465–78.

Hash, C. T., Bhasker Raj, A. G., Lindup, S., Sharma, A., Beniwal, C. R., Folkertsma, R. T., Mahalakshmi, V., Zerbinid, E. and Blummeld, M. 2003. Opportunities for marker-assisted selection (MAS) to improve the feed quality of crop residues in pearl millet and sorghum. *Field Crops Res.*, 84:79–88.

Inuyama, S., Musick, J. T. and Dusek, D. A. 1976. Effect of plant water deficit at various growth stages on growth, grain yield and leaf water potential of irrigated grain sorghum. *Proc. Crop Sci. Soc. Jpn.* 45:298–307.

Islam, M. A., Du, H., Ning, J., Ye, H. and Xiong, L. 2009. Characterization of Glossy1-homologous genes in rice involved in leaf wax accumulation and drought resistance. *Plant Mol. Biol.*, 70:443–56.

Jenks, M. A., Rich, P. J., Rhodes, D., Ashworth, E. N., Axtell, J. D. and Ding, C. K. 2000. Leaf sheath cuticular waxes on bloomless and sparse bloom mutants of *Sorghum bicolor*. *Phytochem.*, 54:577–84.

Kadam, N. N., Xiao, G., Melgar, R. J., Bahuguna, R. N., Quinones, C., Tamilselvan, A., Prasad, P. V. V. and Jagadish, K. S. V. 2014. Agronomic and physiological responses to high temperature, drought, and elevated CO_2 interactions in cereals. *Adv. Agron.*, 127:111–56.

Kagawa, T. and Wada, M. 2002. Blue light-induced chloroplast relocation. *Plant Cell Physiol.*, 43:367–71.

Kebede, H., Subudhi, P. K., Rosenow, D. T. and Nguyen, H. T. 2001. Quantitative trait loci influencing drought tolerance in grain sorghum (*Sorghum bicolor* L. Moench). *Theor. Appl. Genet.*, 103:266–76.

Knoll, J. and Ejeta, G. 2008. Marker assisted selection for early-season cold tolerance in sorghum: QTL validation across populations and environments. *Theor. Appl. Genet.*, 116:541–53.

Knoll, J., Gunaratna, G. and Ejeta, G. 2008. QTL analysis of early-season cold tolerance in sorghum. *Theor. Appl. Genet.*, 116:577–87.

Kubien, D. S., von Caemmerer, S., Furbank, R. T. and Sage, R. F. 2003. C_4 photosynthesis at low temperature. A study using transgenic plants with reduced amounts of Rubisco. *Plant Physiol.*, 132:1577–85.

Kumar, A. A. 2016. Botany, taxonomy and breeding. In: S. Rakshit and Y. H. Wang (Eds), *The Sorghum Genome, Compendium of Plant Genomes*. Springer International Publishing, Cham, Switzerland, pp. 27–45.

Leport, L., Turner, N. C., French, R. J., Barr, M. D., Duda, R. and Davies, S. L. 2006. Physiological responses of chickpea genotypes to terminal drought in a Mediterranean-type environment. *Eur. J. Agron.*, 11:279–91.

Lewis, R. B., Hiler, E. A. and Jordan, W. R. 1974. Susceptibility of grain sorghum to water deficit at three growth stages. *Agron. J.*, 66:589–91.

Lindahl, M., Spetea, C., Hundal, T., Oppenheim, A. B., Adam, Z. and Andersson, B. 2000. The thylakoid FtsH protease plays a role in the light-induced turnover of the photosystem II D1 protein. *Plant Cell*, 12:419–31.

Major, D. J. 1980. Photoperiod response characteristics controlling flowering of nine crop species. *Canadian J. Plant Sci.*, 60:777–84.

Major, D. J. and Kiniry, J. R. 1991. Predicting daylength effects on phenological processes. In: T. Hodges (Ed.), *Predicting Crop Phenology*. CRC Press, Boca Raton, FL, pp. 15–28.

Maulana, F. and Tesso, T. T. 2013. Cold temperature episode at seedling and flowering stages reduces growth and yield components in sorghum. *Crop Sci.*, 53:564–74.

Mickelbart, M. V., Peel, G., Joly, R. J., Rhodes, D., Ejeta, G. and Goldsbrough, P. B. 2003. Development and characterization of near-isogenic lines of sorghum segregating for glycinebetaine accumulation. *Physiol. Plant.*, 118:253–61.

Miller, E. C. 1916. Comparative study of the root systems and leaf areas of corn and sorghum. *J. Agric. Res.*, 6:311–32.

Mitaru, B. N., Mgonja, M. A., Rwomushana, I. and Opio, F. 2012. Integrated sorghum and millet sector for increased economic growth and improved livelihoods in Eastern and Central Africa. In: *Proceedings of the ECARSAM Stakeholders Conference*, 20–22 November 2006, Dar es Salaam, Tanzania. ASARECA (Association for Strengthening Agricultural Research in Eastern and Central Africa), Entebbe, Uganda.

Mo, G., Nie, D., Kirkham, M. B., He, H., Ballou, L. K., Caldwell, F. W. and Kanemasu, E. T. 1992. Root and shoot weight in a tallgrass prairie under elevated carbon dioxide. *Environ. Exp. Bot.*, 32:193–201.

Munekage, Y., Hashimoto, M., Miyake, C., Tomizawa, K., Endo, T., Tasaka, M. and Shikanai, T. 2004. Cyclic electron flow around photosystem I is essential for photosynthesis. *Nature* 429:579–82.

Murdy, D. S., Tabo, R. and Ajayi, O. 1994. Botanical parts of the inflorescence of a sorghum plant. In: *Sorghum Hybrid Seed Production and Management*. International Crops Research Institute for the Semi-Arid Tropics, Patancheru, Andhra Pradesh, India, pp. 4–12.

Nakayama, F. S. and van Bavel, C. H. M. 1963. Root activity distribution patterns of sorghum and soil moisture conditions. *Agron. J.*, 55:271–4.

Ortiz, D., Hu, J. and Fernandez, M. G. S. 2017. Genetic architecture of photosynthesis in *Sorghum bicolor* under non-stress and cold stress conditions, *J. Exp. Bot.*, 68:4545–57.

Ottman, M. J., Kimball, B. A., Pinter, P. J., Wall, G. W., Vanderlip, R. L., Leavitt, S. W., LaMorte, R. L., Matthias, A. D. and Brooks, T. J. 2001. Elevated CO_2 increases sorghum biomass under drought conditions. *New Phytol.*, 150:261–73.

Owensby, C. E., Ham, J. M., Knapp, A., Rice, C. W., Coyne, P. I. and Auen, L. M. 1996. Ecosystem-level responses of tallgrass prairie to elevated CO_2. In: G. W. Koch and H. A. Mooney (Eds), *Carbon Dioxide and Terrestrial Ecosystems*. Academic Press, San Diego, CA, pp. 147–62.

Paul, C. 1990. *Sorghum Agronomy*. International Crops Research Institute for the Semi-Arid Tropics, Patancheru, Andhra Pradesh, India.

Pimentel, C., Davey, P. A., Juvik, J. A. and Long, S. P. 2005. Gene loci in maize influencing susceptibility to chilling dependent photoinhibition of photosynthesis. *Photosynthesis Res.*, 85:319–26.

Prasad, P. V. V., Boote, K. J. and Allen Jr., L. H. 2006. Adverse high temperature effects on pollen viability, seed-set, seed yield and harvest index of grain-sorghum [*Sorghum bicolor* (L.) Moench] are more severe at elevated carbon dioxide due to high tissue temperature. *Agr. Forest. Meteorol.* 139:237–51.

Prasad, P. V. V., Pisipati, S., Mutava, R. N. and Tuinstra, M. R. 2008. Sensitivity of grain sorghum to high temperature stress during reproductive development. *Crop Sci.*, 48:1911–17.

Prasad, P. V. V., Djanaguiraman, M., Perumal, R. and Ciampitti, I. A. 2015. Impact of high temperature stress on floret fertility and individual grain weight of grain sorghum: Sensitive stages and thresholds for temperature and duration. *Front. Plant Sci.*, 6:820

Premachandra, G. S., Hahn, D. T., Rhodes, D. and Joly, R. J. 1995. Leaf water relations and solute accumulation in two grain sorghum lines exhibiting contrasting drought tolerance. *J. Exp. Bot.*, 46:1833–41.

Prior, S. A., Runion, G. B., Rogers, H. H., Torbert, H. A. and Reeves, D. W. 2005. Elevated atmospheric CO_2 effects on biomass production and soil carbon in conventional and conservation cropping systems. *Global Change Biol.*, 11:657–65.

Rao, S. P., Belum, V. S., Nagaraj, N. and Upadhyaya, H. D. 2015. Sorghum production for diversified uses. In: Y. H. Wang, H. D. Upadhyaya and C. Kole (Eds), *Genetics, Genomics and Breeding of Sorghum*. CRC press and Taylor and Francis Group, Boca Raton, FL, pp. 1–27.

Rathinasabapathi, B., Gage, D. A., Mackill, D. J. and Hanson, A. D. 1993. Cultivated and wild rices do not accumulate glycinebetaine due to deficiencies in two biosynthetic steps. *Crop Sci.*, 33:534–8.

Reddy, B. V. S., Kumar, A. A., Ramesh, S. and Reddy, P. S. 2011. Sorghum genetic enhancement for climate change adaptation. In: S. S. Yadav, R. Redden, J. L. Hatfield, H. Lotze-Campen and A. J. W. Hall (Eds), *Crop Adaptation to Climate Change*, 1st Edition. Wiley-Blackwell, Oxford, UK, pp. 896–927.

Saini, H. S. 1997. Effects of water stress on male gametophyte development in plants. *Sex Plant Reprod.*, 10:67–73.

Sanchez, A. C., Subudhi, P. K., Rosenow, D. T. and Nguyen, H. T. 2002. Mapping QTLs associated with drought resistance in sorghum (*Sorghum bicolor* L. Moench). *Plant Mol. Biol.*, 48:713–26.

Singh, V., Nguyen, C. T., van Oosterom, E. J., Chapman, S. C., Jordan, D. R. and Hammer, G. L. 2015. Sorghum genotypes differ in high temperature responses for seed set. *Field Crops Res.*, 171:32–40.

Stephens, J. C., Miller, F. R. and Rosenow, D. T. 1967. Conversion of alien sorghums to early combine genotypes. *Crop Sci.*, 7:396.

Stone, L. R. and Schlegel, A. J. 2006. Yield-water supply relationship of grain sorghum and winter wheat. *Agron. J.*, 98:1359–66.

Tuinstra, M. R., Grote, E. M., Goldbrough, P. B. and Ejeta, G. 1996. Identification of quantitative trait loci associated with pre-flowering drought tolerance in sorghum. *Crop Sci.*, 36:1337–44.

Vanderlip, R. L. and Reeves, H. E. 1972. Growth stages of sorghum. *Agron. J.*, 64:13–16.

Wall, G. W., Brooks, T. J., Adam, N. R., Cousins, A. B., Kimball, B. A., Pinter Jr., P. J., La Morte, R. L., Triggs, J., Ottman, M. J., Leavitt, S. W., Matthias, A. D., Williams, D. G. and Webber, A. N. 2001. Elevated atmospheric CO_2 improved sorghum plant water status by ameliorating the adverse effects of drought. *New Phytol.*, 152:231–48.

Wang, D., Portis Jr., A. R., Moose, S. P. and Long, S. P. 2008. Cool C_4 photosynthesis: Pyruvate Pi dikinase expression and activity corresponds to the exceptional cold tolerance of carbon assimilation in *Miscanthus × giganteus*. *Plant Physiol.*, 148:557–67.

Xu, W., Subudhi, P. K., Crasta, O. R., Rosenow, D. T., Mullet, J. E. and Nguyen, H. T. 2000. Molecular mapping of QTLs conferring staygreen in grain sorghum (*Sorghum bicolor* L. Moench). *Genome*, 43:461–9.

Chapter 4

Sorghum as a food grain

Barbara J. Stoecker, Oklahoma State University, USA; Kebede Abegaz, Hawassa University, Ethiopia; and Yewelsew Abebe, Alive and Thrive, Ethiopia

1 Introduction

Sorghum (*Sorghum bicolor* L. Moench) is produced in many countries in the tropical, semi-tropical and arid regions of the world. It grows well in many different environments; is naturally drought tolerant; and is versatile as a food, feed and fuel (Food and Agriculture Organization, 1995).

Grain sorghum is the fifth most widely produced cereal crop in the world after rice, wheat, corn and barley, and is a dietary staple for more than half a billion people in more than 30 countries (ICRISAT, 2017). Sorghum is used as a principal food in areas of Africa and Asia as well as being used as a raw material for alcoholic drinks and beverages.

The diversity of sorghum grains is related to sorghum's intrinsic properties, which are derived from starch, protein and phytochemical components. This diversity can be valuable for developing new food products using germination, malting, fermentation, flaking, dry popping and roasting, mixing or extrusion techniques (Vargas-Solórzano et al., 2014).

http://dx.doi.org/10.19103/AS.2017.0015.22

2 Protein and starch of sorghum

The sorghum kernel is composed of three distinct parts: outer protective layer (pericarp), storage tissue (endosperm) and embryo (germ). Some sorghum cultivars have a pigmented sub-coat (testa) located between the pericarp and the endosperm (Earp et al., 2004). The pericarp contains 5% to 8% of the grain protein, while the endosperm contains approximately 81% of sorghum protein. The germ contains approximately 15% of the protein in sorghum (reviewed by Waniska and Rooney (2000)). Eighty per cent of grain nitrogen was found in the endosperm, 16% in the germ and approximately 3% in the pericarp (Taylor and Schüssler, 1986).

The distribution of starch granules within parts of the sorghum kernel varies with grain type and cultivar. In general, the pericarp and embryo together represent a small percentage of the kernel and contain little starch. Most of the starch resides in the endosperm. Starch is packaged as semi-crystalline granules within the endosperm, with the shape, size and content of amylose and amylopectin varying with the cultivar.

2.1 Characteristics of sorghum proteins

In normal sorghum cultivars, most of the proteins in the endosperm are prolamines (soluble in alcohol–water mixtures) as well as some limited amounts of glutelins (soluble in dilute acid and dilute alkali) (Taylor and Schüssler, 1986). The protein in the germ of the kernel is rich in albumins (water-soluble) and globulins (soluble in dilute salt solution) which contain lysine and other essential amino acids (Taylor and Schüssler, 1986). On the basis of solubility in different solvents, the proteins of sorghum grain are divided into: albumins (water-soluble), globulins (salt-soluble), kafirins (prolamines, aqueous alcohol-soluble), cross-linked kafirins (aqueous alcohol + reducing agent-soluble), cross-linked glutelins (detergent + reducing agent + alkaline pH-soluble) and unextracted structural protein residue (Jambunathan et al., 1975; Landry and Moureaux, 1980).

The nutritional value of sorghum is diminished because of low digestibility of grain protein and starch. Digestibility varies depending on variety, even in sorghum lines from the same pedigree (Wong et al., 2009). The indigestibility of sorghum lines is attributed to (a) greater abundance of disulphide-bonded proteins; (b) diverse nature of the protein matrix, protein body packaging and its interaction with starch; and (c) differences in the non-waxy starch and accompanying granule-bound starch synthase (Wong et al., 2010; Wong et al., 2009). According to Wong and colleagues (2009), each of these factors should be considered in efforts to enhance the nutritional value of sorghum grain.

Other factors that contribute to the problem of sorghum digestibility include the interaction of protein with non-protein components (polyphenols, starch, non-starch polysaccharides, phytates and lipids); the nature of the proteins themselves and their organization within the grain (Duodu et al., 2003; Belton et al., 2006; Ezeogu et al., 2005; Ezeogu et al., 2008). The low amount of protein relative to starch affects the functional properties of starch, such as gelatinization and digestion rate, to a greater extent in sorghum than in other cereals (Chandrashekar and Kirleis, 1988; Duodu et al., 2002; Ezeogu et al., 2008; Ezeogu et al., 2005).

There has been an increasing interest in potential new food applications of sorghum, particularly due to the resilience of sorghum against the high temperature and drought conditions that may arise due to climate change (Taylor et al., 2006). However, application of sorghum for food and feed purposes is limited by the relatively poor nutritive value of sorghum grain in comparison with other cereals. One of the reasons for relatively low nutritive value of sorghum grain is resistance of its seed storage proteins (kafirins) to protease digestion (Oria et al., 1995; Duodu et al., 2003).

Oria et al. (2000) suggested that the lower digestibility of proteins in the sorghum endosperm is due to strong disulphide bonds formed by beta and gamma kafirins, which result in enzyme-resistant structure on the periphery of the protein body. The disulphide bonds in the protein matrix also limit the expansion of starch granules and access of amylases (Ezeogu et al., 2008) during starch hydrolysis. Beta and gamma kafirins form extended web-like structures with starch embedded within, after cooking. Formation of these extended structures reduces digestion of both protein and starch.

Kafirins are the major storage proteins of sorghum and their characteristics have been reviewed (Belton et al., 2006; De Mesa-Stonestreet et al., 2010). Generally, the kafirins are located in spherical protein bodies, and both the protein bodies and starch granules are embedded in a glutelin matrix (De Mesa-Stonestreet et al., 2010). Pepsin-indigestible residues come mainly from α-kafirin in uncooked sorghum, while cooking sorghum leads additionally to more disulphide-bonded proteins with reduced digestibility (Duodu et al., 2002; Duodu et al., 2003). The α-kafirins are the last proteins to be digested in the intestine and, because they make up a large amount of the protein in sorghum, create low overall digestibility (Duodu et al., 2003).

Because lysine is the first limiting amino acid in sorghum, seeds from ten genotypes were analysed for protein and amino acid composition. In addition to a commercial genotype, nine high-lysine genotypes developed by the International Crops Research Institute for the Semi-Arid Tropics (ICRISAT) were analysed. Total protein concentration varied by nearly 3-fold, while lysine concentration varied by more than 6-fold among the 10 genotypes (Vendemiatti et al., 2008).

A point mutation has been identified recently that is associated with a highly digestible, lysine-enhanced sorghum phenotype (Wu et al., 2013). Improvements are still needed in several traits including kernel hardness, but the use of a molecular marker may facilitate nutritional improvements (Wu et al., 2013).

2.2 Characteristics of sorghum starch

The most abundant component (about 70%) of the dry grain is starch (Benmoussa et al., 2015; Kotarski et al., 1992; Rooney and Pflugfelder, 1986), which is composed of amylose and amylopectin. Amylose is a linear polymer of glucose, linked through α-1,4-glycosidic bonds. Amylopectin is a branched polymer of glucose, composed of linear α-1,4-glycosidic linkages with a branch point made of α-1,6-glycosidic linkages. In non-waxy sorghum grain, starch is typically composed of a mixture of 30% amylose and 70% amylopectin (Rooney and Pflugfelder, 1986). The low-amylose sorghum grains exhibit endosperm breakage patterns resembling those of wax; for this reason they are termed as 'waxy'. As a result, starch of waxy and non-waxy grains has different gelatinization and digestibility properties that lead to their use for different products.

Waxy and heterowaxy sorghums have a decreasing amount of amylose in their starchy fraction. The advantage of starch in the form of amylopectin is that it gelatinizes at lower temperatures and more completely than do normal sorghum lines containing high amylose content (Wu et al., 2010). The content of amylose and total starch, together with protein digestibility, accounted for 94% of variation in digestibility (Wong et al., 2010). Specialty sorghums such as waxy, heterowaxy and tannin (high antioxidant) hybrids offer benefits for nutrition and/or processing (Rooney and Awika, 2005).

From a processing and utilization standpoint, amylopectin starch lowers gelatinization temperatures. High-protein digestible and waxy sorghum lines take less time to ferment dough for making Ethiopian *injera* (personal communication, Dr Kebede Abegaz).

The methods to improve the nutritional quality and organoleptic properties of cereal-based foods include genetic improvement, amino acid fortification and supplementation or complementation with protein-rich sources as well as processing technologies, which include milling, malting, fermentation or sprouting (Chavan and Kadam, 1989a; Chavan and Kadam, 1989b). Fermentation in sorghum has been reported as a good option for increased digestibility of sorghum proteins (Belton and Taylor, 2004), decreased antinutritional factors such as tannins and phytic acid (Mohamed Nour et al., 2011; Osman, 2004), improved amino acid balance and increased vitamin content (El Tinay et al., 1979). The nutritional disadvantages of sorghum tannins lie primarily in their ability to form poorly digestible complexes with dietary protein (Butler et al., 1984).

3 Effects of cooking and processing on digestibility and nutrient availability

Food processing effects can increase or decrease digestibility of sorghum-based foods. For example, sorghum proteins become less digestible after wet cooking (Eggum et al., 1983) but may be more digestible after extrusion. Thus, the choice of food processing and preparation techniques affects nutritional quality of the product for human consumption.

Traditional mixed diets in developing countries often depend on minimally refined cereals and legumes as their major sources of protein and energy. Because subsistence farming families consume their own grains, strategies that increase digestibility of macronutrients as well as bioavailability of micronutrients are needed at the household level. Germination (sprouting), fermentation and soaking are among the food processing methods used at the household level in developing countries (Gibson et al., 2006). In laboratory trials, germination increased *in vitro* protein digestibility in sorghum grains from Sudan (Mohamed Nour et al., 2010); also, α-amylase activity of germinated sorghum increased several fold compared to non-germinated grain (Dicko et al., 2006a). Fermentation reduced viscosity of sorghum-based complementary foods (Tizazu et al., 2010) which would allow infants or young children to increase their food intake.

3.1 Digestibility of macronutrients

Kafirins, the storage proteins in sorghum, have low digestibility *in vitro* (Duodu et al., 2003; de Morais Cardoso et al., 2017). The proteins in uncooked sorghum were only 73–88% digested by pepsin, trypsin-chymotrypsin or the combination of these enzymes *in vitro* (Hamaker et al., 1986), and cooking the sorghum further reduced protein digestibility (Eggum et al., 1983; Hamaker et al., 1986; Oria et al., 1995; Duodu et al., 2003). Protein cross-linking has been proposed to be the most detrimental factor affecting protein digestibility (Duodu et al., 2003). Sorghum bran high in tannins (PA) reduced *in vitro* protein digestibility compared to white sorghum bran (Dunn et al., 2015). Likewise, sorghum proanthocyanidin (PA) extracts reacted with starch, particularly with amylose, and decreased *in vitro* digestibility. The PA extracts increased the net resistant starch (RS) by approximately two-fold compared to a monomeric polyphenol extract (Barros et al., 2012). Digestibility of flour made from the endosperm is improved compared to that of wholegrain flour (Duodu et al., 2002).

New biofortified sorghum lines with combined waxy starch and high protein digestibility have been developed. In addition to being more digestible, these new types of sorghums may have better dough properties than traditional sorghum lines (Elhassan et al., 2015).

Extrusion of decorticated sorghum increased pepsin digestibility of cooked flour porridges to 79% compared to the digestibility (61%) of cooked porridges made from raw flour (Hamaker et al., 1994). *In vitro* digestibility also was positively associated with extrusion barrel temperature in a sorghum–maize composite flour (Licata et al., 2014). In a study comparing two sorghum genotypes (red, tannin and white, non-tannin), the *in vitro* protein digestibility of the extruded red tannin type (43–58%) was significantly lower than the digestibility (69–73%) of the white type (Mkandawire et al., 2015)

Sorghum contains RS, but the quantity of RS is very dependent on genotype. The amount of RS, defined as starch that is not hydrolysed by incubation with digestive enzymes, varied among 49 sorghum genotypes, from 0.3 to 65.7 g/100 g sorghum flour on a dry weight basis (Teixeira et al., 2016).

Effects of type of heat treatment on RS were tested in two sorghum genotypes considered appropriate for human food. With wet cooking, less than 7% of RS was retained in sorghum grain or flour, while roasting (dry heat) techniques retained as much as 97% of the RS (Teixeira et al., 2016). Starch also has 15–25% reduced digestibility in cooked sorghum flours compared to cooked maize flour. Predigestion of the sorghum flour with pepsin increased starch digestibility, indicating that the sorghum proteins contributed to the slow digestion of the starch (Zhang and Hamaker, 1998). In extruded sorghum–maize composite flour, levels of slowly digestible starch increased with higher proportions of sorghum in the mix and decreased with higher extrusion barrel temperature (Licata et al., 2014).

3.2 Availability of micronutrients and phytochemicals

Although consumption of whole grains is advocated on the basis of health benefits, plant-based diets also have the potential to reduce micronutrient availability from the diet. Traditional preparation techniques including soaking, fermenting and germinating may reduce some antinutritional factors (Gibson et al., 2006). Phytates reduce absorption of minerals such as Zn, Fe and Ca (Gibson et al., 2006). Sorghum may contain condensed tannins and high concentrations of antinutritional factors such as phytates (Sarwar Gilani et al., 2012).

Soaking unmilled sorghum for 1 hour as a strategy to reduce phytate resulted in reductions of only 12–19%, while soaking milled sorghum for only 1 hour prompted 39–57% reduction in phytate from the original concentration in grain. However, soaking also caused substantial losses in nutritionally important elements including calcium, iron and zinc; so it was concluded that soaking was not a viable household method to improve mineral availability from sorghum (Kruger et al., 2014).

Both processing methods and selection of cultivar can affect iron availability in cell culture (Proietti et al., 2014). Cooking had little effect on iron and zinc availability estimates, but fermentation decreased the concentration of iron-binding phenols and phytate which should enhance mineral availability (Proietti et al., 2013). Adding wheat phytase and mushroom polyphenol oxidase to sorghum gruels during fermentation reduced the concentration of phytate and phenolic compounds and increased the concentration of accessible iron (Towo et al., 2006).

Fermentation of several cereal blends, including wheat–red sorghum and teff–white sorghum, was investigated at the household level for making *injera*, a traditional

Ethiopian pancake. Barley malt was added for making *injera* in the highland areas, while malt was not added to the teff–white sorghum by the lowland households. Phytic acid (IP6) was completely hydrolysed in the wheat–red sorghum blend, while only 28% IP6 hydrolysis was observed in the lowland households where barley malt was not added for the fermentation (Baye et al., 2013). However, phytate was not the only inhibitor of iron bioaccessibility; hydrolysis of fibre produced a positive effect independent of phytate hydrolysis (Baye et al., 2015).

Cultivars of sorghum that have pigmented testa contain condensed tannins (Dykes et al., 2005; Awika et al., 2005b). Both wet cooking and extrusion reduced the condensed tannins in sorghum porridges, but losses were greater in the ready-to-eat extruded porridges (Dlamini et al., 2009). Dry heat and extrusion cooking reduced phenolic compounds in sorghum by different degrees. Losses of 3-deoxyanthocyanidins, total flavones and flavanones, and total PAs and their oligomers and polymers were substantially greater with extrusion cooking than with dry heating in a conventional oven (Cardoso et al., 2015). However, PA monomer concentrations and dimers were increased three-fold by extrusion similar to earlier reports from Awika and colleagues (2003). Generally, food processing operations were found to reduce phenolic compounds (Taylor and Duodu, 2015).

High hydrostatic pressure was used to promote dough structure formation of 60:40 composites of wheat and sorghum flour; the high-pressure bread received significantly higher overall acceptability scores than the non-pressured conventional (gluten-added) preparation. With the high-pressure preparation, bioaccessible phenols were reduced by nearly 50%; however, antiradical activity was increased from 28 to 36%, while digestible starch was increased by approximately 10% (Angioloni and Collar, 2012).

4 Protein quality of sorghum for human use

The relatively low digestibility of sorghum proteins and the particularly low concentrations of lysine are not sufficient to meet the needs for rapid growth of infants and young children. Complementary feeding (foods provided in addition to breast milk) and amino acid recommendations for infant and young child feeding (IYCF) are discussed in the following section. Different traits that may influence digestibility of kafirins and compensation for the lysine deficit by mixing sorghum with legumes such as beans or cowpeas have been investigated.

4.1 Sorghum and protein/amino acid needs for infant and young child feeding (IYCF)

An expert consultation for FAO provided recommendations on protein quality for different age groups based on digestible indispensable amino acid scores (DIAASs) (Food and Agriculture Organization, 2013). These scores are to be based on true ileal digestibility of each amino acid preferably determined in humans. Lysine is the first limiting amino acid in sorghum for humans. For young children, aged 6 months to 3 years, the recommended digestible lysine concentration is 57 mg/g protein, while for older children, adolescents and adults the recommendation is 48 mg/g protein. According to the USDA National Nutrient Database, lysine concentration of wholegrain sorghum flour (NDB#20648) is 20.6 mg lysine/g protein (United States Department of Agriculture, 2015). This figure does not include possible reductions in chemical integrity of lysine with processing (i.e. Maillard reaction) or substances such as phytates, tannins or kafirins which may limit utilization of the lysine (Food and Agriculture Organization, 2013).

4.2 Composites of sorghum and legumes (pulses) for complementary foods for IYCF

Because the essential amino acid concentration of sorghum is low, and because legumes (pulses) have higher protein and lysine levels than grains and pseudograins, various sorghum and legume blends have been evaluated for complementary feeding. A guide to the recommendations from the World Health Organization for energy density, protein and micronutrients in complementary foods is referenced (Dewey and Brown, 2003).

Ileal amino acid digestibility values were obtained from eight cereal grains using growing pigs. Eight grains were evaluated (maize (2 types), barley, oats, rice, rye, sorghum and wheat), and the DIAASs were substantially lower for sorghum than for the other grains, indicating a need for more amino acid supplementation to reach the recommended reference patterns for optimal growth (Cervantes-Pahm et al., 2014).

Adding 30% cowpea to sorghum flour nearly doubled the lysine scores compared with sorghum-only foods. *In vitro* protein digestibility of the foods was increased by 13–62% (Anyango et al., 2011b). Composite sorghum meal:marama bean flour (70:30) porridge had significantly higher protein and fat content than sorghum meal alone. Lysine content was 3–4 times higher in the bean-supplemented porridges (Kayitesi et al., 2012). A sorghum:sugar bean (*Phaseolus vulgaris*) composite (40.7:59.3) flour was bioassayed for 22 days in rapidly growing young rats. Growth with the composite flour diet was as fast as with a soy concentrate diet (designed to meet all needs for growth); additionally, growth with the composite diet was much more rapid than with the diet formulated with sorghum flour as the only source of protein and energy (Jackson et al., 2013).

A 1:1 composite of sorghum flour with defatted soy flour was used to make biscuits for school children. The protein content was doubled compared to sorghum alone and the chemically available lysine content was increased by 261% (Serrem et al., 2011).

An eight-month feeding trial was conducted in India in which school children aged 9–12 years were randomly assigned to receive 100% rice or 60% sorghum and 40% rice for their daily cereal intake. Growth rate was higher in girls fed with sorghum and in boys fed with rice. Haemoglobin, albumin, retinol binding protein, ferritin and serum iron concentrations were improved in the sorghum group for both boys and girls. Average cereal intake of boys was 285 g/day and that of girls was 250 g/day (Prasad Mp et al., 2016). Children of this age have lower lysine requirements per gram of protein than children under two years of age who grow very rapidly (Food and Agriculture Organization, 2013), so additional studies are needed to verify availability of protein and specific amino acids from sorghum for complementary feeding.

5 Sensory and functional characteristics of food products containing sorghum

Although wholegrain sorghum has several beneficial health attributes, it frequently is combined with legumes to compensate for low concentrations of available lysine or with other grains to improve functional characteristics of the prepared products (Ragaee and

Abdel-Aal, 2006; Del Giudice et al., 2008). Absence of gluten is an advantage for persons with gluten sensitivities but can be a challenge in terms of rheological properties of dough (Bugusu et al., 2001). Combining legumes with sorghum produces flavour and texture changes (Anyango et al., 2011a). Furthermore, availability of micronutrients and phenolics, as well as slowly digestible starch, can be enhanced or decreased by processing.

5.1 Sorghum gruels (thin porridges) and thick porridges

Moussa and colleagues developed pregelatinized flours from three food-grade sorghum cultivars to be used for instant thin or thick porridges using continuous processor technology. Thin or thick instant porridges were prepared by 1:8 or 1:2 reconstitution in hot water and compared with porridges cooked traditionally from flours from the same cultivars. Sensory panellists in Niger ranked 'overall liking' of the instant product from one cultivar as significantly better than traditionally cooked porridges; instant porridges from the other two cultivars were liked at least as well as the control porridges. Compared to control porridges, the instant thick porridges had stronger gels and the thin products had lower viscosity. The authors concluded that the instant sorghum flour had potential for commercialization and that the lower viscosity of the thin porridge might allow preparation of a higher energy density food for children (Moussa et al., 2011).

Sorghum has been combined with a variety of legumes and other products to make gruels that meet the needs of infants and young children between the ages of 6 and 23 months. Both tannin-type or non-tannin-type sorghum:cowpea (70:30) composites had reduced viscosity in a fermented thin porridge, *uji*, but increased firmness in an unfermented stiff porridge, *ugali*. There was considerable variation in sensory characteristics due to sorghum cultivar. A sensory panel evaluated the *ugali* and identified a significant cowpea flavour as well as differences between the two sorghum cultivars (Anyango et al., 2011a).

Composite sorghum meal:marama bean flour (70:30) porridge had significantly higher protein and fat content than sorghum meal alone (Kayitesi et al., 2012). Although the sensory panel preferred the sorghum porridge, the composite porridge from heated full-fat marama beans was also acceptable (Kayitesi et al., 2010).

In Botswana, children and their parents evaluated the sensory attributes of a porridge made from an extruded sorghum:sugar bean (*Phaseolus vulgaris*) composite (40.7:59.3) flour. The customary sorghum porridge was preferred, but more than 75% of children rated the composite porridge as highly acceptable and over 50% of adult panellists liked the sensory attributes of the sorghum:bean mixture (Jackson et al., 2013). An amaranth–sorghum complementary food formulation was recently reported from Kenya. Steeping and germination of grains reduced tannins and phytates in the complementary food mix; the 90:10 amaranth:sorghum mixture had a significantly higher energy content than other ratios tested and met the protein recommendations for complementary foods (Okoth et al., 2017).

5.2 Breads and biscuits/cookies

Products that depend on gluten for certain viscoelastic qualities are a challenge with sorghum flour which does not contain gluten; often, composite flours or novel mixing techniques are used to obtain familiar functional characteristics (Bugusu et al., 2001). However, flour from high-digestible, high-lysine sorghum produced higher bread loaf volumes in sorghum:wheat composites than did normal sorghum:wheat composites (Goodall et al., 2012).

Sorghum cultivar properties, particularly the quantity of tannins, affected stiffness of a fermented flatbread, *injera*, made from a sorghum:cowpea (70:30) flour. Stiffness of the *injera* using tannin-containing sorghum increased, while stiffness of the non-tannin sorghum decreased (Anyango et al., 2011a). Australian consumers rated overall acceptability of flat breads made with 30% white sorghum flour significantly higher than the 100% wheat flour flat bread (Yousif et al., 2012).

A 1:1 composite of sorghum flour with defatted soy flour was used to make biscuits for school children. The protein content was doubled compared to sorghum alone and the chemically available lysine content was increased by 261%. School children in Pretoria, South Africa, gave hedonic ratings higher than 80% over the four-day evaluation period (Serrem et al., 2011).

5.3 Pasta and other products

Pasta was made with 100% durum wheat semolina (DWS) or red or white sorghum composite flour at 20, 30 or 40%, respectively, with wheat flour. *In vitro* starch digestion was lowered at all incorporation levels. The 20 and 30% blends had acceptable palatability (Khan et al., 2014).

Sorghum-based *tempe* was made with a 70:15:15 composite of sorghum:leguminous pulse:oil seed. After soaking components overnight and draining and steaming the sorghum meal, the sorghum, pulse and oil seed were combined and boiled for 15 minutes. After draining, the residues were inoculated with *Rhizopus oligosporus* and incubated at 37 °C for 24–30 hours to form *tempe*. Lysine remained the limiting amino acid in the mixtures, but the amino acid score increased to 63–69 in the various *tempe* products (Mugula and Lyimo, 2000).

Sorghum can be popped or extruded for ready-to-eat snacks. Such ready-to-eat foods are particularly popular among persons with celiac disease (CD), but the bland flavour of sorghum makes such snacks also appealing to the general population (Asif et al., 2010).

5.4 Alcoholic and non-alcoholic beverages

Gowe, a non-alcoholic cereal-based beverage, is produced from malted cereal flour and then cooked. Producers surveyed in Benin were found to be producing *gowe* from sorghum, maize or composites with maize/sorghum ratios varying from 1 to 3. Consumers were from all age and sociocultural groups (Adinsi et al., 2014).

For production of lactic acid-fermented beverages, malting parameters for sorghum were investigated. Germination time had the most effect on production of high-quality malt (Hassani et al., 2014). At least some barley malt can be substituted by sorghum malt for brewing beers (Taylor et al., 2006).

6 Health benefits of sorghum

Sorghum has a number of intriguing potential health benefits. The grain can be eaten by persons with CD who must avoid gluten. Different sorghum cultivars contain variable amounts of RS; starch which is slowly digested may help to moderate blood glucose peaks and assist with appetite regulation. Sorghum is rich in fibre and bioactive components. Antioxidants in sorghum may reduce inflammation and oxidative stress, two factors known to be associated with various chronic diseases in developed nations. Antiproliferative effects of sorghum extracts have been demonstrated in cell lines, and some components of sorghum may have beneficial effects on the gut microbiome. High-quality trials in animal models and humans are needed to assess effects on health outcomes of increased use of sorghum as a dietary component. Recent reviews on potential health benefits of sorghum have been published (Simnadis et al., 2016; Stefoska-Needham et al., 2015; Taylor and Duodu, 2015; Taylor et al., 2014; Althwab et al., 2015; Del Giudice et al., 2008; Rooney and Awika, 2005) and several individual studies are cited below.

6.1 Sorghum for persons with celiac disease

Sorghum is considered to be a safe cereal for people with CD (Asif et al., 2010). Persons with CD have an immune response to the glutamine- and proline-rich storage proteins found in wheat, rye and barley and possibly a similar protein in oats (Wieser and Koehler, 2008). Collectively these proteins are referred to as glutens, although the structures from the different grains are not identical.

CD affects the small intestine by reducing or flattening normal villi which thus impairs nutrient absorption (Lionetti and Catassi, 2011). The body mounts both innate and adaptive immune responses to gluten which contribute to this destruction of the intestinal mucosa (Wieser and Koehler, 2008). Currently, the only practical solution for confirmed CD is a lifelong avoidance of dietary gluten which means avoiding any products containing wheat, rye, barley and at least some cultivars of oats (Silano et al., 2014; Rubio-Tapia et al., 2013; Ballabio et al., 2011).

Most studies of gluten sensitivity have been *in vitro*. However, a few small studies have been conducted with sorghum extracts in human tissues. From eight patients with active CD and four non-CD controls, endoscopic biopsies were obtained. When tissue sections from these biopsies were challenged *in vitro* with sorghum protein digests, the expression of inflammatory markers from all samples was similar to that produced by the medium alone. Challenge with gliadin (a component of gluten) extract increased antigen production by more than 100% (Ciacci et al., 2007). Two confirmed CD patients also consumed 150 g per day of bread and cookies made from sorghum for five days with no adverse serological tests or symptoms (Ciacci et al., 2007).

Based on testing for immunoglobulin A (IgA) transglutaminase antibodies in serum samples from 7,798 persons from the National Health and Nutrition Examination Survey (NHANES 2009–10), the overall prevalence of CD in the United States was reported to be 0.71% or about 1 in 141. Prevalence tended to be higher in non-Hispanic whites, while minorities had a lower prevalence (Rubio-Tapia et al., 2012). There are regional differences in Europe, but overall prevalence remains at approximately 1%. Diagnosis of CD in other areas of the world is limited, at least partially due to lack of awareness (Catassi et al., 2015).

Computer simulations of the sorghum genome suggest that it does not contain the peptides that are toxic for patients with CD (Pontieri et al., 2013). Likewise, evaluation of prolamines from various varieties of sorghum demonstrated the absence of toxic peptides, indicating that consumption of sorghum should be safe for persons with CD (Pontieri et al., 2013).

The Codex Standard for Foods for Special Dietary Use for Persons Intolerant to Gluten was last amended in 2015 and specifies that gluten-free foods must consist of ingredients which do not contain more than 20 mg gluten/kg in total as sold to the consumer (Food and Agriculture Organization and World Health Organization, 1979). Thus, sorghum meets the Codex standards as a gluten-free food (Pontieri et al., 2013; Althwab et al., 2015; Rosell et al., 2014).

6.2 Delayed carbohydrate absorption from sorghum products

Glycemic index (GI) is a measure of the effect of consuming carbohydrate on postprandial glucose. The GI is calculated as the area under the two-hour blood glucose curve when a portion of a test food containing 50 g of carbohydrate is compared to a reference food containing 50 g of carbohydrate (Wolever, 2003). Methods of cooking and processing, digestibility, amounts of fibre, and antinutrients such as phytates and polyphenols are among the factors in foods that may affect the GI (Moraes et al., 2015).

A few trials of carbohydrate absorption from sorghum products have been conducted in humans. In India, sorghum-based versus wheat- or rice-based foods were compared in ten healthy volunteers; sorghum-based semolina *upma*, *poha* and pasta, but not *roti*, had significantly lower GI than the comparable food made from wheat or rice (*poha*) (Rajendra Prasad et al., 2015). Poquette and colleagues (2014) compared blood glucose and insulin responses to a whole-wheat flour muffin (control) and a grain sorghum muffin in 10 healthy men in a randomized crossover design. Both muffins contained 50 g of starch. Area under the curve for both plasma glucose and insulin was significantly reduced after consuming the sorghum muffin compared to the whole-wheat muffin.

Slowly digesting starch has been investigated *in vitro* in several trials. Amoako and Awika (2016) reacted partially gelatinized starch from normal and waxy maize with condensed tannin extracts to assess the tannin effects on the formation of RS. In 30% ethanol, the tannin extract increased slow-digesting starch from 100 to 274 mg/g in normal maize but not in the waxy maize. The authors suggested that condensed tannins could be advantageous in terms of lowering the caloric impact of foods. Pasta made with 20 to 40% red sorghum flour or white sorghum flour substituted for durum wheat semolina (DWS) had higher RS than the pasta made with 100% DWS. Cooking did not change the RS starch content of the pasta (Khan et al., 2013). Other *in vitro* studies also showed that sorghum PAs reacted with starch and substantially decreased its digestibility (Barros et al., 2012). A starch hydrolysis index estimated *in vitro* was significantly lower in whole sorghum flour, decorticated sorghum flour and sorghum bran than in white bread (Moraes et al., 2015).

6.3 High fibre concentration in sorghum

Dietary fibre in bran from decorticated sorghum was 36–45%. The first two sequentially removed bran fractions also were 3–6 fold higher in phenols and antioxidant activity compared to the whole grain (Awika et al., 2005a). Sorghums containing tannins may reduce caloric availability (Awika and Rooney, 2004). Furthermore, fibre with its high concentrations of tannins and phenolics may have beneficial effects on the gut microbiome (Stefoska-Needham et al., 2015; de Morais Cardoso et al., 2017).

6.4 Antioxidant compounds in sorghum

Sorghum has an unusually high concentration of phytochemicals with antioxidant activity (Awika and Rooney, 2004; Ragaee et al., 2006), but there is wide variability among sorghum cultivars. These phytochemicals are classified as phenolic acids and flavonoids.

The phenolic acids are benzoic acid and cinnamic acid; the flavonoids include tannins, PAs, anthocyanins and 3-deoxyanthocyanins (Gu et al., 2003; Awika and Rooney, 2004; Dykes and Rooney, 2006; Dykes et al., 2005; Althwab et al., 2015).

In a randomized crossover clinical trial, 20 healthy participants consumed test meals of control pasta, 30% red sorghum pasta or 30% white sorghum pasta at one- to two-week intervals. Blood samples at baseline and 2 hours postprandial were collected with each type of pasta. Plasma total polyphenols and antioxidant capacity were increased, and protein carbonyl level was significantly lower following consumption of the red sorghum pasta. White sorghum pasta did not produce similar results (Khan et al., 2015).

Brans from black sorghum had anthocyanin concentrations equivalent to pigmented fruits and vegetables (Awika et al., 2005b). Brown (tannin-containing) and black sorghums had a least 10 times higher antioxidant activity than white sorghum or red wheat brans. Black sorghums had the highest 3-deoxyanthocyanin content (Awika et al., 2005a); anthocyanin in sorghum brans was 3–4× higher than whole grain (Awika et al., 2004). Black sorghum hybrids also contained high concentrations of 3-deoxyanthocyanidins but they were not as high as the original black cultivars (Dykes et al., 2013). The 3-deoxyanthocyanidin concentrations were strongly correlated to the pericarp colour (Dykes et al., 2013). These compounds are stable and may have use as natural food colourants as well as having antioxidant activity (Dykes et al., 2013; Awika et al., 2005b). As a point of reference, major sources of PAs in the American diet are apples, chocolate and grapes (Gu et al., 2004).

Food processing and cooking generally decreased the antioxidant compounds in sorghum (Dunn et al., 2015). For example, procyanidins (also called condensed tannins) in two varieties of brown sorghum differed by variety but were mostly high-molecular weight polymers. Incorporating the sorghum brans into cookies and bread significantly reduced the procyanidins, particularly the higher molecular weight ones (Awika et al., 2003). Ragee and coworkers have reviewed effects of milling and thermal processing on phenolic compounds in sorghum and other cereal grains (Ragaee et al., 2014).

Cardoso and colleagues (2015) compared the effects of extrusion and dry heat in a conventional oven on these antioxidant compounds in sorghum. Dry heat increased the antioxidant capacity in all genotypes; extrusion reduced antioxidant capacity in two of three genotypes. Extrusion also decreased the degree of polymerization (Awika et al., 2003), and traditionally cooked porridges and sorghum grain had higher concentrations of condensed tannin oligomers and polymers than extruded sorghum porridges (Dlamini et al., 2009).

6.5 Antiproliferative effects of sorghum

Sorghum extracts containing 3-deoxyanthocyanins (3-DXA) were tested for their ability to induce phase II enzymes in murine hepatoma cells using the NAD(P)H:quinone oxidoreductase (NQO) assay. NQO is considered to be a good marker for protection of cells against carcinogens. Black sorghum extracts promoted much higher NQO activity than red or white sorghum extracts (Yang et al., 2009). Extracts of all three types of sorghum demonstrated antiproliferative activity against HT-29 colon cancer cells with the black sorghum extract having higher activity than the extracts from the red and white sorghums (Yang et al., 2009). Another study demonstrated that antiproliferative activity in oesophageal OE33 and colon HT-29 adenocarcinoma cells correlated with antioxidant activity but that antiproliferative activity and NQO-specific inducer ratios were not correlated (Awika et al., 2009).

Phytoestrogens from sorghum may contribute to beneficial effects of whole grains on human health. Activation of oestrogen receptor-β (ER- β) has been hypothesized to be one mechanism by which sorghum may be associated with reduced risk for colon cancer. Building on a previous study (Yang et al., 2012), two oestrogenic flavonoids from sorghum were evaluated for their ability to inhibit ER-mediated cell growth in non-malignant young adult mouse colonocytes. Results confirmed that the two compounds, apigenin, a flavone, and naringenin, a flavanone, were synergistic, requiring concentrations only 1/10 and 1/200 as high, respectively, in combination as alone for equivalence to physiological concentrations of oestradiol (Yang et al., 2015). The researchers concluded that composition of flavonoids rather than the content of polyphenolics determined ER activity (Yang et al., 2015). The study also illustrated the importance of evaluation of potential synergies of compounds as they occur naturally in foods rather than studying isolated compounds.

7 Prospects and challenges for sorghum as a food grain

Sorghum's drought resistance and thus its contribution to food security make it a very important crop, particularly in the context of climate change (Rufino et al., 2013). Breeding programmes are developing sorghum cultivars that address some of the difficulties related to its wider use as human food, and various studies are illustrating the inherent health benefits of some components of sorghum.

Challenges in the use of sorghum for human food depend somewhat on the target consumers. Digestibility of foods prepared from sorghum is affected by the cooking method. Traditional wet cooking, such as might be used to make porridge in a subsistence farming household, reduces digestibility. Extrusion can increase digestibility of sorghum proteins; however, extruded foods are more available to families with resources beyond those of the subsistence farmers. RS in sorghum means that it may be more difficult for humans to fully extract the potential energy from carbohydrate. If food is limiting, this is a disadvantage; if one considers the worldwide epidemic of obesity, a food with less easily utilized carbohydrate could be an advantage. Slower release of carbohydrate also should be beneficial for moderation of blood glucose peaks.

The condensed tannins and high levels of phenolics in sorghum provide an excellent source of dietary antioxidants, but some of the compounds may reduce the availability of minerals. The lack of gluten in sorghum makes it an important food for persons with CD but generally reduces the acceptability of typical leavened bread products. Several investigators (Dicko et al., 2006b; Taylor et al., 2006; Tuinstra, 2008) have reviewed the diverse uses of sorghum for foods and beverages in both developed and developing countries.

7.1 Increasing yields and market development

In developing countries, sorghum yields are generally low, and individual farmers have small quantities of the grain available to sell. Better agronomic practices and stable market prices would promote sorghum production. Hybrid sorghums produce higher yields but may have longer growing periods. New food-grade sorghum varieties have been developed, particularly to meet the increased demand for sorghum-based foods in the developed world (Tuinstra, 2008). Certain cultivars have higher lysine content and more digestible protein which makes them more appropriate for use for feeding children.

Heightened interest in gluten-free products drives a niche market for sorghum (De Mesa-Stonestreet et al., 2010). If gluten sensitivity is as high as 1% worldwide, the need for appropriate baked goods will create an enormous demand. Because most currently diagnosed gluten intolerance occurs in developed countries, sorghum producers will need to meet those marketing standards to effectively compete for the higher value markets.

Focus groups conducted in Kansas suggested that health aspects of sorghum appealed most to consumers; however, conjoint analysis and a follow-up Internet survey demonstrated that sensory characteristics were most likely to be the basis for purchase of sorghum-containing granola bars and breakfast cereals (Vázquez-Araújo et al., 2012). Thus, except for persons with CD, sensory considerations may outweigh health considerations for driving market demand.

7.2 Nutritional considerations

Tuinstra has reviewed key qualities of food-grade sorghum cultivars (Tuinstra, 2008). Increased lysine content would enhance the protein quality of sorghum and would particularly help in meeting the requirements for rapid growth of infants and young children (Food and Agriculture Organization, 2013). Protein quality has been improved by suppressing different kafirin subclasses. Modifications in protein body structure may have allowed better digestive enzyme access (Da Silva et al., 2011). However, some food-grade hybrids with better protein quality may be more susceptible to mould and weathering (Tuinstra, 2008).

In a pot experiment, increasing levels of microminerals in the soil resulted in higher levels of increased microminerals in the sorghum (Ahmed et al., 2014). Biofortified sorghums are being developed with enhanced nutritional traits including high levels of iron and zinc. These minerals may also be more bioavailable through phytate reduction (Taylor et al., 2014).

Retention of potentially beneficial phytochemicals in sorghum will be important as new cultivars are developed. Furthermore, to assess effects of sorghum on prevention of chronic disease, high-quality animal and human trials are needed, and the synergy demonstrated in a food matrix among these phytochemicals must be considered.

7.3 Mycotoxin contamination

The African Union's Partnership for Aflatoxin Control in Africa (PACA) is expanding the recognition of food and food safety concerns across the continent. While aflatoxin contamination of groundnuts and maize has received most attention, recent reports have identified contamination of sorghum products as well. Mycotoxin contamination has potential for having serious effects on human health including increased risk for hepatocellular carcinoma and association with stunting in children. The widespread exposure to aflatoxin in sub-Saharan Africa has been reviewed (Gnonlonfin et al., 2013).

Aflatoxin B_1 and ochratoxin A were identified in 6.1% and 21.8%, respectively, of 78 sorghum samples tested in Ethiopia, with the highest concentrations found in samples drawn from underground pits (Ayalew et al., 2006). The mean concentration of aflatoxin B_1 and G_1 in contaminated Ethiopian sorghum samples was found to be higher than the European standard (Chala et al., 2014). Various other mycotoxins were also identified in sorghum and other cereal samples (Ayalew et al., 2006; Chala et al., 2014), indicating a need for improved strategies for cereal grain storage in rural households. A report from Sudan also indicated that some sorghum samples contained aflatoxin B_1 concentrations above the European Union regulatory limits (Elbashir and Ali, 2014).

A multicentre study in India in which 1606 grain sorghum samples were tested in the rainy season showed 73% of the samples to be positive for aflatoxin B_1, but only 0.75% of the total samples contained aflatoxin above the safety limit specified by the Codex Alimentarius Commission (Ratnavathi et al., 2012).

In samples collected during the humid season in Malawi, all sorghum malt and beer prepared with sorghum malt were contaminated with aflatoxins, while 15% of sorghum grain samples and 43% of samples of *thobwa*, a non-alcoholic drink, likewise were contaminated. Only five samples of the traditional beer were collected, but using the VICAM AflaTest immunoaffinity column, the mean aflatoxin concentration in the beer was 22 µg/L, while the Codex Alimentarius Commission has set the maximum level in ready-to-eat foods at 10 ug/kg (Matumba et al., 2011). A diverse array of mycotoxins were identified in 60 sorghum samples from Tunisia and the potential for infants to be high-risk consumers from consumption of complementary foods was discussed (Oueslati et al., 2014).

Both *Aspergillus* and *Fusarium* species have been identified in sorghum samples collected from small-scale farmers in Ethiopia. Ninety grain samples were selected from threshing floors and collections were repeated 5–6 months later from underground storage pits. Aflatoxin B_1 levels ranged from non-detectable to 33 µg/kg grain, while total fumonisin concentrations ranged between 907 and 2041 µg/kg grain. Lower concentrations of both mycotoxins were detected in freshly harvested sorghum grain than in stored grain (Taye et al., 2016).

8 Conclusion

Sorghum is the fifth most abundantly produced cereal crop in the world and its drought resistance is particularly important in the context of climate change. Digestibility of sorghum is lower than that of other major cereal grains, and lysine concentrations of currently utilized sorghum cultivars are too low for optimal growth of infants and young children unless used in combination with pulses or legumes. Furthermore, lack of gluten interferes with baking properties of sorghum flour. However, certain food processing techniques such as extrusion, germination and fermentation may increase the availability of proteins and starch from sorghum. Sorghum has several potential health benefits. The lack of gluten proteins in sorghum prevents adverse reactions in CD patients. The amounts of RS in sorghum can be an advantage in terms of producing a lower postprandial glucose peak. Although high dietary fibre in sorghum may be detrimental for young children who need to consume foods with high nutrient density, the high fibre content may be beneficial in terms of the worldwide obesity epidemic. Furthermore, the high concentration of phytochemicals with antioxidant activity in sorghum may have application in terms of reduction of chronic disease risk. Additional research is needed to improve quality of food-grade sorghum cultivars. Attention to mycotoxin control and to storage and transport of the grain will be critical for improving food quality of subsistence farm families and for developing sustainable international markets for sorghum.

9 Where to look for further information

The characteristics of the protein and starch matrix and steps for improving functionality of foods prepared from sorghum have been reviewed by several scientists (Rooney and

Awika, 2005; Taylor et al., 2014; Stefoska-Needham et al., 2015; Sarwar Gilani et al., 2012). Sorghum contains numerous phytochemicals and potential health benefits of these compounds also have been reviewed (de Morais Cardoso et al., 2017; Awika and Rooney, 2004; Althwab et al., 2015; Taylor and Duodu, 2015).

Human clinical trials, and even animal trials, that assess impacts of sorghum on health outcomes are extremely limited. A recent systematic review has examined 19 studies of health effects associated with sorghum consumption (Simnadis et al., 2016). This very small number of trials for multiple health considerations emphasizes the need for additional studies in humans.

10 References

Adinsi, L., Vieira-Dalode, G., Akissoe, N., Anihouvi, V., Mestres, C., Jacobs, A., Dlamini, N., Pallet, D. and Hounhouigan, D. J. 2014. Processing and quality attributes of gowe: a malted and fermented cereal-based beverage from Benin. *Food Chain (2046-1887)*, 4, 171–83.

Ahmed, S. O., Abdalla, A. W. H., Inoue, T., Ping, A. and Babiker, E. E. 2014. Nutritional quality of grains of sorghum cultivar grown under different levels of micronutrients fertilization. *Food Chemistry*, 159, 374–80.

Althwab, S., Carr, T. P., Weller, C. L., Dweikat, I. M. and Schlegel, V. 2015. Advances in grain sorghum and its co-products as a human health promoting dietary system. *Food Research International*, 77, 349–59.

Amoako, D. B. and Awika, J. M. 2016. Polymeric tannins significantly alter properties and in vitro digestibility of partially gelatinized intact starch granule. *Food Chemistry*, 208, 10–17.

Angioloni, A. and Collar, C. 2012. Effects of pressure treatment of hydrated oat, finger millet and sorghum flours on the quality and nutritional properties of composite wheat breads. *Journal of Cereal Science*, 56, 713–19.

Anyango, J. O., Taylor, J. R. N. and De Kock, H. T. L. 2011a. Evaluation of the functional quality of cowpea-fortified traditional African sorghum foods using instrumental and descriptive sensory analysis [electronic resource]. *LWT - Food Science and Technology*, 44, 2126–33.

Anyango, J. O., Taylor, J. R. N. and De Kock, H. T. L. 2011b. Impact of cowpea addition on the Protein Digestibility Corrected Amino Acid Score and other protein quality parameters of traditional African foods made from non-tannin and tannin sorghum [electronic resource]. *Food Chemistry*, 124, 775–80.

Asif, M., Mack, C. A., Riaz, M. N., Rooney, L. W. and Acosta-Sanchez, D. 2010. Uses of sorghum grain in gluten-free products. *Cereal Foods World*, 55, 285–91.

Awika, J. M. and Rooney, L. W. 2004. Sorghum phytochemicals and their potential impact on human health. *Phytochemistry*, 65, 1199–1221.

Awika, J. M., Rooney, L. W. and Mcdonough, C. M. 2005a. Decorticating sorghum to concentrate healthy phytochemicals. *Journal of Agricultural and Food Chemistry*, 53, 6230–4.

Awika, J. M.Rooney, L. W., Prior, R. L., Dykes, L. and Gu, L. 2003. Processing of sorghum (Sorghum bicolor) and sorghum products alters procyanidin oligomer and polymer distribution and content. *Journal of Agricultural and Food Chemistry*, 51, 5516–21.

Awika, J. M., Rooney, L. W. and Waniska, R. D. 2005b. Anthocyanins from black sorghum and their antioxidant properties. *Food Chemistry*, 90, 293–301.

Awika, J. M., Waniska, R. D. and Rooney, L. W. 2004. Properties of 3-deoxyanthocyanins from sorghum. *Journal of Agricultural and Food Chemistry*, 52, 4388–94.

Awika, J. M., Yang, L., Browning, J. D. and Faraj, A. 2009. Comparative antioxidant, antiproliferative and phase II enzyme inducing potential of sorghum (*Sorghum bicolor*) varieties. *Food Science and Technology*, 42, 1041–6.

Ayalew, A., Fehrmann, H., Lepschy, J., Beck, R. and Abate, D. 2006. Natural occurrence of mycotoxins in staple cereals from Ethiopia. *Mycopathologia*, 162, 57–63.

Ballabio, C., Lionetti, E., Catassi, C., Restani, P., PEÑAS, E., Redaelli, R., Manferdelli, S., Uberti, F., Boggini, G. and Vacca, E. 2011. Molecular characterisation of 36 oat varieties and in vitro assessment of their suitability for coeliacs' diet [electronic resource]. *Journal of Cereal Science*, 54, 110–15.

Barros, F., Awika, J. M. and Rooney, L. W. 2012. Interaction of tannins and other sorghum phenolic compounds with starch and effects on in vitro starch digestibility. *Journal of Agricultural and Food Chemistry*, 60, 11609–17.

Baye, K., Guyot, J.-P., Icard-Vernière, C., Rochette, I. and Mouquet-Rivier, C. 2015. Enzymatic degradation of phytate, polyphenols and dietary fibers in Ethiopian injera flours: Effect on iron bioaccessibility. *Food Chemistry*, 174, 60–7.

Baye, K., Mouquet-Rivier, C., Icard-Vernière, C., Rochette, I. and Guyot, J.-P. 2013. Influence of flour blend composition on fermentation kinetics and phytate hydrolysis of sourdough used to make injera. *Food Chemistry*, 138, 430–6.

Belton, P. S., Shewry, P. R., Halford, N. G. and Delgadillo, I. 2006. Kafirin structure and functionality [electronic resource]. *Journal of Cereal Science*, 44, 272–86.

Belton, P. S. and Taylor, J. R. N. 2004. Sorghum and millets: protein sources for Africa. *Trends in Food Science and Technology*, 15, 94–8.

Benmoussa, M., Chandrashekar, A., Ejeta, G. and Hamaker, B. R. 2015. Cellular response to the high protein digestibility/high-lysine (hdhl) sorghum mutation. *Plant Science: An International Journal of Experimental Plant Biology*, 241, 70–7.

Bugusu, B. A., Hamaker, B. R. and Campanella, O. 2001. Improvement of sorghum-wheat composite dough rheological properties and breadmaking quality through zein addition. *Cereal Chemistry*, 78, 31–5.

Butler, L. G., Blytt, H. J., Lebryk, D. G. and Riedl, D. J. 1984. Interaction of proteins with sorghum tannin: Mechanism, specificity and significance. *Journal of the American Oil Chemists' Society*, 61, 916–20.

Cardoso, L. D. M., Pinheiro, S. S., De Carvalho, C. W. P., Queiroz, V. A. V., De Menezes, C. B., Moreira, A. V. B., De Barros, F. A. R., Awika, J. M., Martino, H. S. D. and Pinheiro-Sant'ana, H. M. 2015. Phenolic compounds profile in sorghum processed by extrusion cooking and dry heat in a conventional oven. *Journal of Cereal Science*, 65, 220–6.

Catassi, C., Gatti, S. and Lionetti, E. 2015. World perspective and celiac disease epidemiology. *Digestive Diseases (Basel, Switzerland)*, 33, 141–6.

Cervantes-Pahm, S. K., Liu, Y. and Stein, H. H. 2014. Digestible indispensable amino acid score and digestible amino acids in eight cereal grains. *British Journal of Nutrition*, 111, 1663–72.

Chala, A., Taye, W., Ayalew, A., Krska, R., Sulyok, M. and Logrieco, A. 2014. Multimycotoxin analysis of sorghum (*Sorghum bicolor* L. Moench) and finger millet (*Eleusine coracana* L. Garten) from Ethiopia. *Food Control*, 45, 29–35.

Chandrashekar, A. and Kirleis, A. W. 1988. Influence of protein on starch gelatinization in sorghum. *Cereal Chemistry*, 65, 457–62.

Chavan, J. K. and Kadam, S. S. 1989a. Nutritional improvement of cereals by fermentation. *Critical Reviews in Food Science and Nutrition*, 28, 349–400.

Chavan, J. K. and Kadam, S. S. 1989b. Nutritional improvement of cereals by sprouting. *Critical Reviews in Food Science and Nutrition*, 28, 401–37.

Ciacci, C., DI Fonzo, N., Pontieri, P., Ioerger, B., Bean, S. R. Massardo, D. R. Caporaso, N., Maiuri, L., Del Giudice, L. and Bucci, C. 2007. Celiac disease: In vitro and in vivo safety and palatability of wheat-free sorghum food products [electronic resource]. *Clinical Nutrition*, 26, 799–805.

Da Silva, L. S., Taylor, J., Taylor, J. R. N., Glassman, K., Jung, R. and Zhao, Z.-Y. 2011. Effect of suppressing the synthesis of different kafirin sub-classes on grain endosperm texture, protein body structure and protein nutritional quality in improved sorghum lines [electronic resource]. *Journal of Cereal Science*, 54, 160–7.

De Mesa-Stonestreet, N. J. Alavi, S. and Bean, S. R. 2010. Sorghum proteins: The concentration, isolation, modification, and food applications of kafirins. *Journal of Food Science*, 75, R90–R104.
De Morais Cardoso, L., Pinheiro, S. S., Martino, H. S. D. and Pinheiro-Sant'ana, H. M. 2017. Sorghum (*Sorghum bicolor* L.): Nutrients, bioactive compounds, and potential impact on human health. *Critical Reviews in Food Science and Nutrition*, 57, 372–90.
Del Giudice, F., Massardo, D. R., Pontieri, P., Maddaluno, L., De Vita, P., Fares, C., Ciacci, C. and Del Giudice, L. 2008. Development of a sorghum chain in the Italian Campania Region: From the field to the celiac patient's table. *Journal of Plant Interactions*, 3, 49–55.
Dewey, K. G. and Brown, K. H. 2003. Update on technical issues concerning complementary feeding of young children in developing countries and implications for intervention programs. *Food and Nutrition Bulletin*, 24, 5–28.
Dicko, M. H., Berkel, W. J. H. V., Voragen, A. G. J., Traoré, A. S., Gruppen, H. and Zouzouho, O. C. 2006a. Effects of germination on the activities of amylases and phenolic enzymes in sorghum varieties grouped according to food end-use properties [electronic resource]. *Journal of the Science of Food and Agriculture*, 86, 953–63.
Dicko, M. H., Gruppen, H., Traore, A. S., Voragen, A. G. J. and Van Berkel, W. J. H. 2006b. Sorghum grain as human food in Africa: Relevance of content of starch and amylase activities. *African Journal of Biotechnology*, 5, 384–95.
Dlamini, N. R., Waniska, R. D., Taylor, J. R. N., Dykes, L. and Rooney, L. W. 2009. Condensed tannins in traditional wet-cooked and modern extrusion-cooked sorghum porridges. *Cereal Chemistry*, 86, 191–6.
Dunn, K. L., Yang, L., Girard, A., Bean, S. and Awika, J. M. 2015. Interaction of sorghum tannins with wheat proteins and effect on in vitro starch and protein digestibility in a baked product matrix. *Journal of Agricultural and Food Chemistry*, 63, 1234–41.
Duodu, K. G., Mills, E. N. C., Belton, P. S., Taylor, J. R. N., Nunes, A., Delgadillo, I. and Parker, M. L. 2002. Effect of grain structure and cooking on sorghum and maize in vitro protein digestibility [electronic resource]. *Journal of Cereal Science*, 35, 161–74.
Duodu, K. G., Taylor, J. R. N., Belton, P. S. and Hamaker, B. R. 2003. Factors affecting sorghum protein digestibility [electronic resource]. *Journal of Cereal Science*, 38, 117–31.
Dykes, L. and Rooney, L. W. 2006. Sorghum and millet phenols and antioxidants. *Journal of Cereal Science*, 44, 236–51.
Dykes, L., Rooney, W. L. and Rooney, L. W. 2013. Evaluation of phenolics and antioxidant activity of black sorghum hybrids. *Journal of Cereal Science*, 58, 278–83.
Dykes, L., Rooney, W. L., Waniska, R. D. and Rooney, L. W. 2005. Phenolic compounds and antioxidant activity of sorghum grains of varying genotypes. *Journal of Agricultural and Food Chemistry*, 53, 6813–18.
Earp, C. F., Mcdonough, C. M., Awika, J. and Rooney, L. W. 2004. Testa development in the caryopsis of *Sorghum bicolor* (L.) Moench. *Journal of Cereal Science*, 39, 303–11.
Eggum, B. O., Munck, L., Axtell, J., Monowar, L. and Knudsen, K. E. B. 1983. Nutritional quality of sorghum and sorghum foods from Sudan. *Journal of Cereal Science*, 1, 127–37.
El Tinay, A. H. EL Hidai, M. and Gadir, A. M. A. 1979. Sorghum fermented Kisra bread. I. Nutritive value of Kisra. *Journal of the Science of Food and Agriculture*, 30, 859–63.
Elbashir, A. A. and Ali, S. E. A. 2014. Aflatoxins, ochratoxins and zearalenone in sorghum and sorghum products in Sudan. *Food Additives and Contaminants: Part B: Surveillance Communications*, 7, 135–40.
Elhassan, M. S. M., Naushad Emmambux, M., Hays, D. B., Peterson, G. C. and Taylor, J. R. N. 2015. Novel biofortified sorghum lines with combined waxy (high amylopectin) starch and high protein digestibility traits: Effects on endosperm and flour properties. *Journal of Cereal Science*, 65, 132–9.
Ezeogu, L. I., Taylor, J. R. N. and Duodu, K. G. 2005. Effects of endosperm texture and cooking conditions on the in vitro starch digestibility of sorghum and maize flours. *Journal of Cereal Science*, 42, 33–44.

Ezeogu, L. I., Taylor, J. R. N., Emmambux, M. N. and Duodu, K. G. 2008. Influence of Cooking conditions on the protein matrix of sorghum and maize endosperm flours. *Cereal Chemistry*, 85, 397–402.

Food and Agriculture Organization. 1995. *Sorghum and Millets in Human Nutrition*, Rome, FAO.

Food and Agriculture Organization. 2013. *Dietary Protein Quality Evaluation in Human Nutrition*, Rome, Food and Agriculture Organization.

Food and Agriculture Organization and World Health Organization. 1979. Standard for Foods for Special Dietary Use for Persons Intolerant to Gluten. CODEX STAN 118-1979, Amended 1983 and 2015. Revision: 2008. Rome: Food and Agriculture Organization.

Gibson, R. S., Perlas, L. and Hotz, C. 2006. Improving the bioavailability of nutrients in plant foods at the household level. *Proceedings of the Nutrition Society*, 65, 160–8.

Gnonlonfin, G. J. B., Hell, K., Adjovi, Y., Fandohan, P., Koudande, D. O., Mensah, G. A., Sanni, A. and Brimer, L. 2013. A review on aflatoxin contamination and its implications in the developing world: A sub-Saharan African perspective. *Critical Reviews in Food Science and Nutrition*, 53, 349–65.

Goodall, M. A., Hamaker, B. R., Ejeta, G. and Campanella, O. H. 2012. Grain of high digestible, high lysine (HDHL) sorghum contains kafirins which enhance the protein network of composite dough and bread [electronic resource]. *Journal of Cereal Science*, 56, 352–7.

Gu, L., Holden, J., Haytowitz, D., Prior, R. L., Kelm, M. A., Hammerstone, J. F. and Beecher, G. 2003. Screening of foods containing proanthocyanidins and their structural characterization using LC-MS/MS and thiolytic degradation. *Journal of Agricultural and Food Chemistry*, 51, 7513–21.

Gu, L., Kelm, M. A., Hammerstone, J. F., Beecher, G., Holden, J., Haytowitz, D., Gebhardt, S. and Prior, R. L. 2004. Concentrations of proanthocyanidins in common foods and estimations of normal consumption. *Journal of Nutrition*, 134, 613–17.

Hamaker, B. R., Axtell, J. D. and Mertz, E. T. 1994. Effect of extrusion on sorghum kafirin solubility. *Cereal Chemistry*, 71, 515–17.

Hamaker, B. R., Axtell, J. D., Mertz, E. T. and Kirleis, A. W. 1986. Effect of cooking on the protein profiles and in vitro digestibility of sorghum and maize. *Journal of Agricultural and Food Chemistry*, 34, 647–9.

Hassani, A., Zarnkow, M. and Becker, T. 2014. Influence of malting conditions on sorghum (*Sorghum bicolor* (L.) Moench) as a raw material for fermented beverages. *Food Science and Technology International*, 20, 453–63.

ICRISAT. 2017. *Sorghum* [Online]. Available: exploreit.icrisat.org/page/sorghum/882 [Accessed January 8 2017].

Jackson, J. C., Weatherspoon, L., Nnyepi, M., Malete, L., Mokgatlhe, L., Lyoka, P. and Bennink, M. 2013. Sorghum bean composite porridge nutritional quality and acceptability. *Nutrition and Food Science*, 43, 453–61.

Jambunathan, R., Mertz, E. T. and Axtell, J. D. 1975. Fractionation of soluble proteins of high-lysine and normal sorghum grain. *Cereal Chemistry*, 52, 119–21.

Kayitesi, E., DE Kock, H. L., Minnaar, A. and Duodu, K. G. 2010. Sensory quality of marama/sorghum composite porridges [electronic resource]. *Journal of the Science of Food and Agriculture*, 90, 2124–32.

Kayitesi, E., DE Kock, H. L., Minnaar, A. and Duodu, K. G. 2012. Nutritional quality and antioxidant activity of marama–sorghum composite flours and porridges. *Food Chemistry*, 131, 837–42.

Khan, I., Yousif, A., Johnson, S. K. and Gamlath, S. 2013. Effect of sorghum flour addition on resistant starch content, phenolic profile and antioxidant capacity of durum wheat pasta. *Food Research International*, 54, 578–86.

Khan, I., Yousif, A. M., Johnson, S. K. and Gamlath, S. 2014. Effect of sorghum flour addition on in vitro starch digestibility, cooking quality, and consumer acceptability of durum wheat pasta. *Journal of Food Science*, 79, S1560–67.

Khan, I., Yousif, A. M., Johnson, S. K. and Gamlath, S. 2015. Acute effect of sorghum flour-containing pasta on plasma total polyphenols, antioxidant capacity and oxidative stress markers in healthy subjects: A randomised controlled trial. *Clinical Nutrition (Edinburgh, Scotland)*, 34, 415–21.

Kotarski, S. F., Waniska, R. D. and Thurn, K. K. 1992. Starch hydrolysis by the ruminal microflora. *Journal of Nutrition*, 122, 178–90.

Kruger, J., Oelofse, A. and Taylor, J. R. N. 2014. Effects of aqueous soaking on the phytate and mineral contents and phytate:mineral ratios of wholegrain normal sorghum and maize and low phytate sorghum. *International Journal of Food Sciences and Nutrition*, 65, 539–46.

Landry, J. and Moureaux, T. 1980. Distribution and amino acid composition of protein groups located in different histological parts of maize grain. *Journal of Agricultural and Food Chemistry*, 28, 1186–91.

Licata, R., Chu, J., Wang, S., Coorey, R., James, A., Zhao, Y. and Johnson, S. 2014. Determination of formulation and processing factors affecting slowly digestible starch, protein digestibility and antioxidant capacity of extruded sorghum-maize composite flour. *International Journal of Food Science and Technology*, 49, 1408–19.

Lionetti, E. and Catassi, C. 2011. New clues in celiac disease epidemiology, pathogenesis, clinical manifestations, and treatment. *International Reviews of Immunology*, 30, 219–31.

Matumba, L., Lakudzala, D. D., Khonga, E. B. and Monjerezi, M. 2011. Aflatoxins in sorghum, sorghum malt and traditional opaque beer in southern Malawi [electronic resource]. *Food Control*, 22, 266–8.

Mkandawire, N. L., Weier, S. A., Weller, C. L., Jackson, D. S. and Rose, D. J. 2015. Composition, in vitro digestibility, and sensory evaluation of extruded whole grain sorghum breakfast cereals. *LWT – Food Science and Technology*, 62, 662–7.

Mohamed Nour, A. A., Ahmed, I. A. M. and Babiker, E. E. 2011. Nutritional evaluation of sorghum flour (*Sorghum bicolor* L. Moench) during processing of injera. *International Journal of Biological, Biomolecular, Agricultural, Food and Biotechnological Engineering*, 5, 99–103.

Mohamed Nour, A. A., Mohamed Ahmed, I. A., Babiker, E. E. and Yagoub, A. E. A. 2010. Investigations on winter season Sudanese sorghum cultivars: Effect of sprouting on the nutritional value. *International Journal of Food Science and Technology*, 45, 884–90.

Moraes, É. A., Marineli, R. D. S., Lenquiste, S. A., Steel, C. J., Menezes, C. B. D., Queiroz, V. A. V. and Maróstica Júnior, M. R. 2015. Sorghum flour fractions: Correlations among polysaccharides, phenolic compounds, antioxidant activity and glycemic index. *Food Chemistry*, 180, 116–23.

Moussa, M., Qin, X., Chen, L.-F., Campanella, O. H. and Hamaker, B. R. 2011. High-quality instant sorghum porridge flours for the West African market using continuous processor cooking. *International Journal of Food Science and Technology*, 46, 2344–50.

Mugula, J. K. and Lyimo, M. 2000. Evaluation of the nutritional quality and acceptability of sorghum-based tempe as potential weaning foods in Tanzania. *International Journal of Food Sciences and Nutrition*, 51, 269–77.

Okoth, J. K., Ochola, S. A., Gikonyo, N. K. and Makokha, A. 2017. Development of a nutrient-dense complementary food using amaranth-sorghum grains. *Food Science and Nutrition*, 5, 86–93.

Oria, M. P., Hamaker, B. R., Axtell, J. D. and Chia-Ping, H. 2000. A highly digestible sorghum mutant cultivar exhibits a unique folded structure of endosperm. *Proceedings of the National Academy of Sciences of the United States of America*, 97, 5065–70.

Oria, M. P., Shull, J. M. and Hamaker, B. R. 1995. Resistance of sorghum alpha-, beta-, and gamma-kafirins to pepsin digestion. *Journal of Agricultural and Food Chemistry*, 43, 2148–53.

Osman, M. A. 2004. Changes in sorghum enzyme inhibitors, phytic acid, tannins and in vitro protein digestibility occurring during Khamir (local bread) fermentation. *Food Chemistry*, 88, 129–34.

Oueslati, S., Blesa, J., Moltó, J. C., Ghorbel, A. and Mañes, J. 2014. Presence of mycotoxins in sorghum and intake estimation in Tunisia. *Food Additives and Contaminants. Part A: Chemistry, Analysis, Control, Exposure and Risk Assessment*, 31, 307–18.

Pontieri, P., Mamone, G., De Caro, S., Tuinstra, M. R. Roemer, E., Okot, J., De Vita, P., Ficco, D. B. M., Alifano, P., Pignone, D., Massardo, D. R. and Del Giudice, L. 2013. Sorghum, a healthy and gluten-free food for celiac patients as demonstrated by genome, biochemical, and immunochemical analyses. *Journal of Agricultural and Food Chemistry*, 61, 2565–71.

Poquette, N. M., Gu, X. and Lee, S.-O. 2014. Grain sorghum muffin reduces glucose and insulin responses in men. *Food and Function*, 5, 894–9.
Prasad M. P. R., Benhur, D., Kommi, K., Madhari, R., Rao, M. V. and Patil, J. V. 2016. Impact of sorghum supplementation on growth and micronutrient status of school going children in southern India – A randomized trial. *Indian Journal of Pediatrics*, 83, 9–14.
Proietti, I., Mantovani, A., Mouquet-Rivier, C. and Guyot, J.-P. 2013. Modulation of chelating factors, trace minerals and their estimated bioavailability in Italian and African sorghum (*Sorghum bicolor* (L.) Moench) porridges. *International Journal of Food Science and Technology*, 48, 1526–32.
Proietti, I., Tait, S., Aureli, F. and Mantovani, A. 2014. Modulation of sorghum biological activities by varieties and two traditional processing methods: an integrated in vitro /modelling approach. *International Journal of Food Science and Technology*, 49, 1593–9.
Ragaee, S. and Abdel-Aal, E.-S. M. 2006. Pasting properties of starch and protein in selected cereals and quality of their food products. *Food Chemistry*, 95, 9–18.
Ragaee, S., Noaman, M. and Abdel-Aal, E. S. M. 2006. Antioxidant activity and nutrient composition of selected cereals for food use. *Food Chemistry*, 98, 32–8.
Ragaee, S., Seetharaman, K. and Abdel-Aal, E.-S. M. 2014. The impact of milling and thermal processing on phenolic compounds in cereal grains. *Critical Reviews in Food Science and Nutrition*, 54, 837–49.
Rajendra Prasad, M. P., Rao, B. D., Kalpana, K., Rao, M. V. and Patil, J. V. 2015. Glycaemic index and glycaemic load of sorghum products. *Journal of the Science of Food and Agriculture*, 95, 1626–30.
Ratnavathi, C. V., Das, I. K., Patil, J. V., Komala, V. V. and Kumar, B. S. V. 2012. Natural occurrence of aflatoxin B1 in sorghum grown in different geographical regions of India [electronic resource]. *Journal of the Science of Food and Agriculture*, 92, 2416–20.
Rooney, L. W. and Awika, J. M. 2005. Overview of products and health benefits of specialty sorghums. *Cereal Foods World*, 50, 109–15.
Rooney, L. W. and Pflugfelder, R. L. 1986. Factors affecting starch digestibility with special emphasis on sorghum and corn. *Journal of Animal Science*, 63, 1607–23.
Rosell, C. M., Barro, F., Sousa, C. and Mena, M. 2014. Cereals for developing gluten-free products and analytical tools for gluten detection. *Journal of Cereal Science*, 59, 354–64.
Rubio-Tapia, A., Hill, I. D., Kelly, C. P., Calderwood, A. H. and Murray, J. A. 2013. ACG clinical guidelines: Diagnosis and management of celiac disease. *American Journal of Gastroenterology*, 108, 656–76.
Rubio-Tapia, A., Ludvigsson, J. F., Brantner, T. L., Murray, J. A. and Everhart, J. E. 2012. The prevalence of celiac disease in the United States. *American Journal of Gastroenterology*, 107, 1538–44.
Rufino, M. C., Jones, P. G., Van Wijk, M. T., Herrero, M., Thornton, P. K., Ng'ang'a, S. K. and Mutie, I. 2013. Transitions in agro-pastoralist systems of East Africa: Impacts on food security and poverty [electronic resource]. *Agriculture, Ecosystems and Environment*, 179, 215–30.
Sarwar Gilani, G., Wu Xiao, C. and Cockell, K. A. 2012. Impact of antinutritional factors in food proteins on the digestibility of protein and the bioavailability of amino acids and on protein quality. *British Journal of Nutrition*, 108 Suppl. 2, S315–32.
Serrem, C. A., De Kock, H. L. and Taylor, J. R. N. 2011. Nutritional quality, sensory quality and consumer acceptability of sorghum and bread wheat biscuits fortified with defatted soy flour. *International Journal of Food Science and Technology*, 46, 74–83.
Silano, M., Penas Pozo, E., Uberti, F., Manferdelli, S., Del Pinto, T., Felli, C., Budelli, A., Vincentini, O. and Restani, P. 2014. Diversity of oat varieties in eliciting the early inflammatory events in celiac disease. *European Journal of Nutrition*, 53, 1177–86.
Simnadis, T. G., Tapsell, L. C. and Beck, E. J. 2016. Effect of sorghum consumption on health outcomes: A systematic review. *Nutrition Reviews*, 74, 690–707.
Stefoska-Needham, A., Beck, E. J., Johnson, S. K. and Tapsell, L. C. 2015. Sorghum: An underutilized cereal whole grain with the potential to assist in the prevention of chronic disease. *Food Reviews International*, 31, 401–37.

Taye, W., Ayalew, A., Chala, A. and Dejene, M. 2016. Aflatoxin B_1 and total fumonisin contamination and their producing fungi in fresh and stored sorghum grain in East Hararghe, Ethiopia. *Food Additives and Contaminants. Part B, Surveillance*, 1–9.

Taylor, J. R. N., Bean, S. R. and Schober, T. J. 2006. Novel food and non-food uses for sorghum and millets [electronic resource]. *Journal of Cereal Science*, 44, 252–71.

Taylor, J. R. N., Belton, P. S., Beta, T. and Duodu, K. G. 2014. Increasing the utilisation of sorghum, millets and pseudocereals: Developments in the science of their phenolic phytochemicals, biofortification and protein functionality. *Journal of Cereal Science*, 59, 257–75.

Taylor, J. R. N. and Duodu, K. G. 2015. Effects of processing sorghum and millets on their phenolic phytochemicals and the implications of this to the health-enhancing properties of sorghum and millet food and beverage products. *Journal of the Science of Food and Agriculture*, 95, 225–37.

Taylor, J. R. N. and Schüssler, L. 1986. The protein compositions of the different anatomical parts of sorghum grain. *Journal of Cereal Science*, 4, 361–9.

Teixeira, N. D. C., Queiroz, V. A. V., Rocha, M. C., Amorim, A. C. P., Soares, T. O., Monteiro, M. A. M., De Menezes, C. B., Schaffert, R. E., Garcia, M. A. V. T. and Junqueira, R. G. 2016. Resistant starch content among several sorghum (*Sorghum bicolor*) genotypes and the effect of heat treatment on resistant starch retention in two genotypes. *Food Chemistry*, 197, 291–6.

Tizazu, S., Urga, K., Abuye, C. and Retta, N. 2010. Improvement of energy and nutrient density of sorghum-based complementary foods using germination. *African Journal of Food, Agriculture, Nutrition and Development*, 10, 2927–42.

Towo, E., Svanberg, U. and Matuschek, E. 2006. Fermentation and enzyme treatment of tannin sorghum gruels: effects on phenolic compounds, phytate and in vitro accessible iron. *Food Chemistry*, 94, 369–76.

Tuinstra, M. R. 2008. Food-grade sorghum varieties and production considerations: A review. *Journal of Plant Interactions*, 3, 69–72.

United States Department of Agriculture. 2015. USDA National Nutrient Database for Standard Reference, 28. Revised slightly May, 2016 ed. Washington, D.C.

Vargas-Solórzano, J. W., Carvalho, C. W. P., Takeiti, C. Y., Ascheri, J. L. R. and Queiroz, V. A. V. 2014. Physicochemical properties of expanded extrudates from colored sorghum genotypes. *Food Research International*, 55, 37–44.

Vázquez-Araújo, L., Chambers, E. T. and Cherdchu, P. 2012. Consumer input for developing human food products made with sorghum grain. *Journal of Food Science*, 77, S384–9.

Vendemiatti, A., Rodrigues Ferreira, R., Humberto Gomes, L., Oliveira Medici, L. and Antunes Azevedo, R. 2008. Nutritional quality of sorghum seeds: Storage proteins and amino acids. *Food Biotechnology*, 22, 377–97.

Waniska, R. D. and Rooney, L. W. 2000. Structure and chemistry of the sorghum caryopsis. In: Smith, C. W. and Frederiksen, R. A. (eds), *Sorghum: Origin, History, Technology, and Production*. New York, NY: John Wiley and Sons, Inc.

Wieser, H. and Koehler, P. 2008. The biochemical basis of celiac disease. *Cereal Chemistry*, 85, 1–13.

Wolever, T. M. 2003. Carbohydrate and the regulation of blood glucose and metabolism. *Nutrition Reviews*, 61, S40–8.

Wong, J. H., Lemaux, P. G., Pedersen, J. F., Buchanan, B. B., Marx, D. B. and Wilson, J. D. 2010. Principal component analysis and biochemical characterization of protein and starch reveal primary targets for improving sorghum grain [electronic resource]. *Plant Science*, 179, 598–611.

Wong, J. H., Wilson, J. D., Hurkman, W. J., Buchanan, B. B., Lemaux, P. G., Vensel, W. H., Cai, N., Lau, T., Pedersen, J. F. and Singh, J. 2009. Digestibility of protein and starch from sorghum (*Sorghum bicolor*) is linked to biochemical and structural features of grain endosperm [electronic resource]. *Journal of Cereal Science*, 49, 73–82.

Wu, X., Peteson, G., Rooney, W., Wang, D., Shi, Y.-C., XU, F., Robbins, A., Jampala, B., Yan, S. and Hays, D. 2010. Ethanol fermentation performance of grain sorghums (*Sorghum bicolor*) with modified endosperm matrices [electronic resource]. *Journal of Agricultural and Food Chemistry*, 58, 9556–62.

Wu, Y., Yuan, L., Guo, X., Holding, D. R. and Messing, J. 2013. Mutation in the seed storage protein kafirin creates a high-value food trait in sorghum. *Nature Communications*, 4, 2217. doi:10.1038/ncomms3217.

Yang, L., Allred, K. F., Dykes, L., Allred, C. D. and Awika, J. M. 2015. Enhanced action of apigenin and naringenin combination on estrogen receptor activation in non-malignant colonocytes: implications on sorghum-derived phytoestrogens. *Food and Function*, 6, 749–55.

Yang, L., Allred, K. F., Geera, B., Allred, C. D. and Awika, J. M. 2012. Sorghum phenolics demonstrate estrogenic action and induce apoptosis in nonmalignant colonocytes. *Nutrition and Cancer*, 64, 419–27.

Yang, L., Awika, J. M. and Browning, J. D. 2009. Sorghum 3-deoxyanthocyanins possess strong phase II enzyme inducer activity and cancer cell growth inhibition properties [electronic resource]. *Journal of Agricultural and Food Chemistry*, 57, 1797–1804.

Yousif, A., Nhepera, D. and Johnson, S. 2012. Influence of sorghum flour addition on flat bread in vitro starch digestibility, antioxidant capacity and consumer acceptability. *Food Chemistry*, 134, 880–7.

Zhang, G. and Hamaker, B. R. 1998. Low alpha-amylase starch digestibility of cooked sorghum flours and the effect of protein. *Cereal Chemistry*, 75, 710–13.

Chapter 5

Sorghum as a forage and energy crop

Scott Staggenborg and Hui Shen, Chromatin Inc., USA

1 Introduction

Sorghum is an important source of grain and fodder, forage and biomass throughout the world. In the United States, grain sorghum is the prominent crop; however, non-grain or forage sorghum plays an important role as a feedstock globally. Forage sorghum is phenotypically diverse with cultivars that are used primarily for silage, cultivars that may or may not include sudangrass that are primarily used for haying and grazing, and more recently, high-yielding and sweet cultivars that have been positioned for use as renewable feedstocks. Sorghum was quickly identified as one of the most appropriate dedicated energy crops. In addition, having scalable and well-understood production practices makes sorghum more acceptable than many lesser developed perennial grasses. Relative high yields make it an excellent choice over grain-based annuals, and because it is generally not used for human food, it is of less concern in the food-versus-fuel debate (Staggenborg et al., 2008). Sorghum has been used as a renewable feedstock to produce cellulosic ethanol, steam via combustion and methane via anaerobic digestion. Traits such as brown midrib (BMR) and brachytic dwarfism, and heat and drought tolerance make sorghum adaptable to many marginal environments.

2 Forage and biomass sorghum types

The economic yield of sorghum can be defined as grain, forage and biomass, and sugars depending on the end use and the genetic composition of a particular hybrid. This section focuses on three groups: forage and the sweet sorghum hybrids, sorghum × sudangrass

http://dx.doi.org/10.19103/AS.2017.0015.24

(So×Su) and sudangrass (Su×Su). So×Su hybrids are known for high biomass yields and rapid regrowth, and Su×Su hybrids are characterized by rapid early season growth, rapid regrowth, fine stems and lower potential for hydrogen cyanide production, when the plant is damaged, wounded or fed on, compared with forage sorghums and So×Su hybrids.

2.1 Plant height and maturity

These three groups can be further subdivided on the basis of hybrid maturity. Hybrid maturity is controlled by seven flowering genes: *Ma1* through *Ma7*, which influence the duration of growth, or days to maturity (Quinby, 1974; Rooney et al., 2007; Mullet et al., 2010). *Ma1* (Murphy et al., 2011), *Ma3* (Childs et al., 1997; Yang et al., 2014), *Ma5* and *Ma6* (Murphy et al., 2014; Mullet and Rooney, 2013) had been cloned. *Ma2*, *Ma4* and *Ma7* genes are not known, although there are suggested candidate genes (or proposed gene loci) available based on QTL mapping studies (Hart et al., 2001; Rooney and Aydin, 1999). *Ma1* is considered to have large effect on maturity because it has direct role in repressing flowering time. Plants with the *ma1* mutant tend to flower early, and thus produce shorter plants with fewer internode numbers. *Ma1* encodes a pseudoresponse regulator protein 37 (Murphy et al., 2011). It represses the expression of *FT* (*FLOWERING LOCUST*) through enhancing the expression of *CO (CONSTANS)* and inhibiting *Ehd1* (*EARLY HEADING DATE 1*) (Doi et al., 2004). *Ma1* has a known allelic series and these mutants were classified into seven groups depending on the mutations *Prr37* gene carries (Murphy et al., 2011; Klein et al., 2015). All these mutants produce photoperiod-insensitive phenotypes under day lengths that are longer than 12 hours.

Manipulation of these genes can result in a forage sorghum, So×Su hybrid, or Su×Su hybrid that either produce a head after some variable vegetative stage (headed) or do not produce a head during a normal growing season in a temperate environment (photoperiod sensitive). Photoperiod-sensitive hybrids are those that require decreasing day length, a particular temperature and a certain photoperiod to trigger floral initiation (Craufurd et al., 1999; Dingkuhn et al., 2008). In temperate environments, this photoperiod response results in floral initiation after a photoperiod of approximately 12 h and 20 min, so panicles are rarely observed or are present late enough to produce no grain. Because these hybrids do not flower until very late in the year, they become very tall and produce very high biomass yields. They are traditionally higher in moisture at harvest because they do not head and trigger a natural dry down process like headed hybrids. From a forage-quality perspective, these hybrids produce no grain and are often considered to have low forage quality.

Sorghum height is mainly controlled by four genes (dwarf): Dw_1, Dw_2, Dw_3 and Dw_4. Genes Dw_1 and Dw_2 are known to affect internode length, Dw_3 affects internode number and Dw_4 appears to affect panicle length (Goud and Vasudev Rao, 1977). Only *Dw1* and *Dw3* genes have been cloned so far. Recent studies suggest *Dw1* gene regulates the length of the internodes. It does not affect leaf size and other plant parts (Klein et al., 2015). The *dw1dw1* can be used to reduce internode length (plant height) without reducing the canopy and leaf development (Klein et al., 2015, Hilley et al., 2016). The *Dw3* gene plays important roles in controlling plant height through regulating total plant height, and flag leaf height and has little impact on panicle exertion (Quinby and Karper 1961; Li et al., 2015). Interestingly, a QTL loci, *qHT7.1*, was found to regulate the interval between flag leaf to apex only (Quinby and Karper 1961). Grain sorghum hybrids have been dwarfed to facilitate mechanized harvest and seldom exceed 2 m in height. Most grain sorghum hybrids developed in the United States are recessive at three height loci (3-dwarf) and their genotypes are generally $dw_1Dw_2dw_3dw_4$ (Quinby and Karper, 1954). One height gene, *dw3*, is known only to reduce internodes on the lower part of the stalk and has little impact on panicle exsertion. From a forage perspective, plant height results in three distinct forage sorghum types: traditional, dual-purpose and brachytic dwarf forage hybrids. It is possible to find brachytic dwarf So×Su hybrids.

2.2 Brown midrib and brachytic dwarf sorghum

Lignin decreases forage digestibility, so efforts to reduce lignin in forage sorghum hybrids have been conducted for decades. BMR mutations (*bmr1* through *bmr19*) were introduced into sorghum by Porter et al. (1978) and Bittinger et al. (1981) via chemical mutagenesis of two grain sorghum lines. Allelism tests of these identified *bmr* mutants result in four allelic groups, *bmr2*, *bmr6*, *bmr12* and *bmr19* (Saballos et al., 2008). Fritz et al. (1990) selected three from this original population that were agronomically acceptable, and they have been the source of three alleles: *bmr6*, *bmr12* and *bmr18*. Since the *bmr19* mutant is not publicly available, and it has limited value for forage and bioenergy applications due to less reduction of lignin content (Saballos et al., 2008), most of the developed *bmr* varieties are based on *bmr6* and *bmr12* mutants (Pedersen et al., 2006a,b,c, 2008). Sorghum *Bmr6* gene encodes cinnamyl alcohol dehydrogenase (Saballos et al., 2009; Sattler et al., 2009), which catalyses the last step in monolignol biosynthesis. Sorghum *Bmr12* gene encodes caffeic acid 3-*O*-methyltransferase (COMT) (Bout and Vermerris, 2003) and sorghum *Bmr2* gene encodes a 4-coumarate coenzyme A ligase (4CL), which catalyses an early step in monolignol biosynthesis (Saballos, et al., 2012). BMR hybrids have noticeable brown pigment in the midrib of the leaves, the stem, the pith and immature panicles (Porter et al., 1978). These hybrids have reduced lignin content and generally higher digestibility, yet they are often low in yield (Porter et al., 1978; Bucholtz et al., 1980; Hanna et al., 1981; Bean and McCollum, 2006; McCuistion et al., 2009, 2010).

Traditional forage sorghums have no recessive dwarf genes, resulting in hybrids that can exceed 4 m in height. While brachytic dwarf forage hybrids have a similar leaf number as traditional forage sorghum hybrids, they have brachytic mutations, which reduce internode length, resulting in a 'stacking' appearance of the leaves. It is believed that because of their higher leaf-to-stem ratios, brachytic dwarf hybrids are more digestible compared with traditional forage sorghum hybrids. Plant height is also important for breeding forage (biomass) sorghum because sorghum plants with increased stature always invite lodging under wind stress. Early BMR hybrids exhibited above average lodging. To overcome this problem, dwarf sorghum mutants were used to breed BMR lines, which will be discussed in the next section. According to Cook (1915), plants with short statues were classified as dwarfs and brachytes. A dwarf is a plant deficient in stature and parts comparing to a normal plant. But a brachytic plant only has a shorter vertical axis, caused by shortening of the internodes, without reduction of other parts such as plant leaf length or size (Cook, 1915). To distinguish these two dwarf types, people sometimes call the brachytic a 'brachytic dwarf'. Plants with longer flowering time, such as photoperiod-sensitive lines, may produce more leaf numbers. However, longer flowering time normally results in taller plant if dwarf genes are not present. As reviewed above, the *dw1* allele can be used to breed brachytic dwarf plant if proper maturity genes are present. The combination of both *ma1ma1dw1dw1* alleles may result in brachytic dwarf plants with less internode numbers. It was reported that *Ma1* and *Dw2* loci are tightly linked on chromosome 6 (Klein et al.,

2008; Multani et al., 2003). The impacts of *ma1ma1dw2dw2dw1dw1* alleles to brachytic dwarf phenotype are not known. Achieving a desirable brachytic dwarf phenotype with larger number of leaves and early maturity through manipulating both the *Ma* genes and *Dw* genes is difficult, as early flowering time will most likely produce fewer leaves due to the developmental transition from vegetative to reproductive stage. A new trait with rapid cell division to produce many, shorter internodes before flowering will benefit early-maturity brachytic dwarf products.

2.3 Water use

Sorghum tolerance to drought and heat stress has been reported by a large number of researchers. Forage sorghum is often compared with corn for improved water-use efficiency and for producing higher yields under water-limiting scenarios (Table 1). Marsalis et al. (2009) reported forage sorghum yields greater or equal to corn silage when the amount of applied irrigation water was reduced to two-thirds of what would have been applied to a fully irrigated corn crop. Howell et al. (2008) reported that water use in BMR sorghum was approximately 73% of that in corn water, as measured by a lysimeter, and Rajan and Maas (2007) reported higher daily water-use rates for corn compared with forage sorghum via eddy covariance and spectral crop coefficient methods in large-scale production fields.

McCuistion et al. (2009) compared the water use and yields of BMR forage sorghum, traditional forage sorghum and photoperiod forage sorghum. Photoperiod forage sorghum had the highest yields and greatest water-use efficiency, 44 kg ha^{-1} dry matter mm^{-1} of water. The water-use efficiency of headed forage sorghums was 21 kg ha^{-1} dry matter mm^{-1} of water. Saeed and El-Nadi (1998) reported water-use efficiencies for forage sorghum ranging from 65 to 86 kg ha^{-1} mm^{-1}, with light, frequent irrigation resulting in the highest water-use efficiencies and dry matter yields. Ottman (2010) reported the greatest water-use efficiency at 75% of evapotranspiration (ET) replacement for forage sorghum. He reported an applied water-use efficiency of 78 kg ha^{-1} mm^{-1}. Hussein and Alva (2014) reported lower water-use efficiency for forage sorghum, 6 kg ha^{-1} mm^{-1}. They did not

Table 1 Silage yields of corn, forage sorghum, and BMR forage sorghum in New Mexico under limited irrigation in 2005 and 2006

	2005	2006	2-yr average
Optimum harvest (60–65% moisture)		Mg ha^{-1}	
Corn	48.7	55.3	51.9
Forage sorghum	61.5	59.0	60.3
BMR-FS	51.6	44.7	48.4
Late harvest (50–60% moisture)			
Corn	57.3	56.3	56.8
Forage sorghum	54.8	57.0	56.1
BMR-FS	46.7	43.5	45.2
SEM	1.1	1.1	1.1

† Data from Marsalis et al., 2009; irrigation: 18–20 inches, or about two-thirds of full corn irrigation amounts.

state the cultivar planted, so it is difficult to properly compare the data due to the diversity of forage sorghum types and yield potentials. Narayanan et al. (2013) reported water-use efficiencies ranging from 34 to 76 kg ha^{-1} mm^{-1} from eight public biomass sorghum varieties. Their results illustrate the importance of biomass yield to improve water-use efficiency. Water use ranged from 218 to 256 kg m^{-2} during 2 years, a relatively narrow range compared with yields that ranged from 796 to 1717 g m^{-2}.

3 Forages as animal feed

Forage sorghum, So×Su, and Su×Su hybrids have traditionally been included in rations for ruminant animals. These animals' unique digestive systems are capable of breaking down the complex carbohydrates into simple sugars, which are easily digested. Traditional harvesting and storage methods include grazing, greenchop, dry hay or silage.

3.1 Direct feeding and dry hay methods

Direct-feeding methods, grazing and greenchop require additional management to reduce the potential of negative effects of prussic acid and nitrate poisoning. Prussic acid levels can be reduced by delaying grazing or greenchop harvest until plant heights exceed 45 cm. Prussic acid is concentrated in the leaf tissue, so delaying direct feeding and greenchop harvest until this height is achieved reduces the leaf-to-stem ratio. Nitrates accumulate in grasses that experience growth-reducing stress, drought being the most common. Reductions in dry matter because of stress result in higher concentrations of nitrates, which accumulate in the plant early in the growing season. Nitrate management in a direct-feeding scenario should be managed by testing the forage before feeding.

Sorghum × sudangrass and Su×Su hybrids are the primary hybrids used to produce dry hay. These hybrids can be harvested multiple times during the summer when adequate solar radiation and wind are available to reduce forage moisture content from 80 to 15% (Hill, 1976). When forage sorghum hybrids are harvested for baled hay, additional crimping is required to break the large internodes and aid in the drying process (Hartley et al., 2011).

3.2 Silage

Sorghum silage is predominately produced from forage sorghum hybrids, although photoperiod-sensitive So×Su hybrids are occasionally used in a single-cut system to produce silage. Forage sorghum hybrids can often be divided into three categories: traditional, dual-purpose and brachytic dwarf. The inclusion of BMR genes is common in all three types.

Dual-purpose hybrids are dwarfed forage hybrids that resemble grain sorghum hybrids. Dual-purpose forage sorghums grow to between 1.5 and 2.5 m tall, depending on growing conditions, and will produce a significant amount of grain (Table 2). Lab analyses of silage from these hybrids will always result in high starch levels (~20%) because of the grain in the silage (Table 2). If, however, the plants are not harvested correctly, this starch availability may mean that the lab results are misleading for a nutritionist (e.g. although the lab report may read 25% starch, only 15% may be available to the animal). In order for animals to digest all of this starch, harvest must occur between the milk and soft-dough stages. As the grain

Table 2 Grain and fodder yield as well as compositional analyses of three different forage sorghum types

Variable	Sorghum Partners NK 300	Sorghum Partners SS405	Sorghum Partners 1990
Harvest moisture†	62.1	65.2	70.0
Yield (Mg ha^{-1} @ 65% moisture)†	50.0	58.7	69.2
Grain (Mg ha^{-1})†	7.9	2.9	0.0
Crude protein (%)‡	9.76 ± 0.54§	8.01 ± 0.31	9.54 ± 1.36
Acid detergent fibre (%)‡	24.96 ± 0.06	34.63 ± 1.52	33.16 ± 3.61
Neutral detergent fibre (%)‡	39.50 ± 1.13	54.43 ± 2.5	54.54 ± 0.62
Total digestible nutrients (%)‡	65.24 ± 2.78	65.75 ± 0.35	61.26 ± 4.45
Starch (%)‡	33.64 ± 5.15	19.79 ± 9.73	5.93 ± 1.92
Crude fat (%)‡	2.25 ± 0.15	2.08 ± 0.52	0.84 ± 0.05
Ash (%)‡	6.68 ± 0.76	7.43 ± 0.41	8.65 ± 0.37
Net energy$_{lactation}$ (Mcal/kg)‡	0.77 ± 0.43	0.65 ± 0.02	0.67 ± 0.08

† From Bean and McCollum 2006.
‡ Chromatin internal data collected from different samples.
§ Standard deviation.

matures (proceeds from hard-dough to maturity), it becomes more difficult to digest because of grain hardness. At this point, the grain must be cracked or ground for the starch to be fully digested by the animal.

Full-season headed forage sorghum hybrids are preferred for sorghum silage production. These hybrids produce high biomass yields, contain some grain and are typically drier at harvest than photoperiod-sensitive forage sorghum hybrids (Table 2) (Bean and McCollum, 2006). High biomass yields are accomplished through the production of long internodes and large leaves. Tall plants are often associated with lodging, so brachytic dwarf hybrids have become more popular in the industry. As mentioned previously, these hybrids have significantly smaller internode lengths, which reduce lodging. Forage hybrids also derive a significant yield from grain production, requiring proper harvesting to optimize silage digestion (Bell et al., 2015). While starch digestibility is optimized between the milk and soft-dough stages, forage moisture at harvest is also critical for optimizing silage quality. It is recommended that forages for silage be harvested at 35% dry matter. Forages harvested at higher moisture contents can result in poor fermentation and excessive effluent drainage from the silage storage structure.

3.3 Forage quality

From a livestock feed perspective, forage composition focuses on nutritive value. Analyses for livestock nutrition emphasize digestibility of the fibre and use methods that resemble an animal rumen. The overall goal is to evaluate the carbohydrates, fat, protein, minerals and vitamins available to animals for digestion and growth. Grass species are typically high in carbohydrates compared with legume forage sources. Neutral detergent fibre (NDF) is an estimation of the structural carbohydrates in forage (hemicellulose, cellulose and lignin), and acid detergent fibre (ADF) is an estimate of the cellulose and lignin. Forage analyses also include *in vitro* digestibility estimation methods such as NDF digestibility and *in vitro* dry matter digestibility (Martin and Barnes, 1980; Hatfield et al., 1994).

Attempts have been made to develop composite values that describe forage quality as a single number. Total digestible nutrients, net energy, relative feed value and relative feed quality are all composite values that have been developed to assess feed quality (Undersander et al., 1993). A range of handbooks exist that illustrate livestock response to sorghum (Brouk and Bean, 2011; Brouk, 2012).

4 Dedicated energy sorghum

Sorghum biomass yields have been evaluated throughout the United States for almost three decades (Table 3). As expected, yields vary considerably on the basis of location

Table 3 Research studies conducted to evaluate sorghum as renewable energy feedstocks

Source	Location	Years	Dry yield Max.	Min.	Mean	Description of entries†
			Mg ha^{-1}			
Caravetta et al., 1990	Lafayette, IN	1987–88	15.9	8.3	11.8	PS
Miller and McBee, 1993	College Station, TX	1986–87	28.0	11.4	17.7	Cultivar comparison
Hallam et al., 2001	Iowa	1988–92	20.7	14.6	16.6	SWS and SS
Bean and McCollum, 2006	Bushland, TX	2000–5	10.7	7.6	9.6	Irrigated PS, multiyear means
Rooney et al., 2007	College Station, TX	1985	8.5	5.1	6.7	FS nitrogen study
Venuto and Kindiger, 2008	El Reno, OK	2004–6	33.1	23.6	29.4	PS and SS hybrids
Marsalis et al., 2009	Clovis, NM	2005–6	21.5	19.2	20.4	FS hybrid
Marsalis et al., 2010	Clovis, NM	2007–8			24.0	FS hybrid
Ottman, 2010	Maricopa, AZ	2009	15.1	7.6	12.1	FS irrigation study
Propheter et al., 2010	KS	2007–8	26.8	12.7	19.7	FS, PS and BMR
Dahlberg et al., 2011	Bushland, TX	2007	22.5	10.5	15.6	Irrigated SS and FS
Tamang et al., 2011	Lubbock, TX	2008–9	20.0	10.8	15.1	Irrigated PS N study
Maughan et al., 2012	Illinois	2009–10	39.9	13.5	24.5	ES N study, sum of two harvests
Rocateli et al., 2012	Shorter, AL	2008–9	30.1	8.0	15.0	PS and FS hybrids
Snider et al., 2012	Alabama and Arkansas	2009–10	61.0	15.0	31.8	PS hybrid

† FS, forage sorghum; PS, photoperiod sensitive; SWS, sweet sorghum; SS, sorghum × sudangrass; ES, energy sorghum.

and year. The lowest yields reported in the peer-reviewed literature were either rainfed or low-irrigation treatments in Texas (Bean and McCollum, 2006; Rooney et al., 2007) or Arizona (Ottman, 2010). The greatest yields with traditional forages were produced in high-yield environments with photoperiod-sensitive hybrids. The highest yield, 61 Mg ha^{-1} dry matter, was reported in Alabama in a narrow-row configuration with photoperiod-sensitive forage hybrid (Snider et al., 2012). Yields above 30 Mg ha^{-1} dry matter were reported with similar hybrids by Miller and McBee (1993), Venuto and Kindiger (2008), and Rocateli et al. (2012). The highest yields for energy sorghum were obtained in Illinois, and that sorghum outyielded traditional forage hybrids by 35% (Maughan et al., 2012).

As with digestibility in ruminant animals, lignin reduces conversion of biomass to sugars and eventually ethanol or other chemicals (Dien et al., 2001). Because lignin is highly resistant to chemical cleaving and fills in the spaces between cellulose and hemicellulose, pre-treatment processes to cleave lignin strands into more digestible segments have been evaluated in attempts to improve sugar recovery from cellulosic feedstocks (Yang and Wyman, 2004; Wyman et al., 2005; Pedersen et al., 2010; Hu et al., 2012). With the only commercial market currently being the livestock forage market, commercial-scale applications of BMR feedstocks into the biochemical markets are not widespread.

A range of *in vitro* lab procedures has been developed to estimate forage digestibility by ruminant animals; however, the classification of the carbohydrates as structural and non-structural differs among methods. As interest in the use of crop biomass as a source of simple sugars increased, it became apparent that different analytical methods would be needed. Although estimations of cellulose and hemicellulose could be calculated from detergent fibre methods (ADF and NDF), these methods do not often correlate with dietary or extraction methods (Wolfrum et al., 2009). Methods developed by the US DOE National Renewable Energy Laboratory (NREL) are the standards used in evaluating biomass samples for structural and non-structural carbohydrates, lignin and ash (Sluiter et al., 2005a,b, 2011).

The composition of sorghum biomass has focused largely on the structural carbohydrates: cellulose, hemicellulose and pectin. All three components are composed primarily of six sugars arranged in various complex and amorphous configurations (Heltd and Piechulla, 2011). Miller and McBee (1993) analysed forage and grain sorghum for both structural and non-structural carbohydrates (Table 4). As with mixed sorghum types, the wide range in starch values often indicates the presence of grain. Their values for hemicellulose and cellulose are similar to those from annual crops. Dien et al. (2009) reported similar data for a sample set that contained sorghum with the *bmr* genes, thus the reason for the low lignin values. Corredor et al. (2009) found that total structural carbohydrates in sorghum ranged from 48 to 80%, with lignin ranging from 11 to 20%. They also reported cellulose concentrations of 24–38% and hemicellulose concentrations of 18–22%. Additional work by Theerarattananoon et al. (2010) reported similar percentages of structural carbohydrates and also illustrated the range of compositions across different forage sorghum types. They reported lignin levels to be as low as 15% in a BMR forage sorghum compared with 18–20% in other forage and corn biomass samples. The BMR forage samples also had some of the highest glucan and xylan levels of the forages tested.

Dahlberg et al. (2011) conducted one of the most extensive evaluations of sorghum cultivars for yield and composition. They selected 22 cultivars that comprised both commercial forage and sweet sorghums (Table 4). The group included forage sorghum hybrids, So×Su hybrids and sweet sorghum varieties. They also included subsets of BMR, non-BMR, photoperiod-sensitive and non-photoperiod-sensitive cultivars. They found that

Table 4 Forage and energy sorghum composition from a range of studies conducted in the USA.

Source		Lignin	Cellulose	Hemicellulose	Starch	Ash	Description
		mg g^{-1}					
Miller and McBee, 1993	Max.	90.0[†]	344.0	251.0	52.0	NR[‡]	Forage sorghum and grain types
	Min.	53.0	226.0	184.0	15.0	NR	
	Mean	70.3	2663.8	204.8	29.7	NR	
Dien et al., 2009	Max.	148.8	256.0	203.0	39.0	67.6	Degrained samples, includes BMR[§] hybrids
	Min.	100.3	240.0	183.0	19.0	51.6	
	Mean	125.7	246.9	191.6	28.9	59.7	
Corredor et al., 2009	Max.	204.7	387.2	224.8	229.1	108.7	PS-BMR, PS-non-BMR, forage BMR and forage sorghum
	Min.	110.6	242.1	123.2	8.4	69.3	
	Mean	163.3	341.2	182.0	96.7	92.7	
Dahlberg et al., 2011	Max.	158.5	434.5	239.2	428.3	117.9	Forage, BMR, sorghum–sudangrass and non-grain types
	Min.	94.9	187.5	158.4	0.0	73.9	
	Mean	123.6	274.3	188.7	154.1	92.1	

† Lignin determination method not listed.
‡ NR, not reported.
§ BMR, brown midrib.

carbohydrate composition varied widely among the different cultivars. Ash ranged from 7.3 to 11.3 g g^{-1}, lignin from 9.9 to 16.3 g g^{-1}, glucans from 18.7 to 35.2 and xylans from 12.6 to 19.9 g g^{-1}.

4.1 Saccharification from sorghum biomass

Biomass to ethanol conversion, as with most plant-based biochemical production process, is driven by glucose and xylose, the two primary C6 and C5 sugars found in plant tissues. Pre-treatments and extraction rates may differ with the method and potentially by feedstock species; however, if good analytical data can be obtained for the concentrations of the C6 and C5 sugars extracted from a given amount of biomass, then the ethanol yields will be predictable. The NREL currently cites the rate for the conversion of C6 and C5 sugars to ethanol as 0.51 kg ethanol/kilogram sugar (USDOE, 2006). These conversion factors take into account the addition of water during hydrolysis. This approach has been used by numerous authors to estimate ethanol yields. Dahlberg et al. (2011) estimated that ethanol yields from forage sorghum and sorghum-sudan hybrids ranged from 336 to 440 L Mg^{-1}, with a mean of 403 L Mg^{-1}. Theerarattananoon et al. (2010) reported sorghum composition data that resulted in ethanol estimates ranging from 436 to 521 L Mg^{-1}. Estimated ethanol yields for wheat, corn, big bluestem and biomass sorghum were 473,

476, 449 and 471 L Mg^{-1}, respectively, based on composition data (Theerarattananoon et al., 2012). As previously mentioned, BMR sorghum hybrids typically have lower lignin concentrations and have been used in the livestock-feeding industry because of higher conversion rates. These characteristics carry over into cellulosic ethanol production. Dahlberg et al. (2011) reported that the highest conversion rate, 420 L Mg^{-1}, was from a BMR hybrid, but low yields across all the BMR hybrids evaluated did not result in higher ethanol yields per hectare compared with the other sorghum cultivars examined. Dien et al. (2009) reported glucose yields to be 34% higher in double mutant *bmr6* and *bmr12* hybrids compared with their near-isogenic wild types. Ethanol yields improved from 21% for *bmr6* to 43% for the double mutant compared with the wild type.

The production of organic acids, primarily acetic, propionic, butyric and lactic acids, from biomass and grain has been under parallel development in the lignocellulosic ethanol production industry but has gained less public attention (Du and Yu, 2002; Zhan et al., 2002; Carole et al., 2004; Causey et al., 2004). These acids serve as precursors to plastics, fibres and polyurethanes, which are currently produced primarily from petroleum substrates. While most of these developmental efforts have focused on a wide range of feedstocks, those with grain and sweet sorghum have met with a great deal of success because these cultivars have starches and free sugars that are easier to access compared with sugars sourced from biomass (Samuel et al., 1980; Richter and Berthold, 1998; Richter and Träger, 1994; Zhan et al., 2003; Wee et al., 2006).

4.2 Thermal conversion of sorghum biomass

The combustion of raw materials for the creation of heat has been a part of human existence for millennia. New technology in the late nineteenth century led to this energy source being harnessed to generate electricity. Coal dominated as the preferred source for heat and electricity generation until the late twentieth and early twenty-first centuries, when atmospheric CO_2 concentrations became a concern. Combustion of biomass became a topic of interest, with wood being the dominant source, and biomass from perennial and annual crops used on a smaller scale. More than 56 million megawatts of electricity are generated annually in the United States from biomass (USEIA, 2012), with wood being the dominant fuel; however, crop biomass and residues also have been used to generate electricity. The University of Minnesota Morris generates electricity with a gasification system using corn residue, wood and perennial grasses as fuel (Univ. Minnesota, Morris, 2013). Sorghum has been successfully combusted to generate electricity in California (Chromatin, 2012).

As a heat-generation fuel, sorghum has high heating values (HHVs) that are comparable to those of other renewable resources (Table 5). Sorghum biomass HHVs have been reported to range from 16.00 to 18.32 MJ kg^{-1}. One of the challenges of using annual herbaceous crop biomass as a fuel source is that they are high in alkali metals (e.g. K and Na) and halides (e.g. Cl), with amounts often significantly exceeding the 2–3% ash that is typically desired (Miles et al., 1995). High levels of these chemicals increase sintering and agglomeration in reactors and reactor beds (Ergudenler and Gahly, 1993; Miles et al., 1996; Jenkins et al., 1998; Nielsen et al., 2000; Fryda et al., 2008). The addition of kaolin, dolomite and ammonium sulphate has been reported to reduce slagging and the corrosive effects of halides such as Cl (Davidsson et al., 2002; Ohman et al., 2004; Davidsson et al., 2008; Kassman et al., 2011).

Probably the most widely known use of sorghum for electricity generation is that of sweet sorghum bagasse at ethanol plants in India, Brazil and China (Prasad et al., 2007;

Table 5 Fuel analyses for several plant biomass and fossil fuel sources from various reports

		Fixed carbon	Volatile material	Ash									
Source	Crop	% (w/w) on dry wt. basis	MJ kg^{-1}	% of ash	C	H	O	N	S	High heating value	SiO in ash	K_2O in ash	Cl in ash or plant material
Fryda et al., 2008	Sweet sorghum bagasse	–	–	3.20	49.50	6.20	40.10	0.90	0.01	18.32	31.60	31.60	5.10
	Arundo donax L.	–	–	2.48	46.50	5.70	44.70	0.50	0.01	17.98	44.20	30.00	–
Hartley et al., 2011	Biomass sorghum	–	–	6.58	45.16	5.61	42.23	0.46	0.00	17.99	–	–	–
Ture et al., 1997	Sweet sorghum	–	–	2.18	44.04	6.26	–	0.22	0.08	16.83	–	–	–
	Corn stover	–	–	11.63	46.64	5.66	39.59	0.67	0.08	18.26	–	–	–
	Wheat straw	–	–	10.22	43.88	5.26	38.75	0.63	0.16	17.36	–	–	–
	Switchgrass	–	–	5.84	47.26	5.58	40.70	0.59	0.09	18.66	–	–	–
Gonzales et al., 2006	Sweet sorghum	10.30	61.30	2.70	34.00	4.50	60.23	0.80	0.02	16.00	–	–	0.45†
	Forest pellet	13.80	76.40	1.00	46.50	6.80	44.77	1.90	0.00	18.40	–	–	0.03†
	Arundo donax L.	11.40	58.40	2.20	40.30	5.30	53.13	0.40	0.07	17.40	–	–	0.80†
Channiwala and Parikh, 2002	Sugarcane bagasse	13.51	83.66	3.20	45.48	5.96	45.21	0.15	–	18.73	–	–	–
	Methane	–	–	–	74.85	25.15	–	–	–	55.35	–	–	–
	Bituminous coal	65.25	33.45	6.30	76.65	4.78	10.87	0.54	0.42	31.00	–	–	
	Municipal solid waste	–	–	12.00	47.60	6.00	32.90	1.20	0.30	19.88	–	–	–
Jenkins and Baxter, 1998	Yard waste	13.59	66.04	20.37	41.54	4.79	31.91	0.85	0.24	16.30	59.65	2.96	0.30
	Mixed paper	7.42	84.25	8.33	47.99	6.63	36.84	0.14	0.07	20.78	28.10	0.16	0.00

† Cl reported as % of plant dry matter.

Reddy et al., 2007; Zhang et al., 2010). Cubuk et al. (2011) reported that co-combustion of sweet sorghum biomass and lignite coal resulted in similar NOx emissions but reduced C emissions. Cubuk and Heperkan (2004) noted that a similar mixture reduced pollutant concentrations in the emissions from lignite coal. Tillman (2000) pointed out that co-firing biomass with coal reduced the emission of NOx, SOx, fossil fuel CO_2 and trace minerals, such as mercury; however, co-firing biomass with coal resulted in reduced boiler efficiency in every test he reviewed.

The combustion of sweet sorghum bagasse can be quickly adopted because sugarcane bagasse has been used for centuries as a power source at sugarcane mills and also because it is readily available and eliminates the need for disposal. This system does not have the fouling and corrosion problems mentioned above because the K, Na and Cl are physically removed from the bagasse during the sugar extraction process along with the sugar. Das et al. (2004) and Jenkins et al. (1996) reported that nearly all (>92%) of the measured K, Na and Cl were removed by leaching the biomass with water.

Sorghum has been evaluated as feedstock for other thermochemical conversation processes. Pyrolysis results from sorghum bagasse and sweet sorghum biomass indicated that sorghum produced similar results as other grasses (Piskorz et al., 1998; Yin et al., 2013). Sorghum has also been studied as a feedstock for slow pyrolysis as well (Cordella et al., 2013). Yue et al. (2017) and Mafu et al. (2016) also evaluated sorghum as feedstock for torrefaction, or the production of biochar. In all cases, sorghum performed in manners similar to other annual grasses.

4.3 Anaerobic digestion of sorghum

Anaerobic digestion as a pathway to convert sorghum to biogas for heat and electricity or for transportation fuel is not widely used in the United States. Anaerobic digestion has been used for decades to reduce emissions from livestock waste systems by stimulating the biodegradable material to degrade and capturing the subsequent methane gas that is produced during degradations (Karellas et al., 2010). The inclusion of higher-carbon feedstocks that are also high in N are more beneficial than the digestion of animal manure alone. Karellas et al. (2010) listed the advantages as (i) more flexibility in siting anaerobic digesters (ADs); (ii) reduction in AD volumes as well as construction and operating costs; (iii) lower mixing rates, which reduce parasitic electricity consumption; (iv) reduced cost of feedstock logistics because of higher-density materials; and (v) production of higher-quality fertilizer in the mineral effluents.

Biogas production rates are highly dependent on the composition of the feedstock. Total solids (TS) and volatile solids (VS) are two important components to be considered when evaluating a potential feedstock (Schievano et al., 2008; Karellas et al., 2010). Schievano et al. (2008 and 2009) reported that only lab measurements of VS and oxygen demand in 20 h (OD20) were needed to predict methane production in an AD. Angelidaki et al. (1999) developed a model to estimate biogas and methane production that accounted for substrate carbohydrate and N concentrations as well as the OD20, whereas Chandra et al. (2012) indicated that a C-to-N ratio between 20 and 30 was optimal for digestion.

Sorghum biomass is high in both TS and VS (Table 6), which is optimal for the fermentation process (Richards et al., 1991). As with animal nutrition, feedstocks with lower lignin levels result in greater digestion of TS (Herrmann et al., 2011). Herrmann et al. also found that ensiling had minimal impact on AD and that most of the dry matter lost during the ensiling process was converted to lactic and acetic acid which are easily consumed in an AD. Also

Table 6 Composition and methane yields from a range of agricultural based feedstocks

		Total solids	Volatile solids	Methane
Source	Feedstock	g kg^{-1}	g kg^{-1} TS	L g VS^{-1}
Schievano et al., 2009	Cattle manure	18	799	0.058†
Schievano et al., 2009	Swine manure	30	602	0.126†
Schievano et al., 2009	Poultry litter	235	680	0.153†
Schievano et al., 2009	Maize silage	300	915	0.330†
Schievano et al., 2009	Sweet sorghum silage	200	905	0.290†
Richards et al., 1991	Sorghum silage	900	945	0.360
Lee et al., 2011	Corn thin stillage	73	64	0.630
Antonopoulou et al., 2008	Sweet sorghum juice	20	20	0.317‡
Miller and McBee, 1993	Forage sorghum	300	958	0.285
Jerger et al., 1987	Sweet sorghum (Rio)	343	945	0.400
Jerger et al., 1987	Grain sorghum (RS 610)	294	894	0.310
Jerger et al., 1987	High energy (ATx623 × Rio)	322	942	0.340

† Values estimated from model.
‡ Values estimated from H_2 yields.

of interest is the use of thin stillage from grain-ethanol plants. Thin stillage is the liquid removed from wet distillers grains after ethanol production. Thin stillage is very low in dry matter (~6%) (Egg et al., 1985). Thin stillage from either corn or sorghum is quite limited in its ability to provide VS or organically available dry matter for digestion (Table 6). Sweet sorghum juice and thin stillage are both low TS and VS, suggesting that both could be used as the liquid substrate in ADs to increase methane production while reducing the water use.

Addition of plant material as AD substrates increases yields over those of manure-only substrates. Zhou et al. (2011) reported a sixfold increase in biogas production when grass silage was digested compared with cow manure alone. Forage and sweet sorghum are feedstocks that produce high yields of biogas because of their elevated content of free sugars and VS (Table 6). High biomass sorghum has been shown to be a good biogas feedstock. Claassen et al. (2004) and Antonopoulou et al. (2008) described systems in which sweet sorghum bagasse was used as a substrate by thermophilic bacteria to produce H_2 gas for biopower. High biomass yield per land area and high digestion rates result in high energy production per unit land area. Mahmood and Honermeier (2012) reported production of 8114 m^3 biogas ha^{-1} and 4333 m^3 methane ha^{-1} from sorghum. Jerger et al. (1987) reported more than 90% conversion of VS from sweet sorghum biomass to methane, with yields near 0.36 L g^{-1} VS.

5 Sweet sorghum

Sweet sorghum cultivars are forage sorghum cultivars that have been selected for their high stem juice yields and elevated sap sugar concentrations. Sweet sorghum cultivars routinely have sugar concentrations above 15% (v/v) (Wu et al., 2010). Compared with grain sorghum, which produces a large panicle of grain containing complex carbohydrates, sweet sorghum stores non-structural carbohydrates in its stems (McBee et al., 1988). Sweet sorghum was originally used in the United States for human consumption as either crystal sugar or molasses.

Sweet sorghum production in the United States has largely been concentrated in the South and occupied enough acres to warrant attempts at improvement (Walton et al., 1938). It was estimated that nearly 190 million L of sweet sorghum syrup was produced in 1920. Cultivar improvement efforts were located at USDA facilities in Meridian, MS, and Weslaco, TX, from 1950 to 1970 (Coleman, 1970; Hipp et al., 1970). As a result, sweet sorghum cultivars were bred for either molasses production or crystal sugar production (Nathan, 1978). Sugar extracted from sweet sorghum, much like that from sugarcane, is composed primarily of sucrose, glucose and fructose. Concentrations of each sugar vary among cultivars, with sucrose concentrations ranging from 65 to 93% (Corn, 2009; Kim and Day, 2011; Teetor et al., 2011; Wu et al., 2011). Glucose typically represents 75% of the remaining sugars, and fructose the balance. With increased interest in renewable fuels, research on sweet sorghum rose again in the early part of the twenty-first century, but largely on the same cultivars released from the 1950s through the 1980s (Miller and Ottman, 2010; Propheter et al., 2010; Wortmann et al., 2010; Teetor et al., 2011; Godsey et al., 2012).

Resent research indicates that sweet sorghum can attain similar fresh weights and fermentable sugars per unit of planted area as sugarcane (Table 7). The highest yields in the United States were reported in the humid, warm locations of Louisiana and South Texas and under irrigation in the desert Southwest. Yields were reduced by nearly 50% when sweet sorghum was grown under rainfed conditions in the southern and northern Great Plains, as reported by Tamang et al. (2011) and Wortmann et al. (2010). Sweet sorghum performance at locations that may be construed as the northern limit for sweet sorghum production in the United States – northeast Kansas (Propheter et al., 2010) and south central Nebraska (Wortmann et al., 2010) – indicate that water supply (stored soil water and in-season rainfall) can influence sweet sorghum yields, but these environments are also capable of producing relatively high yields.

Across all of the locations analysed, juice concentration was less affected, which could have resulted from the situation that a majority of the data reported is from M81E, an open-pollinated sweet sorghum variety developed in 1981 at the USDA facility in Meridian, MS (Broadhead et al., 1981). It is expected that sugarcane sugar extraction technology will be used to extract sugar from the sweet sorghum stalks. This assumption allows for theoretical juice and sugar yields to be calculated on the basis of moisture content and sugar concentration. In many instances, sugar concentration is measured in degrees Brix, which is a measure of dissolved solids in a solution. For our analyses, sugar yields were estimated with Eq. 1, in which sugar and juice are measured in Mg ha^{-1} (Corn, 2009)

$$\text{Sugar yield} = 0.95 \times \text{juice yield} \times 0.873 \times \frac{\text{brix}}{100} \quad (1)$$

This relationship assumes that the expected sugar extraction rate from sweet sorghum stalks is approximately 95% (Bennett and Anex, 2009). The 0.873 is the percentage of fermentable sugars present (in degrees Brix) and is based on measurements made by Corn (2009).

Using a conversion rate for sugar to ethanol of 584 L Mg^{-1} sugar for sweet sorghum juice (Shapouri et al., 2006), 478 L Mg^{-1} biomass for sweet sorghum residual biomass (bagasse)

Table 7 Sweet sorghum biomass yield, juice yields, and juice composition from a range of studies conducted in the USA

		Dry yields					Ethanol[†]			
	Site	Biomass	Stover[‡]	Grain[‡]	Juice[§]	Juice concentration	Juice	Bagasse	Grain	Total
Source			Mg ha^{-1}			Brix		L ha^{-1}		
Propheter et al., 2010	KS	30.4	28.8	1.7	53.4	14.8	4390.1	13742.5	686.4	18819.0
Miller and Ottman, 2010	AZ	26.0	24.6	1.4	45.7		2472.2	11753.5	587.1	14812.7
Teetor et al., 2011	AZ	39.1	37.0	2.1	68.7	12.4	3324.5	17671.1	882.6	21878.2
Kim and Day, 2011	LA	30.0	28.4	1.6	52.7	10.5	3066.6	13561.7	677.4	17305.6
Corn, 2009	TX	43.7	40.2	3.5	74.7	13.7	3368.4	19228.3	1456.0	24052.8
Tamang et al., 2011	TX	14.9	14.1	0.8	26.2	11.3	1643.7	6735.6	336.4	8715.7
Tew et al., 2008	LA	28.6	27.1	1.6	50.3		4900.0	12942.4	646.4	18488.8
Wortmann et al., 2010	NE	9.8	9.2	0.5	17.2	9.9	1087.7	4418.8	220.7	5727.3
Wortmann et al., 2010	NE	15.9	15.1	0.9	28.0	11.2	2485.4	7208.0	360.0	10053.5

[†] Ethanol yields estimated from juice yields or dry yields using the following conversion rates: juice = 584.8 L ethanol per Mg sugar; bagasse = 478 L ethanol per Mg dry matter; grain = 416 L ethanol per Mg dry matter; sugar yield ha^{-1} = 0.95 × juice yield × 0.873 × (brix/100).
[‡] Grain and stover yields may have been estimated based on total biomass yields, subtraction and an assumed harvest index of 0.05.
[§] Theoretical juice yield estimated as the difference between fresh weights and dry weights. If not reported, moisture content was assumed to be 65%.

(Dahlberg et al., 2011), and 416 L ethanol Mg^{-1} for sweet sorghum grain (Shapouri et al., 2006) resulted in ethanol estimates ranging from 5727 to more than 24 000 L ha^{-1} (Table 7). On average, juice accounts for close to 20%, bagasse 77% and grain 3% of these ethanol values. These estimates clearly illustrate the value of large amounts of bagasse as a feedstock. The fate of bagasse will be dictated by the value of the end energy product, which can include ethanol, electricity or steam. This value proposition will also be dictated by the local infrastructure and energy needs.

Sweet sorghum hybrids and varieties are often overlooked as potential livestock feed. Their high yields of biomass and sugar make them ideal for forage products. In fact, many forage sorghum hybrids used for silage have high sugar contents but low juice yields. Adewakun et al. (1989) found that Brandes sweet sorghum resulted in similar average daily weight gain as corn silage and greater average daily weight gain compared with fescue hay. Organic matter digestibility, digestibility of crude protein and ether extracts, and gross energy digestibility were higher in steers fed sweet sorghum compared with corn silage. Caswell et al. (1983) suggested that the greater amounts of sugar and ether extracts in sweet sorghum provide more readily available energy for animal performance. Lance et al. (1964) reported similar milk yields from sweet sorghum feed compared with corn silage. Stefaniak et al. (2012) noted that sweet sorghum hybrids were higher in starch than other forage sorghum groups, suggesting that this could influence animal performance.

One advantage of sweet sorghum as a livestock feed is that the chemical producer is primarily interested in the sugar-containing juice. The by-product of this process would be the bagasse. Most studies have shown that there is very little loss of dry matter as a result of the juicing process and, in fact, that the biomass is macerated to into very fine pieces, potentially improving digestibility (Propheter et al., 2010). Being a by-product, bagasse also has the potential to reduce feed costs for the livestock producer and an additional income source for the chemical manufacturer.

6 Summary

Sorghum is one of the most genetically diverse crops used in agriculture. Forage sorghum is no exception, with the diversity increasing with the inclusion of So×Su and Su×Su hybrids. Current forage sorghum end users have the benefit of new traits, such as BMR and brachytic dwarfism, to improve digestibility and harvested yield. To date, most of these forages are used in the diets of ruminant livestock. In the past decade, an interest in using sorghum for renewable energy feedstocks has given rise to biomass sorghum, which comprises selections from forage sorghum hybrids with a focus on biomass yield and less on biomass quality. Attention to sorghum as a renewable feedstock also stimulated a renewed interest in sweet sorghum, which was developed by selecting forage sorghum hybrids with high stem-sugar content and juice yields. These hybrids have been positioned to extend the sugarcane-growing season in some environments and have also been found to produce high levels of biogas in ADs. Biomass sorghum has also been proposed as a combustion feedstock, but slagging caused by high levels of K and Cl requires additional technology to be deployed to successfully use biomass sorghum as a heat source. Forage and biomass sorghum have proven to be reliable renewable food and fibre sources, especially in marginal environments in which water is limiting.

7 Where to look for further information

Improving sorghum for any use mentioned in this chapter can be enhanced by improving the crop's adaptability to environmental stresses. Since sorghum is typically grown in marginal environments, developing more drought- and salt-tolerant sorghum inbred lines and hybrids will benefit sorghum breeding programs and end users alike. Effort by the Department of Energy through the ARPA-E TERRA program have sought to identify germplasm that is more drought and salt tolerant. This program is also focused on developing rapid phenotyping sensor arrays and platforms to decrease the time required to manually measure and evaluate such germplasm.

Since many of the characteristics that improve sorghum as a livestock feed component also improve its efficiency in renewable energy programs, improvements in yields and quality are necessary. Reducing lignin or improving digestibility while maintaining agronomic traits will result in desirable products for all sorghum based industries. Increasing seed size and starch digestibility are also important traits that will increase silage digestibility and should be a focal point of all forage sorghum breeding programs. Harvestable yield has long been a stigma in tall forage sorghum. Lodging is always the concern with these high yielding sorghum hybrids and improving standability while maintaining forage digestibility should also be a focal point in improving forage and biomass sorghum.

8 References

Adewakun, L. O., A. O. Famuyiwa, A. Felix and T. A. Omole. 1989. Growth performance, feed intake and nutrient digestibility by beef calves fed sweet sorghum silage, corn silage, and fescue hay. *J. Anim. Sci.* 67: 1341–9.

Angelidaki, I., L. Ellegaard and B. K. Ahring. 1999. A comprehensive model of anaerobic bioconversion of complex substrates to biogas. *Biotechnol. Bioeng.* 63: 363–72. doi:10.1002/(SICI)1097-0290(19990505)63:3<363::AID-BIT13>3.0.CO;2-Z.

Antonopoulou, G., H. N. Gavala, I. V. Skiadas, K. Angelopoulos and G. Lyberatos. 2008. Biofuels generation from sweet sorghum: Fermentative hydrogen production and anaerobic digestion of the remaining biomass. *Bioresour. Technol.* 99: 110–19. doi:10.1016/j.biortech.2006.11.048.

Bean, B. and T. McCollum. 2006. Summary of six years of forage sorghum variety trials. Publ. SCS-2006-04. Texas A&M AgriLife, Texas A&M Univ.: College Station.

Bell, J., Q. Xue, T. McCullom, P. Sirmon, T. Brown and D. Pietsch. 2015. 2014 Texas Panhandle Sorghum Silage Trial. Texas A&M AgriLife, Texas A&M Univ.: College Station.

Bennett, A. S. and R. P. Anex. 2009. Production, transportation and milling costs of sweet sorghum as a feedstock for centralized bioethanol production in the upper Midwest. *Bioresour. Technol.* 100: 1595–607. doi:10.1016/j.biortech.2008.09.023.

Bittinger, T. S., R. P. Cantrell and J. D. Axtell. 1981. Allelism tests of the brown-midrib mutants of sorghum. *J. Hered.* 72: 147–8.

Bout, S. and W. Vermerris. 2003. A candidate-gene approach to clone the sorghum Brown midrib gene encoding caffeic acid O-methyltransferase. *Mol. Genet. Genomics* 269: 205–14.

Broadhead, D. M., K. C. Freeman and N. Zummo. 1981. M 81E: A New Variety of Sweet Sorghum. Info. Sheet 1309. Mississippi Agric. and Forestry Exp. Stn.: Starkville.

Brouk, M. J. 2012. *Sorghum in Beef Production Feeding Guide*. United Sorghum Checkoff Program: Lubbock, TX.

Brouk, M. J. and B. Bean. 2011. *Sorghum in Dairy Cattle Production Feeding Guide*. United Sorghum Checkoff Program: Lubbock, TX.

Bucholtz, D. L., R. P. Cantrell, J. D. Axtell and V. L. Lechtenberg. 1980. Lignin biochemistry of normal and brown midrib mutant sorghum. *J. Agric. Food Chem.* 28: 1239–41. doi:10.1021/jf60232a045.

Caravetta, G. J. and K. D. Johnson. 1990. Within-row spacing influences on diverse sorghum genotypes: II. Dry matter yield and forage quality. *Agron. J.* 82: 210–15.

Carole, T. M., J. Pellegrino and M. D. Paster. 2004. Opportunities in the industrial biobased products industry. *Appl. Biochem. Biotechnol.* 115: 871–85. doi:10.1385/ABAB:115:1-3:0871.

Caswell, L. F., R. S. Kalmbacher and F. G. Martin. 1983. Yield and silage fermentation characteristics of corn, sweet sorghum and grain sorghums. *Proc. Soil Crop Sci. Soc. Fla.* 42: 139–42.

Causey, T. B., K. T. Shanmugam, L. P. Yomano and L. O. Ingram. 2004. Engineering *Escherichia* coli for efficient conversion of glucose to pyruvate. *Proc. Natl. Acad. Sci. U. S. A.* 101: 2235–40. doi:10.1073/pnas.0308171100.

Chandra, R., H. Takeuchi and T. Hasegawa. 2012. Methane production from lignocellulosic agricultural crop wastes: A review in context to second generation of biofuel production. *Renew. Sustain. Energy Rev.* 16: 1462–76. doi:10.1016/j.rser.2011.11.035.

Childs K. L., F. R. Miller, M. M. Cordonnier-Pratt, L. H. Pratt, P. W. Morgan and J. E. Mullet. 1997. Plant Physiol. 113: 611–19. doi:http://dx.doi.org/10.1104/pp.113.2.611.

Chromatin. 2012. Constellation energy and chromatin announce partnership to test sorghum biomass as fuel to generate power. Press release, Chromatin, Inc., 21 September 2011. http://www.chromatininc.com/news/Constellation-Energy-and-Chromatin-Announce-Partnership-to-Test-Sorghum-Biomass-as-Fuel-to-Generate-Power (Accessed 9 May 2013).

Claassen, P. A. M., T. de Vrije and M. A. W. Budde. 2004. Biological hydrogen production from sweet sorghum by thermophilic bacteria. *Proceedings of the 2nd World Conference on Biomass for Energy, Industry, and Climate Protection, Rome Italy*. 10–14 May. ETA-Renewable Energies, Florence, Italy, and WIP, Munich, Germany, pp. 1522–5.

Coleman, O. H. 1970. Syrup and sugar from sweet sorghum. In J. S. Wall and W. M. Ross (Eds), *Sorghum Production and Utilization*. AVI Publishing: Westport, CT, pp. 416–41.

Cook, O. F. 1915. Brachysm, a hereditary deformity of cotton and other plants. *J. Agric.Res.* 3: 387–99.

Cordella, M., C. Berrueco, F. Santarelli, N. Paterson, R. Kandiyoti and M. Millan, 2013. Yields and ageing of the liquids obtained by slow pyrolysis of sorghum, switchgrass and corn stalks. *J. Anal. Appl. Pyrolysis*, 104: 316–24. ISSN: 0165-2370, http://dx.doi.org/10.1016/j.jaap.2013.07.001.

Corn, R. J. 2009. Heterosis and composition of sweet sorghum. Ph.D. diss. Texas A&M Univ.: College Station. http://hdl.handle.net/1969.1/ETD-TAMU-2009-12-7409.

Corredor, D. Y., J. M. Salazar, K. L. Hohn, S. Bean, B. Bean and D. Wang. 2009. Evaluation and characterization of forage sorghum as feedstock for fermentable sugar production. *Appl. Biochem. Biotechnol.* 158: 164–79. doi:10.1007/s12010-008-8340-y.

Craufurd, P. Q., V. Mahalakshmi, F. R. Bidinger, S. Z. Mukuru, J. Chantereau, P. A. Omanga, A. Qi, E. H. Roberts, R. H. Ellis, R. J. Summerfield and G. L. Hammer. 1999. Adaptation of sorghum: Characterisation of genotypic flowering responses to temperature and photoperiod. *Theor. Appl. Genet.* 99: 900–11. doi:10.1007/s001220051311.

Cubuk, M. H. and H. A. Heperkan. 2004. Investigation of pollutant formation of sweet sorghum–lignite (Orhaneli) mixtures in fluidised beds. *Biomass Bioenergy* 27: 277–87. doi:10.1016/j.biombioe.2004.02.001.

Cubuk, M. H., D. B. Ozkan and O. Emanet. 2011. NOx formation of co-combustion of sweet sorghum–ignite (Orhaneli) mixtures in fluidized beds. In H. Gökçekus, U. Türker and J. W. LaMoreaux (Eds), *Survival and Sustainability*. Springer: Berlin, pp. 931–41.

Dahlberg, J., E. Wolfrum, B. W. Bean and W. L. Rooney. 2011. Compositional and agronomic evaluation of sorghum biomass as a potential feedstock for renewable fuels. *J. Biobased Mat. Bioenergy* 5: 507–13. doi:10.1166/jbmb.2011.1171.

Das, P., A. Ganesh and P. Wanikar. 2004. Influence of pretreatment for deashing of sugarcane bagasse on pyrolysis products. *Biomass Bioenergy* 27: 445–57.

Davidsson, K. O., L. E. Åmand, B. M. Steenari, A. L. Elled, D. Eskilsson and B. Leckner. 2008. Countermeasures against alkali-related problems during combustion of biomass in a circulating fluidized bed boiler. *Chem. Eng. Sci.* 63: 5314–29. doi:10.1016/j.ces.2008.07.012.

Davidsson, K. O., J. G. Korsgren, J. B. B. Perrersson and U. Jaglid. 2002. The effects of fuel washing techniques on alkali release from biomass. *Fuel* 81: 137–42. doi:10.1016/S0016-2361(01)00132-6.

Dien, B. S., G. Sarath, J. F. Pedersen, S. E. Sattler, H. Chen, D. L. Funnell-Harris, N. N. Nichols and M. A. Cotta. 2009. Improved sugar conversion and ethanol yield for forage sorghum (*Sorghum bicolor* L. Moench) lines with reduced lignin contents. *BioEnergy Res.* 2: 153–64. doi:10.1007/s12155-009-9041-2.

Dien, B. S., N. N. Nichols and R. J. Bothast. 2001. Recombinant *Escherichia coli* engineered for production of L-lactic acid from hexose and pentose sugars. *J. Ind. Microbiol. Biotechnol.* 27: 259–64. doi:10.1038/sj.jim.7000195.

Dingkuhn, M., M. Kouressy, M. Vaksmann, B. Blerget and J. Chantereau. 2008. A model of sorghum photoperiodism using the concept of threshold-lowering during prolonged appetence. *Eur. J. Agron.* 28: 74–89. doi:10.1016/j.eja.2007.05.005.

Doi K, T. Izawa and T. Fuse. 2004. Ehd1, a B-type response regulator in rice, confers short-day promotion of flowering and controls FT-like gene expression independently of Hd1. *Genes Dev.* 18(8): 926–36. doi:10.1101/gad.1189604.

Du, G. and J. Yu. 2002. Green technology for conversion of food scraps to biodegradable thermoplastic polyhydroxyalkanoates. *Environ. Sci. Technol.* 36: 5511–16. doi:10.1021/es011110o.

Egg, R. P., J. M. Sweeten, and C. G. Coble. 1985. Grain sorghum stillage recycling: Effect on ethanol yield and stillage quality. *Biotechnol. Bioeng.* 27: 1735–8.

Ergudenler, A. and A. E. Gahly. 1993. Agglomeration of silica sand in a fluidized bed gasifier operating on straw. *Biomass Bioenergy* 4: 135–47. doi:10.1016/0961-9534(93)90034-2.

Fritz, J. O., K. J. Moore and E. H. Jaster. 1990. Digestion kinetics and cell wall composition of brown midrib sorghum × sudangrass morphological components. *Crop Sci.* 30: 213–19. doi:10.2135/cropsci1990.0011183X003000010046x.

Fryda, L. E., K. D. Panopoulos and E. Kakaras. 2008. Agglomeration in fluidised bed gasification of biomass. *Powder Technol.* 181: 307–20.

Godsey, C. B., J. Linneman, D. Bellmer and R. Hunke. 2012. Developing row spacing and plant density recommendations for rainfed sweet sorghum production in the Southern Plains. *Agron. J.* 104: 280–6. doi:10.2134/agronj2011.0289.

Goud, J. V. and M. Vasudev Rao. 1977. Inheritance of height in sorghum. *Genet. Agrar.* 31: 39–51.

Hallam, A., I. C. Anderson and D. R. Buxton. 2001. Comparative economic analysis of perennial, annual, and intercrops for biomass production. *Biomass Bioenergy* 21: 407–24.

Hanna, W. W., W. G. Monson and T. P. Gaines. 1981. In vitro dry matter digestibility, total sugars, and lignin measurements on normal and brown midrib (*bmr*) sorghums at various stages of development. *Agron. J.* 73: 1050–2. doi:10.2134/agronj1981.00021962007300060034x.

Hart, G., K. Schertz, Y. Peng and N. Syed. 2001. Genetic mapping of *Sorghum bicolor* (L.) Moench QTLs that control variation in tillering and other morphological characters. *Theor. Appl. Genet.* 103: 1232–42. doi:10.1007/s001220100582.

Hatfield, R. D., J. G. Jung, J. Ralph, D. R. Buxton and P. J. Weimer. 1994. A comparison of the insoluble residues produced by the Klason lignin and acid detergent lignin procedures. *J. Sci. Food Agric.* 65: 51–8. doi:10.1002/jsfa.2740650109.

Hartley, B. E., J. D. Gibson, J. A. Thomasson and S. W. Searcy. 2011. Moisture loss and ash characterization of high-tonnage sorghum. Paper presented at ASABE Annual International Meeting, Louisville, KY. 7–10 August. Paper 1111528.

Heldt, H. W. and B. Piechulla. 2011. *Plant Biochemistry*. 4th ed. Academic Press: Burlington, MA.

Herrmann, C., M. Heiermann and C. Idler. 2011. Effects of ensiling, silage additives and storage period on methane formation of biogas crops. *Bioresour. Technol.* 102: 5153–61. doi:10.1016/j.biortech.2011.01.012.

Hill, J. D. 1976. Predicting the natural drying of hay. *Agric. Meteorol.* 17: 195–204. doi:10.1016/0002-1571(76)90055-8.

Hilley J., S. Truong, S. Olson, D. Morishige and J. Mullet. 2016. Identification of Dw1, a regulator of sorghum stem internode length. *PLoS One* 11(3): e0151271. doi:10.1371/journal.pone.0151271.

Hipp, B. W., W. R. Cowley, C. J. Gerard and B. A. Smith. 1970. Influence of solar radiation and date of planting on yield of sweet sorghum. *Crop Sci.* 10: 91–2. doi:10.2135/cropsci1970.0011183X001000010033x.

Howell, T. A, S. R. Evett, J. A. Tolk, K. S. Copeland, P. D. Colaizzi and P. H. Gowda. 2008. Evapotranspiration of corn and forage sorghum for silage. *Wetting Front.* 10(1): 3–11

Hu, G., C. Cateto, Y. Pu, R. Samuel and A. J. Ragauskas. 2012. Structural characterization of switchgrass lignin after ethanol organosolv pretreatment. *Energy Fuels* 26: 740–5. doi:10.1021/ef201477p.

Hussein, M. M. and A. K. Alva. 2014. Growth, yield, and water use efficiency of forage sorghum as affected by N, P, K fertilizer and deficit irrigation. *Am. J. Plant Sci.* 5: 2134–40. doi:10.4236/ajps.2014.513225.

Jenkins, B. M., R. R. Bakker and J. B. Wei. 1996. On the properties of washed straw. *Biomass Bioenergy* 10: 177–200. doi:10.1016/0961-9534(95)00058-5.

Jenkins, B. M., L. L. Baxter, T. R. Miles Jr. and T. R. Miles. 1998. Combustion properties of biomass. *Fuel Process. Technol.* 54: 17–46. doi:10.1016/S0378-3820(97)00059-3.

Jerger, D. E., D. P. Chynoweth and H. R. Isaacson. 1987. Anaerobic digestion of sorghum biomass. *Biomass* 14: 99–113.

Karellas, S., I. Boukis and G. Kontopoulos. 2010. Development of an investment decision tool for biogas production from agricultural waste. *Renew. Sustain. Energy Rev.* 14: 1273–82. doi:10.1016/j.rser.2009.12.002.

Kassman, H., M. Brostrom, M. Berg and L. Amand. 2011. Measures to reduce chlorine in deposits: Application in a large-scale circulating fluidised bed boiler firing biomass. *Fuel* 90: 1325–34. doi:10.1016/j.fuel.2010.12.005.

Kemanian, A. R., C. O. Stockle, D. R. Huggins and L. M. Viega. 2007. A simple method to estimate harvest index in grain crops. *Field Crops Res.* 103: 208–16. doi:10.1016/j.fcr.2007.06.007.

Kim, M. and D. F. Day. 2011. Composition of sugar cane, energy cane, and sweet sorghum suitable for ethanol production at Louisiana sugar mills. *J. Ind. Microbiol. Biotechnol.* 38: 803–7. doi:10.1007/s10295-010-0812-8.

Klein, R. R., F. R. Miller, D. V. Dugas, P.J. Brown, M. Burrell, P.E. Klein. 2015. Allelic variants in the PRR37 gene and the human-mediated dispersal and diversification of sorghum. *Theor. Appl. Genet.* 128: 1669. doi:10.1007/s00122-015-2523-z.

Klein R. R., J. E. Mullet, D. R. Jordan, F. R. Miller, W. L. Rooney and M. A. Menz. 2008. The effect of tropical sorghum conversion and inbred development on genome diversity as revealed by high-resolution genotyping. *Crop Sci.* 48: 12–26. doi:10.2135/cropsci2007.06.0319tpg.

Lance, R. D., D. C. Foss, C. R. Kruegar, B. R. Baumgardt and R. R. Niedermeir. 1964. Evaluation of corn and sorghum silages on the basis of milk production and digestibility. *J. Dairy Sci.* 47: 254–7. doi:10.3168/jds.S0022-0302(64)88635-5.

Lee, P, J. Bae, J. Kim and W. Chen. 2011. Mesophilic anaerobic digestion of corn thin stillage: A technical and energetic assessment of the corn-to-ethanol industry integrated with anaerobic digestion. *J. Chem. Technol. Biotechnol.* 86: 1514–20.

Li X., X. Li, E. Fridman, T. T. Tesso and J. Yu. 2015. Dissecting repulsion linkage in the dwarfing gene Dw3 region for sorghum plant height provides insights into heterosis. *Proc. Natl. Acad. Sci. U. S. A.* 112: 11823–8. doi:10.1073/pnas.1509229112.

Mafu, L. D., H. W. J. P. Neomagus, R. C. Everson, M. Carrier, C. A. Strydom and J. R. Bunt. 2016. Structural and chemical modifications of typical South African biomasses during torrefaction. *Bioresour. Technol.* 202: 192–7. ISSN: 0960-8524, http://dx.doi.org/10.1016/j.biortech.2015.12.007 (http://www.sciencedirect.com/science/article/pii/S0960852415016375).

Mahmood, A. and B. Honermeier. 2012. Chemical composition and methane yield of sorghum cultivars with contrasting row spacing. *Field Crops Res.* 128: 27–33. doi:10.1016/j.fcr.2011.12.010.

Marsalis, M. A., S. V. Angadi and F. E. Contreras-Govea. 2010. Dry matter yield and nutritive value of corn, forage sorghum, and BMR forage sorghum at different plant populations and nitrogen rates. *Field Crops Res.* 116: 52–7. doi:10.1016/j.fcr.2009.11.009.

Marsalis, M. A., S. V. Angadi, F. E. Contreras-Govea and R. E. Kirksey. 2009. Harvest Timing and Byproduct Addition Effects on Corn and Forage Sorghum Silage Grown Under Water Stress. Bull. 799. New Mexico State Univ.: Las Cruces.

Martin, G. C. and R. F. Barnes. 1980. Prediction of energy digestibility of forages with in vitro rumen fermentation and fungal enzyme systems. In W. G. Pigden, C. C. Balch and M. Graham (Eds), *Standardization of Analytical Methodology for Feeds*. Publ. IDR-134e. Intl. Res. Dev. Cent.: Ottawa, ON.

Maughan, M., T. Voigt, A. Parrish, G. Bollero, W. Rooney and D. K. Lee. 2012. Forage and energy sorghum response to nitrogen fertilization in central and southern Illinois. *Agron. J.* 104: 1032–40. doi:10.2134/agronj2011.0408.

McBee, G. G., R. A. Creelman and F. R. Miller. 1988. Ethanol yield and energy potential of stems from a spectrum of sorghum biomass types. *Biomass* 17: 203–11. doi:10.1016/0144-4565(88)90114-X.

McCuistion, K. C., B. W. Bean and F. T. McCollum. 2009. Yield and water-use efficiency response to irrigation level of brown midrib, non-brown midrib, and photoperiod-sensitive forage sorghum cultivars. *Forage Grazinglands* 7. doi:10.1094/FG-2009-0909-01-RS.

McCuistion, K. C., B. W. Bean and F. T. McCollum. 2010. Nutritional composition response to yield differences in brown midrib, non-brown midrib, and photoperiod sensitive forage sorghum cultivars. *Forage Grazinglands* 8. doi:10.1094/FG-2010-0428-01-RS

Miles, T. R., T. R. Miles Jr., L. L. Baxter, R. W. Bryers, B. M. Jenkins and L. L. Oden. 1995. Alkali Deposits Found in Biomass Power Plants. A Preliminary Investigation of Their Extent and Nature. Publ. NREL/TP-433–8142. Vol. 2. Sandia Natl. Lab.: Albuquerque, NM.

Miles, T. R., T. R. Miles Jr., L. L. Baxter, R. W. Bryers, B. M. Jenkins and L. L. Oden. 1996. Boiler deposits from firing biomass fuels. *Biomass Bioenergy* 10: 125–38. doi:10.1016/0961-9534(95)00067-4.

Miller, A. N. and M. J. Ottman. 2010. Irrigation frequency effects on growth and ethanol yield in sweet sorghum. *Agron. J.* 102: 60–70. doi:10.2134/agronj2009.0191.

Miller, F. R. and G. G. McBee. 1993. Genetics and management of physiological systems of sorghum for biomass production. *Biomass Bioenergy* 5: 41–9. doi:10.1016/0961-9534(93)90006-P.

Mullet, J. E., W. L. Rooney, P. E. Klein, D. Morishige, R. Murphy and J. A. Brady. 2010. Discovery and utilization of sorghum genes (*ma5/ma6*). US Patent application 20100024065 A1. Published: 28 January.

Mullet J. E. and W. L. Rooney. 2013. Method for production of sorghum hybrids with selected flowering times. US Patent 20130298274 A1.

Multani, D. S., S. P. Briggs, M. A. Chamberlin, J. J. Blakeslee, A. S. Murphy and G. S. Johal. 2003. Loss of an MDR Transporter in Compact Stalks of Maize br2 and Sorghum dw3 Mutants. *Science* 302(5642): 81–84. doi:10.1126/science.1086072.

Murphy R. L., R. R. Klein, D. T. Morishige, J. A. Brady, W. L. Rooney, F. R. Miller, D. V. Dugas, P. E. Klein and J. E. Mullet. Coincident light and clock regulation of pseudoresponse regulator protein 37 (PRR37) controls photoperiodic flowering in sorghum. *PNAS* 108(39): 16469–74. published ahead of print 19 September 2011, doi:10.1073/pnas.1106212108.

Murphy, R. L., D. T. Morishige, J. A. Brady, W. L. Rooney, S. Yang, P. E. Klein and J. E. Mullet. 2014. Ghd7 (Ma6) represses sorghum flowering in long days: Ghd7 alleles enhance biomass accumulation and grain production. *Plant Genome* 7. doi:10.3835/plantgenome2013.11.0040; doi:10.3835/plantgenome2013.11.0040.

Narayanan, S., R. M. Aiken, P. V. V. Prasad, Z. Xin and J. Yu. 2013. Water and radiation use efficiencies in sorghum. *Crop Sci.* 105:649–56.

Nathan, R. A. (Ed.). 1978. Fuels from Sugar Crops: Systems Study for Sugarcane, Sweet Sorghum, and Sugar Beets. Tech. Info. Cent. US DOE 71D-22781. US DOE: Washington, DC.

Nielsen, H. P., F. J. Frandsen, K. Dam-Johansen and L. L. Baxter. 2000. The implications of chlorine-associated corrosion on the operation of biomass-fired boilers. *Prog. Energy Combust. Sci.* 26:283–98. doi:10.1016/S0360-1285(00)00003-4.

Ohman, M., D. Bostrom, A. Nordin and H. Hedman. 2004. Effect of kaolin and limestone addition on slag formation during combustion of wood fuels. *Energy Fuels* 18: 1370–6. doi:10.1021/ef040025+.

Ottman, M. J. 2010. Water use efficiency of forage sorghum grown with sub-optimal irrigation. In *Forage and Grain Report*. Univ. of Ariz. Coop. Ext. Publ. AZ1526. Univ. of Arizona: Tucson, pp. 26–9. http://cals.arizona.edu/pubs/crops/az1526/az1526d.pdf.

Pedersen, J. F., D. L. Funnell, J. J. Toy, A. L. Oliver and R. J. Grant. 2006a. Registration of -Atlas *bmr12*' forage sorghum. *Crop Sci*. 46: 478.

Pedersen, J. F., D. L. Funnell, J. J. Toy, A. L. Oliver and R. J. Grant. 2006b. Registration of seven forage sorghum genetic stocks near-isogenic for the brown midrib genes *bmr6* and *bmr12*. *Crop Sci*. 46: 490–1.

Pedersen, J. F., D. L. Funnell, J. J. Toy, A. L. Oliver and R. J. Grant. 2006c. Registration of 12 grain sorghum genetic stocks near-isogenic for the brown midrib genes *bmr6* and *bmr12*. *Crop Sci*. 46: 491–2.

Pedersen, J. F., J. J. Toy, D. L. Funnell, S. E Sattler, A. L. Oliver and R. A. Grant. 2008. Registration of BN611, A/BN612, and RN613 sorghum genetic stocks with stacked *bmr6* and *bmr12* genes. *J. Plant Reg*. 2: 258–62.

Pedersen, M., A. Vikso-Nielsen and A. S. Meyer. 2010. Monosaccharide yields and lignin removal from wheat straw in response to catalyst type and pH during mild thermal pretreatment. *Process Biochem*. 45: 1181–6. doi:10.1016/j.procbio.2010.03.020.

Piskorz J., P. Majerski, D. Radlein, D. S. Scott and A. V. Bridgwater, 1998. Fast pyrolysis of sweet sorghum and sweet sorghum bagasse. *J. Anal. Appl. Pyrolysis* 46:15–29. ISSN: 0165-2370, http://dx.doi.org/10.1016/S0165-2370(98)00067-9.

Porter, K. S., J. D. Axtell, V. L. Lechtenberg and V. F. Colenbrander. 1978. Phenotype, fiber composition, and in vitro dry matter disappearance of chemically induced brown midrib (*bmr*) mutants of sorghum. *Crop Sci*. 18: 205–8. doi:10.2135/cropsci1978.0011183X00180 0020002x.

Prasad, S., A. Singh, N. Jain and H. C. Joshi. 2007. Ethanol production from sweet sorghum syrup for utilization as automotive fuel in India. *Energy Fuels* 21: 2415–20. doi:10.1021/ef060328z.

Propheter, J. L., S. A. Staggenborg, X. Wu and D. Wang. 2010. Performance of annual and perennial biofuel crops: Yield during the first two years. *Agron. J*. 102: 806–14. doi:10.2134/ agronj2009.0301.

Quinby, J. R. 1974. The genetic control of flowing and growth in sorghum. *Adv. Agron*. 25: 125–62. doi:10.1016/S0065-2113(08)60780-4.

Quinby J. R. and R. E. Karper. 1961. Inheritance of duration of growth in the Milo Group of Sorghum1. *Crop Sci*. 1: 8–10. doi:10.2135/cropsci1961.0011183X000100010004x.

Quinby, J. R. and R. E. Karper. 1954. Inheritance of height in sorghum. *Agron. J*. 46: 211–16. doi:10.2134/agronj1954.00021962004600050007x.

Rajan, N. and S. Maas. 2007. Comparative evaluation of actual crop water use of forage sorghum and corn for silage. Report as of FY2007 for 2007TX271B. Dep. of the Interior, USGS. http://water. usgs.gov/wrri/07grants/2007TX271B.html (accessed 1 June 2015).

Reddy, B. V. S., A. Ashok Kumar and S. Ramesh. 2007. Sweet sorghum: A water saving bio-energy crop. Paper presented at the International conference on Linkages between Energy and Water Management for Agriculture in Developing Countries, ICRISAT campus, Hyderabad, India. 29–30 January.

Richards, B. K., R. J. Cummings, W. J. Jewell and F. G. Herndon. 1991. High solid anaerobic methane fermentation of sorghum and cellulose. *Biomass Bioenergy* 1: 47–53. doi:10.1016/0961-9534(91)90051-D.

Richter, K. and C. Berthold. 1998. Biotechnological conversion of sugar and starchy crops into lactic acid. *J. Agric. Eng. Res*. 71: 181–91. doi:10.1006/jaer.1998.0314.

Richter, K. and A. Träger. 1994. L(+)-lactic acid from sweet sorghum by submerged and solid-state fermentations. *Acta Biotechnol*. 14: 367–78. doi:10.1002/abio.370140409.

Rocateli, A. C., R. L. Raper, K. S. Balkcom, F. J. Arriaga and D. I. Bransby. 2012. Biomass sorghum production and components under different irrigation/tillage systems for the southeastern U.S. Ind. *Crops Prods*. 36:589–98. doi:10.1016/j.indcrop.2011.11.007.

Rooney, W. L. and S. Aydin. 1999. Genetic control of a photoperiod-sensitive response in sorghum bicolor (L.) moench. *Crop Sci.* 39: 397–400. doi:10.2135/cropsci1999.0011183X003900020016x.

Rooney, W. L., J. Blumenthall, B. Bean and J. E. Mullet. 2007. Designing sorghum as a dedicated bioenergy feedstock. *Biofuels Bioprod. Biorefin.* 1: 147–57. doi:10.1002/bbb.15.

Saballos, A., G. Ejeta, E. Sanchez, C. Kang and W. Vermerris. 2009. A genomewide analysis of the cinnamyl alcohol dehydrogenase family in sorghum [*Sorghum bicolor* (L.) Moench] identifies SbCAD2 as the Brown midrib6 gene. *Genetics* 181: 783–95.

Saballos, A., S. E. Sattler, E. Sanchez, T. P. Foster, Z. Xin and et al. 2012. Brown midrib2 (*bmr2*) encodes the major 4-coumarate: Coenzyme A ligase involved in lignin biosynthesis in sorghum (*Sorghum bicolor* (L.) Moench). *Plant J.* 70: 818–30.

Saballos, A., W. Vermerris, L. Rivera and G. Ejeta, 2008 Allelic association, chemical characterization and saccharification properties of brown midrib mutants of sorghum (*Sorghum bicolor* (L.) Moench). *Bioenerg. Res.* 1:193–204. doi:10.1007/s12155-008-9025-7.

Saeed, I. A. M. and A. H. El-Nadi. 1998. Forage sorghum yield and water use efficiency under variable irrigation. *Irrig. Sci.* 18: 67–71. doi:10.1007/s002710050046.

Samuel, W. A., Y. Y. Lee and W. B. Anthony. 1980. Lactic acid fermentation of crude sorghum extract. *Biotechnol. Bioeng.* 22: 757–77. doi:10.1002/bit.260220404.

Sattler, S. E., A. J. Saathoff, E. J. Haas, N. A. Palmer, D. L. Funnell-Harris and et al. 2009. A nonsense mutation in a cinnamyl alcohol dehydrogenase gene is responsible for the sorghum brown midrib6 phenotype. *Plant Physiol.* 150: 584–95.

Schievano, A., M. Pognani, G. D'Imporzano and F. Adani. 2008. Predicting anaerobic biogasification potential of ingestates and digestates of a full-scale biogas plant using chemical and biological parameters. *Bioresour. Technol.* 99: 8112–17. doi:10.1016/j.biortech.2008.03.030.

Schievano, A., B. Scaglia, G. D'Imporzano, L. Malagutti, A. Gozzi and F. Adani. 2009. Prediction of biogas potentials using quick laboratory analyses: Upgrading previous models for application to heterogeneous organic matrices. *Bioresour. Technol.* 100:5777–82. doi:10.1016/j.biortech.2009.05.075.

Shapouri, H., M. Salassi and J. Nelson. 2006. *The Economic Feasibility of Ethanol Production from Sugar in the United States*. USDA: Washington, DC. http://www.usda.gov/oce/reports/energy/EthanolSugarFeasibilityReport3.pdf (Accessed 6 March 2016).

Sluiter, A., B. Hames, R. Ruiz, C. Scarlata, J. Sluiter, and D. Templeton. 2005a. Determination of Ash in Biomass, Laboratory Analytical Procedure. Tech. Rep. 510-42622. National Renewable Energy Lab: Golden, CO.

Sluiter, A., B. Hames, R. Ruiz, C. Scarlata, J. Sluiter, D. Templeton, and D. Crocker. 2011. Determination of Structural Carbohydrates and Lignin in Biomass, Laboratory Analytica Procedure. Tech. Rep. 510-42618. National Renewable Energy Lab: Golden, CO.

Sluiter, A., R. Ruiz, C. Scarlata, J. Sluiter and D. Templeton. 2005b. Determination of Extractives in Biomass, Laboratory Analytical Procedure. Tech. Rep. 510-42619. National Renewable Energy Lab: Golden, CO.

Snider, J. L., R. L. Raper and E. B. Schwab. 2012. The effect of row spacing and seeding rate on biomass production and plant stand characteristics on non-irrigated photoperiod-sensitive sorghum (*Sorghum bicolor* (L.) Moench). *Ind. Crops Prod.* 37: 527–535. doi:10.1016/j.indcrop.2011.07.032.

Staggenborg, S. A., K. C. Dhuyvetter and W. B. Gordon. 2008. Grain sorghum and corn comparisons: Yield, economic, and environmental responses. *Agron. J.* 100: 1600–4. doi:10.2134/agronj2008.0129.

Stefaniak, T. R., J. A. Dahlberg, B. W. Bean, N. Dighe, E. J. Wolfrum and W. L. Rooney. 2012. Variation in biomass composition components among forage, biomass, sorghum-sudangrass, and sweet sorghum types. *Crop Sci.* 52: 1949–54. doi:10.2135/cropsci2011.10.0534.

Tamang, P. L., K. F. Bronson, A. Malapati, R. Schwartz, J. Johnson and J. Moore-Kucera. 2011. Nitrogen requirement for ethanol production from sweet sorghum and photoperiod sensitive sorghums in the Southern High Plains. *Agron. J.* 103: 431–40. doi:10.2134/agronj2010.0288.

Teetor, V. H., D. V. Duclos, E. T. Wittenberg, K. M. Young, J. Chawhuaymak, M. R. Riley and D. T. Ray. 2011. Effects of planting date on sugar and ethanol yield of sweet sorghum grown in Arizona. *Ind. Crops Prod.* 34: 1293–300. doi:10.1016/j.indcrop.2010.09.010

Tew, T. L., R. M Cobill and E. P. Richard. 2008. Evaluation of sweet sorghum and sorghum × sudangrass hybrids as feedstocks for ethanol production. *Bioenerg. Res.* 1: 147–52.

Theerarattananoon, K., X. Wu, S. Staggenborg, R. Propheter, R. Madl and D. Wang. 2010. Evaluation and characterization of sorghum biomass as feedstock for sugar production. *Trans. ASABE* 53: 509–25. doi:10.13031/2013.29561.

Theerarattananoon, K., F. Xu, J. Wilson, S. Staggenborg, L. McKinney, P. Vadlani, Z. Pei and D. Wang. 2012. Effect of pelleting conditions on chemical composition and sugar yield of corn stover, big bluestem, wheat straw and sorghum stalk pellets. *Bioprocess Biosyst. Eng.* 35: 615–23. doi:10.1007/s00449-011-0642-8.

Tillman, D. A. 2000. Biomass cofiring: The technology, the experience, the combustion consequences. *Biomass Bioenergy* 19: 365–84. doi:10.1016/S0961-9534(00)00049-0.

Undersander, D., D. R. Mertens and N. Theix. 1993. *Forage Analyses Procedures*. Natl. Forage Testing Assoc.: Omaha, NE.

University of Minnesota, Morris. 2013. *Renewable Energy*. University of Minnesota: Morris. https://www.morris.umn.edu/sustainability/renewable/(Accessed 1 April 2016).

US Department of Energy (USDOE). 2006. Theoretical Ethanol Yield Calculator. USDOE: Washington, DC. http://www1.eere.energy.gov/biomass/ethanol_yield_calculator.html (accessed 30 April 2013).

US Energy Information Administration (USEIA). 2012. *Electric Power Monthly*. USDOE: Washington, DC, http://www.eia.gov/electricity/annual/archive/03482012.pdf (Accessed 30 March 2016).

Venuto, B. and B. Kindiger. 2008. Forage and biomass feedstock production from hybrid forage sorghum and sorghum-sudangrass hybrids. *Grassl. Sci.* 54: 189–96. doi:10.1111/j.1744-697X.2008.00123.x.

Walton, C. F. Jr., E. K. Ventre and S. Byall. 1938. Farm Production of Sorgo Sirup. Farmers' Bull. 1791. USDA: Washington, DC.

Wee, J. Y., J. N. Kim and H. W. Ryu. 2006. Biotechnological production of lactic acid and its recent applications. *Food Technol. Biotechnol.* 44(2): 163–172.

Wolfrum, E. J., A. J. Lorenz, and N. deLeon. 2009. Correlating detergent fiber analyses and dietary fiber analysis data for corn stover collected by NIRS. *Cellulose* 16(4): 577–85. DOI:10.1007/s10570-009-9318-9.

Wortmann, C. S., A. J. Liska, R. B. Ferguson, D. J. Lyon, R. N. Klein and I. Dweikat. 2010. Dryland performance of sweet sorghum and grain crops for biofuel in Nebraska. *Agron. J.* 102: 319–26. doi:10.2134/agronj2009.0271.

Wu, L., M. Arakane, M. Ike, M. Wada, T. Takai, M. Gau and K. Tokuyasu. 2011. Low temperature alkali pretreatment for improving enzymatic digestibility of sweet sorghum bagasse for ethanol production. *Bioresour. Technol.* 102:4793–9. doi:10.1016/j.biortech.2011.01.023.

Wu, X. R., S. A. Staggenborg, J. L. Propheter, W. L. Rooney, J. Yu and D. Wang. 2010. Features of sweet sorghum juice and their performance in ethanol fermentation. *Ind. Crops Prod.* 31: 164–70. doi:10.1016/j.indcrop.2009.10.006.

Wyman, C. E., B. E. Dale, R. T. Elander, M. Holtzapple, M. R. Ladisch and Y. Y. Lee. 2005. Comparative sugar recovery data from laboratory scale application of leading pretreatment technologies to corn stover. *Bioresour. Technol.* 96: 2026–32. doi:10.1016/j.biortech.2005.01.018.

Yang, B. and C. E. Wyman. 2004. Effect of xylan and lignin removal by batch and flowthrough pretreatment on the enzymatic digestibility of corn stover cellulose. *Biotechnol. Bioeng.* 86(1): 88–98. doi:10.1002/bit.20043.

Yang S, R. L. Murphy, D. T. Morishige, P. E. Klein and W. L. Rooney. 2014. Sorghum Phytochrome B Inhibits Flowering in Long Days by Activating Expressionof SbPRR37 and SbGHD7, Repressors of SbEHD1, SbCN8 and SbCN12. *PLoS One* 9(8): e105352. doi:10.1371/journal.pone.0105352.

Yin R., R. Liu, Y. Mei, W. Fei and X. Sun, 2013. Characterization of bio-oil and bio-char obtained from sweet sorghum bagasse fast pyrolysis with fractional condensers. *Fuel* 112: 96–104. ISSN: 0016-2361, http://dx.doi.org/10.1016/j.fuel.2013.04.090.

Yue, Y., H. Singh, B. Singh and S. Mani. 2017. Torrefaction of sorghum biomass to improve fuel properties. *Bioresour. Technol.* 232: 372–9. ISSN 0960-8524, http://dx.doi.org/10.1016/j.biortech.2017.02.060. (http://www.sciencedirect.com/science/article/pii/S0960852417301839).

Zhan, X., D. Wang, X. S. Sun, S. Kim and D. Y. C. Fung. 2002. Lactic acid production from extrusion cooked grain sorghum. *Trans. ASAE* 46:589–93.

Zhan, X., D. Wang, M. R. Tuinstra, S. Bean, P. A. Seib and X. S. Sun. 2003 Ethanol and lactic acid production as affected by sorghum genotype and location. *Ind. Crops Prod.* 18: 245–55. doi:10.1016/S0926-6690(03)00075-X.

Zhang, C., G. Xie, S. Li, L. Ge and T. He. 2010. The productive potentials of sweet sorghum ethanol in China. *Appl. Energy* 87: 2360–8. doi:10.1016/j.apenergy.2009.12.017.

Zhou, H., D. Löffler and M. Kranert. 2011. Model-based predictions of anaerobic digestion of agricultural substrates for biogas production. *Bioresour. Technol.* 102: 10819–28. doi:10.1016/j.biortech.2011.09.014.

Part 2

Sorghum production and improvement across the world

Introducing new technologies and market strategies for sorghum producers in developing countries: the Sahel case

John H. Sanders, Purdue University, USA; Botorou Ouendeba, former Director of the 3N Program, Niger; Ababacar Ndoye, former Director of the Institute of Food Technology, Senegal; and Niaba Témé, Institute of the Agricultural Economy (IER), Mali

1 Introduction

The dilemma of raising yields for a crop produced principally for home consumption is that yields, input use and expected prices are all low but interrelated, requiring different strategies to change. Soils in semi-arid regions of sub-Saharan Africa are poor. This requires fertilzers which farmers cannot always afford. Low yields mean little surplus marketed so that the farmers deal with the lowest level of the marketing chain, sell at the depressed seasonal price right after harvest, typically selling a low-quality product with 10–20% impurities. Changing the marketing strategy requires increased storage capacity, better quality control and the ability to amass and sell larger quantities to higher levels of the marketing chain. This requires both improvements in technology and institutional change (e.g. farmers' associations for cooperative marketing).

http://dx.doi.org/10.19103/AS.2017.0015.20

The secondary markets for sorghum will also become more important as with income growth, people will eat less cereal and more quality foods including poultry. So the conditions for sorghum to substitute for maize in the feed ration are also considered.

A final issue to consider is the need for research to respond to new problems as they arise. As an example, an improved variety proved 'softer' than local varieties and this required changing the processing of the local staple food 'tô' so that it retained the consistency required by consumers.

This chapter reports on the field research of a programme combining agronomic and economic components over the period 2004–16. The programme initially included four Sahelian countries and then focused on Mali as it moved from the pilot to the scaling-up process. Specifically, we are concerned with introducing new sorghum technologies and improving marketing strategies. Also a focus is needed on the development of an institution to facilitate this process and the ability to respond to evolving problems.

2 Sorghum performance and potential: Mali and Burkina Faso

Over the last decade, aggregate maize yields have doubled from 1.5 to 2.8 t/ha in Mali. Sorghum yields have stagnated, occasionally making it to one ton in good rainfall years but often remaining around 750 kg/ha (Fig. 1). In Burkina Faso, maize yields began the period lower than in Mali at 1.25 t/ha. Yields did not increase as much as in Mali reaching less than 2 t/ha (Fig. 2). With the introduction of Bt cotton in Burkina, returns to cotton have not fallen as much as in Mali so there has been less shift of farmer concentration to maize in the cotton zone (Coulibaly et al., 2015).

There are various reasons for the superior performance of maize as compared with sorghum in both countries. There has been a more rapid introduction of both improved

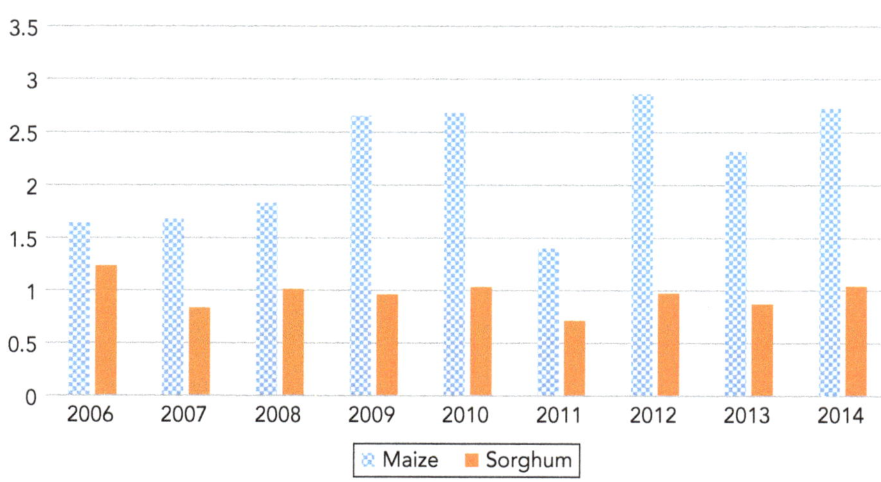

Figure 1 Maize and sorghum yields (t/ha) in Mali, 2006–14. Source: ADA (Directorate of National Agriculture), 2015.

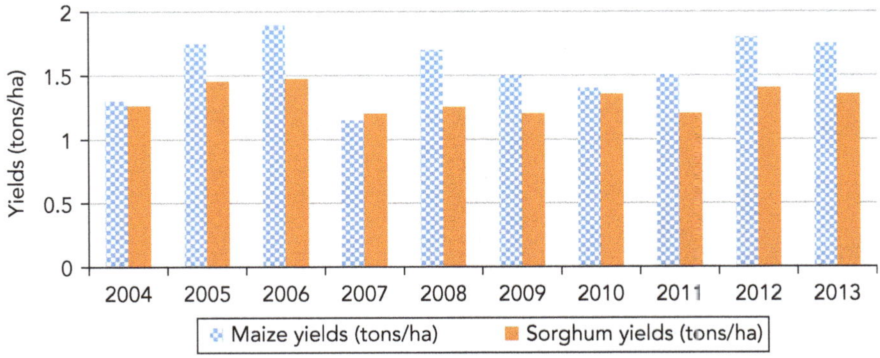

Figure 2 Maize and sorghum yields (t/ha) in Burkina Faso, 2004–13. Source: Département d'Agriculture, 2015.

maize cultivars and hybrids in both countries. Moreover, inorganic fertilizer use and more recently extensive use of fertilizer subsidies have favoured the yield and production increases for maize. There has been support from both IITA (International Institute for Technology in Africa) and CIMMYT (International Maize and Wheat Improvement Center) on introducing new maize cultivars and other improved technologies. Moreover, maize production is concentrated in the higher rainfall regions on better soils in a high input using region. In contrast, sorghum is found in most regions of both countries and is a very important cereal cultivar in lower rainfall and poorer soil fertility regions.

Considering production, maize went from 600 000 t to almost two million t over the decade, more than tripling in Mali (Fig. 3). In Burkina, maize production also tripled from 500 000 to 1.5 million t over the decade (Fig. 4). In Mali, sorghum production ranged from one million to 1.5 million t. Whereas in Burkina, sorghum production ranged from 1.5 million to almost two million t. So the big gains were with maize productivity, while sorghum productivity remained stagnant but increased area and weather had a large effect on the sorghum output.

Both sorghum and maize are used for the traditional food product in rural and, to a lesser extent, urban areas, 'tô' a porridge with the consistency of 'polenta'. 'Tô' is eaten with sauces of various types with the historic preference for most regions being the 'tô' from millet. Sorghum is substituted for millet in poor weather as millet is produced on lighter, more marginal soils than sorghum. Also, sorghum 'tô' is often preferred in regions where the predominance of heavier soils favours sorghum production. When other cereals such as maize with its high yields in the cotton–maize–sorghum systems, characterizing southern regions of Mali and Burkina Faso, become cheaper, they are substituted for the millet or sorghum in the 'tô'.

So, what is the potential for sorghum to increase yields? Sorghum yields in Mali and Burkina are compared with the historic performance of yield increases of sorghum and maize in the United States. Sorghum yields were tripled in the United States in less than 20 years (1955–72) from 1.25 to 3.5 t/ha (Miller and Kebede, 1987). Maize yield increases started slowly during the Depression but ultimately reached twice the sorghum yields by 1987 (Duvick, 1987).

Maize is grown on better soils with higher rainfall and generally higher input levels than those of sorghum. The basis of these yield increases for both cereals was improved

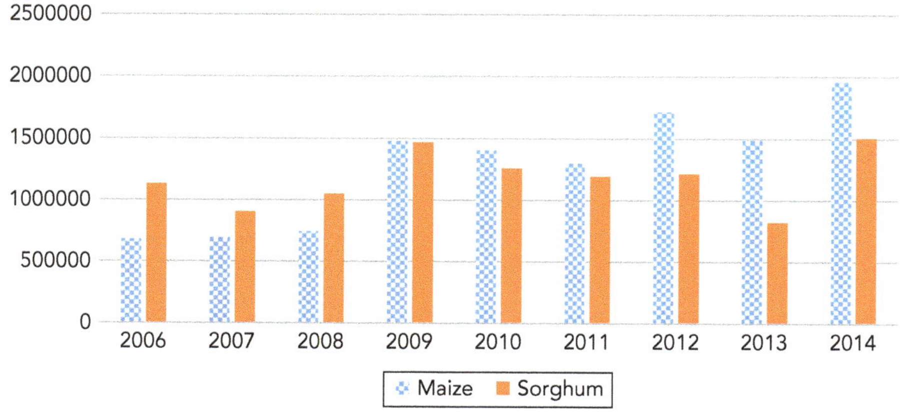

Figure 3 Maize and sorghum production (t) in Mali, 2006–14. Source: ADA (Directorate of National Agriculture), 2015.

cultivars, mainly hybrids, increasing fertilization, improved agronomic practices and some irrigation (Miller and Kebede, 1987; Duvick, 1987).

The bottom line is that maize has achieved much higher yields in both the Sahelian countries and in the United States. However, there is a very large yield gap between actual Malian and Burkina yields and potential sorghum yields as indicated by experiment station yields in Mali and Burkina or with the US historic yield performance of sorghum. Hence, there is a substantial potential to close this sorghum yield gap on farmers' fields in Mali and Burkina.

Moreover, sorghum has an increasing comparative advantage as cereal production moves out of the prime agricultural regions into lower, more variable rainfall regions. Even in the high input using countries, as the United States, sorghum continues to be a principal cereal cultivar in the drier regions (western regions of Texas, Kansas and Nebraska).

So what needs to happen for new sorghum technology to become rapidly adopted by farmers?

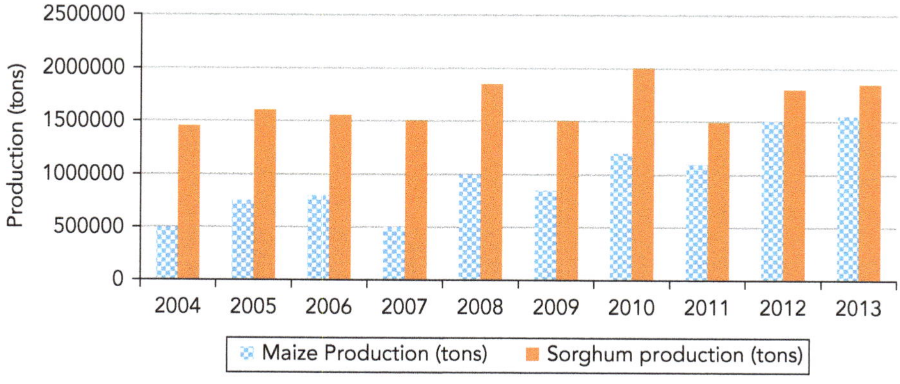

Figure 4 Maize and sorghum production (t) in Burkina Faso, 2004–13.

3 The strategy: breaking out of subsistence

With 'subsistence', crops farmers use few inputs and receive low prices. Yields are low and variable principally due to rainfall variability. Prices are low because farmers sell at the seasonal low point of prices shortly after the harvest (Fig. 5) and the buyers are at the lowest level of the marketing chain. Many collectors pick up at the farms or in the villages, so their costs, and hence margins, are high. Farmers want to sell at harvest as they have a seasonal demand for money at harvest. Pressing harvest time expenditures are school fees, weddings and naming ceremonies (equivalent to baptisms for Muslims), labour payments including presents for wives for their work in the field, deferred health and other pressing household expenses. Farmers also need to store cereals for home consumption during the year and to use small quantities of cereals as cash in exchange at the regular village or town markets.

Getting higher yields is not difficult with moderate levels of inorganic fertilizer, new cultivars responsive to fertilizer and improved agronomy. But it does require farmer training and the availability and purchase of improved seed and fertilizer (for the 2006–07 and 2007–08 crop years, see Abdoulaye et al. (2008) and Baquedano et al. (2009)).

The specific recommendations were initially two 50-kg sacks of NPK (15-15-15) and one sack of Urea (46-0-0) per ha. Improved, fertilizer-responsive cultivars that had been regionally tested were utilized. Agronomic improvements of use of organic fertilizer, water control with ridges or tied ridges, higher densities, thinning and appropriate timing of planting and weeding were recommended. Later the inorganic fertilizer recommendation

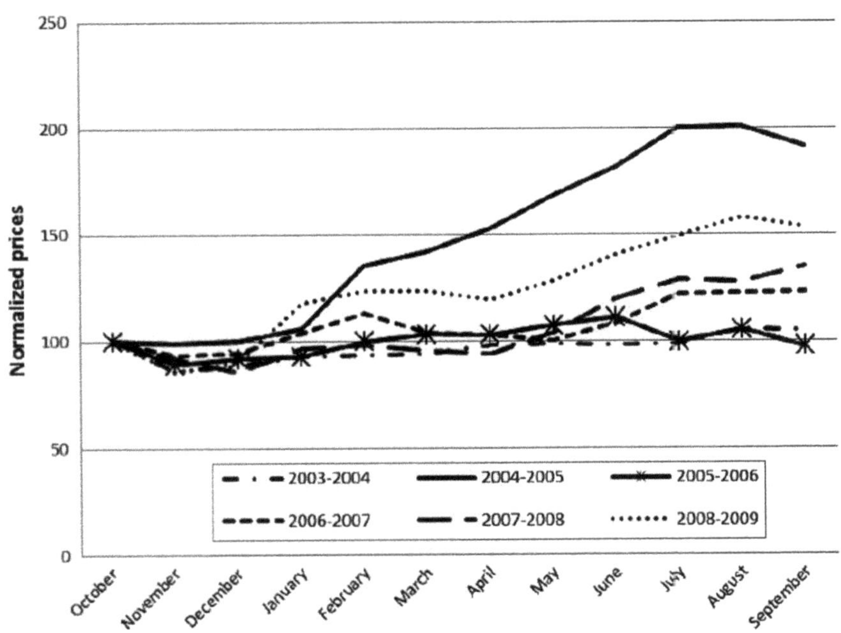

Figure 5 Sorghum prices in Koutiala retail markets, 2003–09. Source: Coulibaly et al., 2015, p. 55.

was changed to one 50-kg sack of DAP (18-48-0) and one sack of urea per ha (manual of agronomic and economic recommendations in Ouendeba et al. (2013)). Comparing the two nutrient levels per ha, the three-sack treatment for above results in 38 kg/ha, 7.5 kg/ha and 7.5 kg/ha of N, P and K, respectively. Whereas the two-sack treatment of higher nutrient DAP combined with urea puts 32 kg/ha of N, 31.5 kg/ha of P and 7.5 kg/ha of K on farmers' fields.

The technology and marketing strategy were followed in farmers' associations organized or adapted for this purpose. Working in farmers' associations also had the benefit of a more rapid diffusion process as farmers learned from other farmers who were successful. The adoption of this system can increase yields by 50–100% (for the 2008–09 and 2010–11 crop years, see Coulibaly (2010) and Coulibaly et al. (2013)).

The critical factors here are that the strategy needs to be profitable and to take risk into account. Focusing first on the risk many have argued for a low input strategy. Clearly, low-income farmers have multiple strategies to reduce risks. They plant on various areas of the topo-sequence (lowlands, slopes, plateau) to reduce the impacts of flooding and drought. They plant various crops and varieties to respond to rainfall variation and insect/disease risks. Family and friend networks are used by making interchanges so that there are contacts in the village and often in the city for help in adverse years. In good years the transfers as of cereals can go to the urban area and in adverse years, grain or money comes from the urban area to the farm.

Nevertheless, many farmers also understand the fundamental risk–return tradeoff: to earn more money it is generally necessary to take more risks. Farmers, like any entrepreneur, need to have the opportunity to make that choice but without putting their basic subsistence at risk. Reducing production risk in this system involves increasing water holding capacity with ridges or tied ridges (this is a technique practised in Africa and Texas of making ridges perpendicular to each other to hold the rainfall).

But there is another important response to risk by increasing returns with marketing innovations. The soils of semi-arid sub-Saharan Africa are notoriously deficient in N and P. Fertilizer is expensive, so many have recommended some types of low input strategies to reduce fertilizer expenditures. However, in dealing with human nutrition, should poor, malnourished people be asked to save money by reducing food expenditures? Similarly, with plants, adequate levels of N and P are essential to increase yields as long as methods can be found to pay for these inputs. This should first lead to a concern with product prices and marketing margins rather than reducing input use.

Working within a farmers' association, the farmer can benefit from product price increases by storage and selling later after the price recovery, from obtaining a price premium by eliminating impurities (dirt, stones, other varieties, etc.) (Toure et al., 2007) and from selling to a higher level of the marketing chain (Abdoulaye and Sanders, 2006; Baquedano et al., 2010; Ndoye et al., 2016).

Farmers can continue to sell some of their sorghum for the required expenditures at harvest and the uses discussed earlier. However, with the higher yields from the new technology, farmers can sell part of the harvest later and to higher levels of the marketing chain.

As a member of a farmers' association with storage facilities, the sale can be put off until the price has recovered or the farmer needs the cash for the inputs for the next production season (Fig. 5). However, over time, the farmers' association must also respond to the processor demand. The processor needs the cereal year around. Assuming that the farmers' association can differentiate their product with a 'clean' sorghum, the buyer will need to pay a premium to buy early.

Food processors and higher levels of the marketing chain want a 'clean cereal' with very low impurities. So the harvesting operations cannot be done on the ground. Moreover, screens are often used at the farm or storage level. There was initially some struggle with processors to get this price premium. Over time, processors have been learning that this is not a zero sum game where if they win, you lose. Rather a long-term relationship is necessary with processors paying for a higher quality product which saves cleaning costs and gives 5–25% more cereal per bag (the levels of impurities from sampling of the cereals at five food processors purchased from eight markets in 2006, see Toure et al. (2007)). Maintaining clean cereal requires more work for farmers and for the farmers' association to guarantee quality. As processors began to recognize their interdependence with suppliers to produce a quality product, the resistance to a price premium in meetings with processors declined and the price premiums they paid began creeping up (field experience in this programme from Senegal and Mali).

In the traditional marketing system (Fig. 6) there are several levels with each earning its own margins. The collector with financing from a regional merchant or wholesaler visits farms at harvest and goes to the village markets later in the year aggregating small quantities. Then the regional merchant sells to the wholesaler either locally or in a major urban area. The wholesaler sells to processors or retailers. Recent additions are the farmers' associations performing the collection and regional merchant functions.

Normally with low yields of poor rainfall years there is little sold even though collectors are offering better prices than in the good rainfall year with abundant supply. The agronomic system with both organic and inorganic fertilizers holds water as well as increases soil fertility thus increasing yields over traditional farmers' practices in poor rainfall years. The yield gains are not as great as in normal and good rainfall years but it is exactly in these poor rainfall years that there is the largest seasonal price increase (Fig. 5).

Besides the timing of the sales the farmers' association also seeks the marketing margins previously received by the lower levels of the marketing chain. A problem frequently occurs as the marketing system resists change. Marketing systems are often 'old boy' networks with personal contacts between levels. So the farmers' associations need to anticipate the potential of cartels between wholesalers or even regionally to avoid changes or to enable the wholesaler to capture the previous margins of the lower levels of the marketing chain, the collectors and the regional merchant. This may require farmers' associations to go outside the regional cartel (arbitrage) as to the urban wholesalers or to processors. More marketing power is possible from the collaboration of farmers' associations to market larger quantities. Often the threat of selling outside the region will be sufficient to obtain a larger share of the marketing margin for farmers' associations.

To summarize, as members in a farmers' association, farmers often have access to storage and can wait for the price recovery. Moreover, by selling a clean, uniform cereal, farmers can demand a higher price. Finally, the farmers' association, by selling larger quantities, can skip stages in the marketing chain by selling directly to either processors or to wholesalers. Just as getting the processors to recognize that a higher price for cereals with reduced impurities did not necessarily mean a higher price for the cereal, the wholesalers and processors can get accustomed to paying higher margins to the farmers' associations rather than to the collectors and regional merchants.

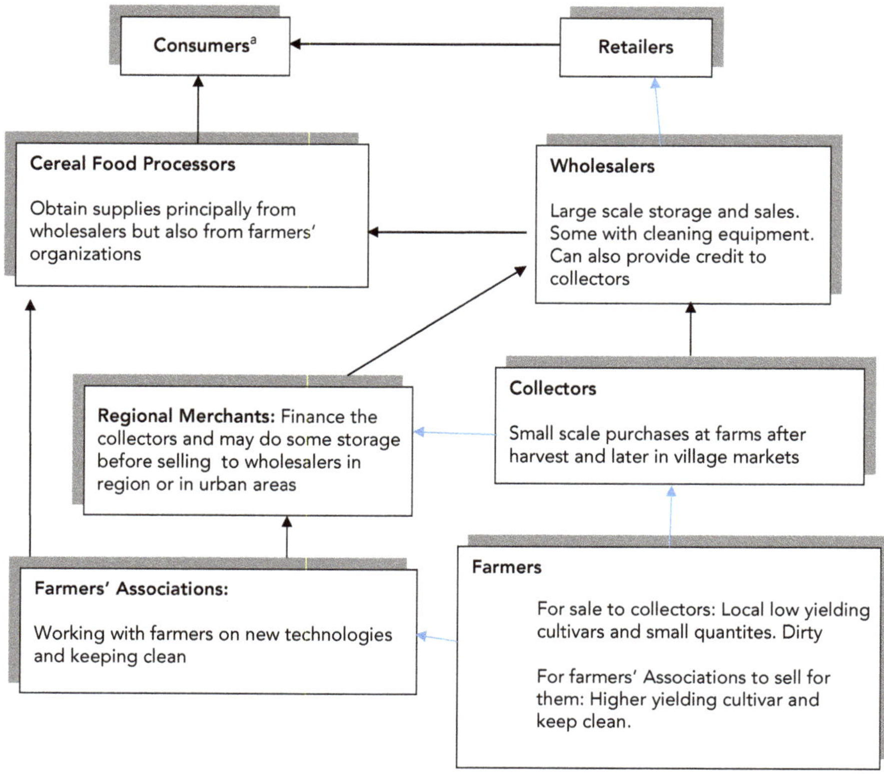

Figure 6 Current sorghum/millet market structure in Mali, 2016. Source: Field work 2015. See Ndoye et al., 2016, p. 103; Jazayeri, 2014. [a] Consumers can buy the unprocessed grain on retail markets and have it dehulled and milled in neighborhood facilities. See the comments in Fig. 9 on changes in these processes.

From different combinations of these activities farmers were able to increase prices by 30–50% (Abdoulaye et al., 2008; Baquedano et al., 2009; Coulibaly, 2010; Coulibaly et al., 2013). This is still considered a poor performance in obtaining price premiums for quality and identifying potential buyers; hence even larger price increases are expected as the negotiating and marketing abilities of the farmers' associations are improved. Moreover, the total revenue of the farmers will increase as they entrust more of their cereals for sale by the association.

4 Pilot project success and scaling-up

In 2007 an especially high-yielding cultivar, Grinkan, was identified (see Fig 7). In 2008 and 2009 average yields from Grinkan in the Garasso, Mali farmers' association were 1.56 t/ha and 1.94 t/ha (our informal field interviews in the farmers' association with 50 members in

Figure 7 Farmer and NGO agent with Grinkan in Koutiala, Mali. 2008.

2008 and 135 members in 2009; Coulibaly reported yields of 1.64 t/ha for 2008; Coulibaly et al., 2015). Sorghum yields in the cotton rotation during this period were 800 kg/ha to 1 t/ha with sorghum often following the highly fertilized cotton in the rotation. This success was the incentive for the USAID-Mali programme to ask one of their agencies (IICEM) specialized in financing and facilitating new technology introduction of minor export crops to undertake the scaling-up of this pilot project.

In the movement from a pilot project to national scale, financing and management become critical constraints. In the pilot there was financing from the project and intensive management (mentoring, financing of inputs, marketing and storage advice). The pilot project financing had to be reimbursed to the farmers' association creating a revolving fund for buying inputs at the start of each crop season. Farmers had to pay for their inputs. Hence, there was no direct subsidy to these programmes. However, there was intensive management for training and monitoring in the pilot project. Since this would not be available as the project scales up to national levels, this was an important indirect subsidy to the pilot project.

As the project moved from experimental to operational phase, the size increased substantially. In the Mopti region[1], the project area increased from approximately 300 ha in 2009 (six farmers' associations) to over 2000 ha with IICEM scaling-up in 2010. With the guarantee of the USAID project by depositing 50% of the loan value in case of default, the national development bank (BNDA) issued loans to the farmers' associations for the fertilizer purchases. The BNDA did not finance the seed purchase but IICEM did facilitate the farmers' associations in obtaining the seed. In the programme in the north (Mopti), there was

1. In the south of Mali, the sorghum variety Grinkan was introduced. In the north (Mopti), the millet cultivar Toroniou was adopted. The rest of the technology and marketing programs were the same. Reporting here on the scaling-up of millet is emphasized as more information was available on this process.

more emphasis on obtaining the variety recommended in the pilot. In the south (Koutiala), this was not the case. However, with regard to IICEM, Grinkan, a soft variety compared with the hard locals had more insect problems in storage. Moreover, the late season rains in 2009 and 2010 made seed quality a problem in the following season given the absence of drying equipment. Both problems could have been resolved if identified earlier. However, obtaining adequate good quality seed will always be a priority in scaling-up rapidly.

With good project management and high repayment rates the guarantee was reduced to one-third the value of the loans in 2011 and was eliminated in 2012. Unfortunately, the programme was disrupted by the war in 2012. When the war in the north forced the banks to close there, USAID-Mali provided financing directly to the farmers' associations in Mopti. USAID-Mali followed the same procedure as that in our pilot project with the farmers required to reimburse the farmers' associations in grain for the credit after the harvest. As of 2016, this financing was still available in the rotating funds of these farmers' organizations. In some regions of the south, the Dutch government provided similar financing with the difference that only half of the funding was required to be repaid.

To give the BNDA more confidence that there were markets for the cereal, IICEM organized contracts between wholesalers and the farmers' associations. The wholesalers pushed for the farmers' associations to sell earlier and to sell more than sufficient to repay the loans. The agreed price was the market price at the time of sale plus a price premium for quality. Since the price varies during the crop season and the price premium for quality was lower than the expected value of the additional cereal available when impurities are removed, the farmers realized the disadvantages of selling either early or more to the wholesaler. Many found ways to avoid both. Nevertheless, the scaling-up was very efficient for furthering the recognition of the importance of the clean cereal, for involving the financial sector and for engaging with higher levels of the marketing chain.

The scaling-up of IICEM also handled well the management problem as the field workers were very active in mentoring the farmers especially stressing the importance of repaying the bank loans. The chronic problem of technology erosion was encountered. As a programme rapidly expands moving to the national level, shortcuts are often taken leading to either lower yields or higher costs or both. This can be due to difficulties of input supply, management time or just the desire of programme administrators to take decisions.

In both the Koutiala region for Grinkan and the Mopti region for Toroniou, the fertilizer recommendations in the scaling-up were frequently not observed and there was insufficient emphasis on seed renewal. As a consequence there were yield declines of 15–20% from the pilot (our field interviews). So the pilot project activity was continued in order to document the importance of getting the input levels right in the scaling-up.

In the second year of the pilot programme in the cotton zone of Burkina Faso, excellent yields for Grinkan were also obtained (Fig. 8) but here the problem of the processing preparation occurred even sooner than in Mali. In the next section, the response to this second-generation problem is explained.

5 Second-generation problems

Since agriculture is a biological system there are continuous changes going on. Often new technology systems will need to respond to these changes. Moreover, breeders or other

Figure 8 Grinkan in Kouakole, Burkina Faso in 2012 with pilot program co-director B. Ouendeba, extension agent and a farmer.

technology developers are not able to anticipate all the possible requirements of the biological system. Hence, it is important for continuing innovation to respond to evolving or unanticipated problems once a successful technology is identified.

This poses a financial problem in developing countries. Generally, the research and extension systems concentrate their expenditures on personnel and depend for operational expenditures on donors and non-governmental organizations (NGOs). So as one project terminates, the national organizations look for new projects and funders and switch their orientation. Who will respond to the emerging technology problems of the previous donor/NGO?

Later there were problems with Grinkan. First, with a softer grain Grinkan was very susceptible to insects in storage. Secondly, women accustomed to the preparation of their 'tô' by soaking overnight did not like the consistency (lack of stickiness) of Grinkan. The insect problem for sale can be handled with PIC (triple plastic sacks which become anaerobic when sealed) sacks or with Phostoxin and polyethylene sacks. The latter is generally preferred for long term seed storage.

The sorghum cultivar Grinkan got off to a fast start in adoption and marketing in Mali until women stated that they did not appreciate the consistency for their principal staple food from it, the 'tô'. The women refused to buy Grinkan in the village markets. This led to the spreading upscale to the larger markets and the demand for Grinkan disappeared. A few farmers continued to produce Grinkan for their own consumption or for their animals.

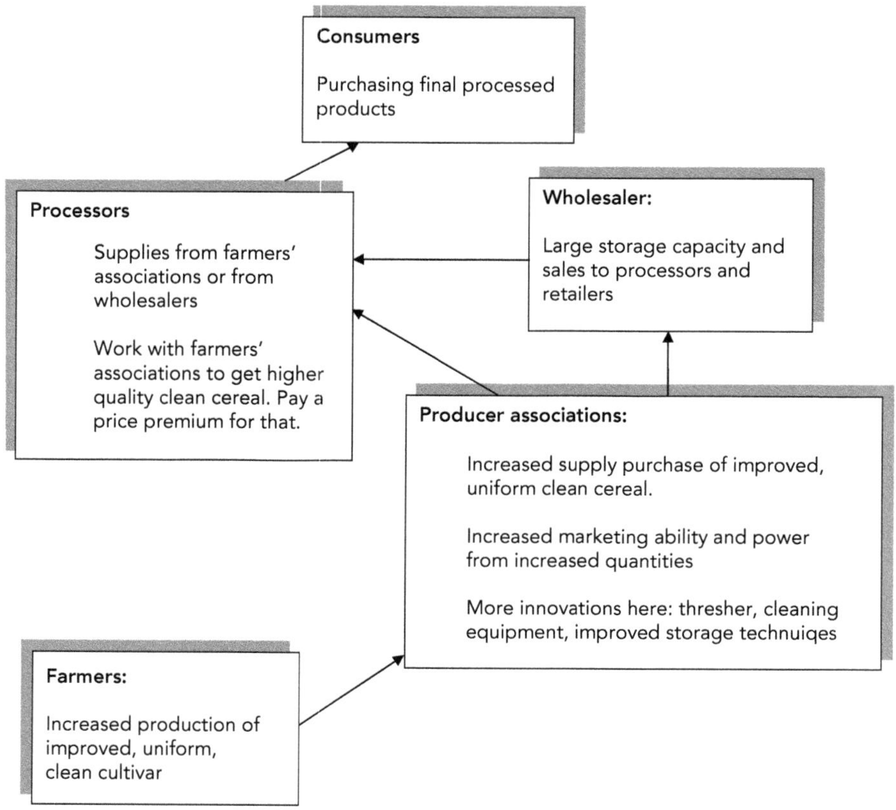

Figure 9 Improved sorghum/millet market structure with strong farmers' associations producing large quantities of clean cereal. Source: Our projections based upon our fieldwork of 2015–16. See Ndoye et al., 2016, p. 104. There is a separate urban retail market where consumers buy the cereal in the market and have it dehulled and milled in local facilities. These facilities often use old equipment leaving metal residues. As costs are reduced in this system and there is more awareness of the health risks of this neighborhood processing, this neighbor milling is expected to decline.

The solution to the processing problem was easy. It was pointed out by a food scientist, Ababacar Ndoye (Sanders et al., 2016) and was discovered independently by some village women. With the soft cultivars such as Grinkan, women needed to avoid the overnight soaking. After adopting this procedure village women reported that they appreciated Grinkan more than their local cultivars (informal conversations in the villages making this adjustment).

Besides the second-generation problems, technology erosion in scaling-up is a continuing problem. So some independent financing to support research and extension over time is important to maintain technology introduction momentum. The dependence for operational funding on outside agencies with different priorities and programmes makes this investment response difficult.

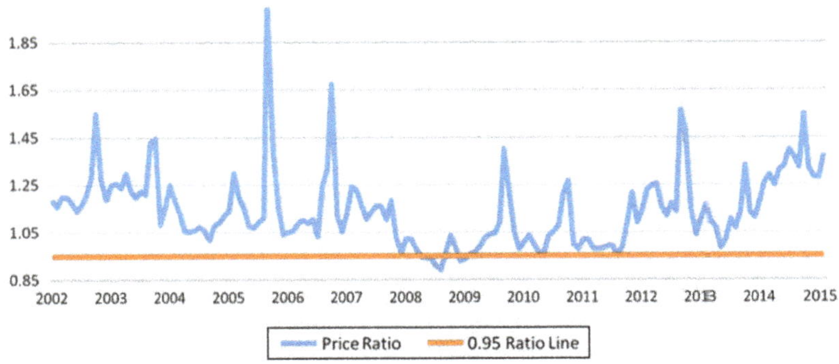

Figure 10 Price ratios of sorghum to maize in Mali, 2002–15. Source: OMA (*Observatoire du Marché Agricole*), 2016. Monthly data. Cited in Sanders et al., 2016, p. 217. Note that it would have been preferable to use wholesale prices as most of the chicken producers bought from wholesalers rather than these retail prices.

Even though sorghum has an advantage over maize in its drought resistance, the deficit and variability of rainfall are continuing concerns in the sorghum-producing regions. However, over this period, excess rainfall causing flooding or occurring in later stages of the crop cycle had more significant negative effects than the rainfall deficits. The seed quality problem from late rainfall in 2010 and 2011 in the absence of drying equipment put an abrupt halt to the rapid expansion of Grinkan in Garasso, where the dramatic yield and production increases of the pilot project had convinced USAID to finance the scaling-up in 2010 (Coulibaly et al., 2015).

To summarize, when the project was scaled up beginning in 2010, area covered was 2000–3000 ha in both regions. The scaling-up programme worked well in the north (Mopti region) where pilot project recommendations were well observed. Even during the war in the north when the banks were closed down, the project was continued with USAID in the north and Dutch grants in the south to finance the revolving funds for the fertilizer purchases of the farmers' associations.

As discussed earlier IICEM had problems in obtaining sufficient quantities of quality Grinkan seed and in observing the fertilizer recommendations of the pilot. But in 2016 Grinkan was still being produced in the region following the pilot recommendations in those villages where the farmers' associations had modified the processing techniques. Another 17 farmers' associations were shown in 2016 how to do this processing and given flash drives of the villages already modifying their processing operations

6 Institutional innovation: farmers' associations as a marketing cooperative

The initial involvement with the farmers' associations was to accelerate diffusion. Farmers tend to follow other farmers and do so at different speeds. They discount outside 'expert' recommendations or even demonstration trials. So, the strategy was to convince approximately 50 farmers in a village association, either new or presently oriented to another crop, to follow the technology recommendations. These recommendations were based upon experience and interactions with the national research organizations over two decades. Credit was provided for the fertilizer to be reimbursed to the farmers' organization. Not all the farmers were expected to follow the recommendations well the first year. But they were expected to follow the best farmers adopting the practices after the first year.

As long as the farmers were happy with the new cultivar this strategy worked very well. In the second year, 50 other farmers were recruited to the programme. The second group would get credit from the programme, while the first group would use the revolving fund. With higher prices being received, the revolving fund tended to increase so they could expand participants or use this for other purposes after providing the fertilizer credits to the first group. A similar process recruiting 50 new members in the third year and new credits was followed and then the programme for this village was ended. This was considered the ideal size for the village and programme. A revolving fund, storage facilities and management experience with both the technology and the marketing would have been built up by then.

The most important function for the farmers' association was to provide the infrastructure, to assure larger quantities marketed and inputs purchased and provide management for the marketing activities. The association needed storage facilities to enable farmers to sell later especially in poor rainfall years when there were the largest seasonal price increases. The farmers' association needed good management so that farmers trusted them to reimburse the farmers for the marketing gains. Also as the 'clean cereal' and larger quantity sales were components of the marketing strategy, a central control of quality and a management ability to identify and contract the best buyers became important. Larger quantities marketed meant that the farmers' associations could deal with higher levels of the marketing chain, the wholesalers or the processors. Moreover, as the number of fertilizers and seed distributors increases, the farmers' associations should be able to get price discounts for quantity purchases.

The marketing chain in the Koutiala region where Grinkan was concentrated is already moving to a new market chain system (Fig. 9). Here regional wholesalers are buying directly from farmers' associations, while there continues to be smaller purchases later in the year in villages on market days. In the future the functions of the collectors and regional merchants will be largely eliminated as the farmers' associations expand storage capacity, improve quality control and develop the ability to identify the best buyers. Then the farmers' associations will become larger and even buy clean cereal from non-members.

Some farmers' associations were selling directly to processors but they will need to guarantee quality to continue that. In one village farmers were doing further processing (flour), which could enable them to deal directly with retailers. Another future role, as management experience is increased, is to use the bank accounts of the farmers' associations, in which the revolving funds are held as leverage to obtain bank lending for other activities of the farmers' organization such as holding the cereal until the 'soudure' or hungry period (the months immediately before the availability of the next harvest).

7 Growth of a secondary market for sorghum

As incomes increase, people consume less grains and tubers and more fruits, vegetables, meat, milk and cheese. One of the principal beneficiaries of these consumption shifts is the intensive production of chicken. Substantial increases in production and consumption are experienced over two to three decades as producers get more efficient at reducing the cost of chicken relative to other meats. So there will be a rapid increase in the demand for feed grains over the next decade. Cereals represent 50–60% of the poultry feed by quantity depending upon the growth stage (Clement et al., 2010; Dowling et al., 2002; Tandiang et al., 2014).

Over the past two decades there have been very large increases in production and productivity (yields) of maize enabling a rapid growth of the poultry industry in both Burkina Faso and Mali. In both countries, the maize production is concentrated in the higher rainfall cotton zone, with 80% of the production in Mali (Mali Best Report, 2015). As the demand for feed increases for the intensive production of poultry, sorghum will have a comparative advantage over maize in the lower and more variable rainfall zones which are north of the cotton zones.

Most poultry producers and their veterinarians or other advisers for feed composition were not aware that sorghums without tannin were available and were approximately equal in feed efficiency to maize (field interviews; Sanders et al., 2016). Hancock has estimated that non-tannin sorghum has 95–97% of the feed efficiency of maize (Hancock conversation, autumn 2015; for more details on comparative feeding of maize or sorghum, see Hancock (2000), Clement et al. (2010), Carmencita et al. (2006), Dowling et al. (2002) and Tandliang et al. (2014)). This implies that with a relative price of sorghum to maize of 0.95, the chicken producers should be indifferent between maize and sorghum and would prefer sorghum at a relative price below 0.95.

So what do the relative prices show? In mid-2008 (Fig. 10) with the sudden increase in maize for biofuel in the United States and the world spike in agricultural prices, there was a brief period in which chicken producers had an incentive to switch. The chicken producers or their veterinarians responsible for the feed composition have to know about the potential of sorghum to substitute, know the relative prices and even more importantly know where to obtain uniform, non-tannin sorghum. Generally, the feed advisers for the poultry producers were not aware of the potential of non-tannin sorghum (field interviews in Mali and Burkina Faso, 2015 and 2016).

Since 2009 the more rapid gains in productivity of maize as compared with sorghum (Figs. 1 and 2) is an explanation for the decline of the substitutability of sorghum with the jump in the relative prices. Also during some of these years there were prohibitions on the export of maize. To the extent these were effective, they reduced the price of maize.

On the high sorghum prices, there are two rainfall effects. In poor rainfall years, millet suffers more on the light soils and then sorghum is needed to substitute for food. Then, millet and sorghum prices increase. In high rainfall years or even periods, sorghum suffers more as it is found on lower lying, heavier soils than millet; hence sorghum supplies are reduced and prices can increase.

In the future (next five years) there will be a definite advantage to sorghum over maize as larger quantities of the cereals are needed for feed and more will need to come from north of the cotton zone (lower rainfall regions). Moreover, sorghum is grown widely in

Mali and hence will not need to match the yield levels of maize for the relative price to decrease. Moderate yield increases of 0.5–1 t/ha over a wide area are expected to be sufficient (see Sanders et al. (2016)). Also there is the human and poultry health factor of maize being more susceptible to fungi in the field than sorghum. This includes the fungus-producing aflatoxin and this could explain the occurrence of sudden chicken deaths reported in field interviews with poultry producers in both Burkina and Mali (Sanders et al., 2016). Both sorghum and maize can experience fungi problems in storage and transportation.

Prospects are good for the increased utilization of sorghum in the poultry ration for two reasons. First, there is a larger yield gap for sorghum than for maize between the experiment station and the present national yields. Secondly, there are many competitive activities in the cotton zone and Bt cotton will enable a comeback of cotton in Mali as it has in Burkina Faso (Baffles, 2011; Coulibaly et al., 2015). This will at least partially reverse the shift to maize that has occurred with falling cotton prices (Coulibaly et al., 2015). Consequently, in the next decade, cereal productivity for feed which is north of the cotton zone will become increasingly important to the poultry industry as well as improving income distribution in the Sahel. This movement north of the increasing cereal supply for poultry feed would favour sorghum over maize.

8 Conclusion

There is substantial potential for yield increases of sorghum in Mali and Burkina Faso. Sorghum is very important not only as a basic staple food, especially when millet yields are poor, but also to support the rapidly expanding poultry industry. The continuing demand growth for feed will require the expansion of cereal production especially outside of the cotton zones.

Increasing cereal yields is well known as resulting first from soil fertility amendments and then adding in varieties (or hybrids) that are fertilizer-responsive, and finally accompanying agronomic inputs. So why has this not been done with sorghum and millet? Unfortunately, low-input concepts have often focused attention on reducing the input costs of these 'subsistence' crops. A more appropriate strategy is to pay for these necessary inputs to increase soil fertility with marketing changes to increase product prices received by farmers (as well as with the economic gains to the new technologies).

With the introduction or expansion of existing farmers' organizations, farmers can benefit from storage and price recovery after the harvest season's price collapse; obtain a price premium from the value added of selling a 'clean', uniform cereal and deal directly with higher levels of the marketing chain. Selling directly to wholesalers gives the farmers' organization the potential to capture the marketing margins of collectors and regional merchants. Moreover, a farmers' organization can achieve more market power by dealing in larger quantities with both the cereal sales and the purchase of inputs.

Since the 1970s, there have been substantial investments in the agricultural research systems of developing countries. These national agricultural research institutions have become increasingly competent at identifying, adapting and developing new cultivars and associated technologies. However, there has often been disappointment about the pace at which these technologies are incorporated by farmers and adopted over large areas. Generally, the research stations argue that this diffusion is the responsibility of the extension service once on-farm or regional trials have been undertaken.

There has not been as much investment in the training and functioning of the national extension services and they have often not fully perceived what the researchers were trying to do and opted to concentrate on encouraging best farmer diffusion rather than accelerating new technology adoption of innovations from the experiment station. Moreover, both national research and extension systems frequently suffer from insufficient operating budgets.

With their budgets focused on paying personnel, both the research and extension, institutions then become dependent upon outside donors and NGOs for operating funds. These NGOs and donors have shifting objectives over time. Since agriculture is a continually evolving biological system, new problems evolve over time or were there already but not identified by the technology developers. Resources need to be found to maintain the gains to innovation as these constraints are identified or new constraints emerge.

Moreover, the scientific and applied issues in moving from the station to the farm and then to the consumer tend to fall between disciplines so they become neglected. This is especially serious for those commodities, which according to conventional wisdom are 'subsistence' crops. Marketing systems and adaptation to consumer requirements are as important to facilitate introduction as the technology practices resulting in the need for the agricultural scientists, food scientists and marketing economists to collaborate.

9 Where to look for further information

On the introduction of hybrid maize and sorghum in the U.S., see Duvick, 1987 and Miller and Kebede, 1987 respectively.

For more information on the project and policy implications and after the data of 2008–2012 reported in section 10, see Abdoulaye and Sanders, 2006; Abdoulaye et al., 2006, 2007, 2008; Ibrahim, 2012, 2014.

For a summary article incorporating sustainability concerns into research and development activities, see Lynam and Herdit, 1989.

10 Appendix: associations and farmer details, 2008–12

Here is a partial listing of associations and estimates of farmers' participation in the pilot projects (Tables A1 and A2). The associations were pressed, not always successfully, to have one farm family per ha so the approximate number of farmers can be easily calculated from the area data. More complete data on area for the period 2005–16 is available by contacting the principal author.

Table A1 Sorghum area (ha) in the program, Mali 2008–12

Project sites	Regions (Segou and Koutiala)	Improved sorghum varieties	2008	2009	2010	2011	2012	Total
Kafara	Koulikoro	Wassa, Nieta	50	50	na	na	na	100
Diola	Koulikoro	Soumba, Niatichama, Grinkan	76	96	na	na	na	172
Garasso, Koutiala	Sikasso	Grinkan	50	150	117	na	na	317
Kaniko, Koutiala	Sikasso	Niatichama, Grinkan	58.5	50	na	na	na	108.5
Zanzoni, Koutiala	Sikasso	Niatichama[b]	–	50	na	na	na	50
Finkoloni, Koutiala	Sikasso	Niatichama	50	na	na	na	na	50
Seed Production Koutiala	Sikasso	Grinkan				20		20
Extension, Koutiala	Sikasso	Grinkan	–	–	–	–	530[c]	530
Kolokani	Kouliokoro	Seguifa	50	110	100	100	100	460
Diankounte Camara	Kayes	Seguifa	–	50	75	na	na	125
Katiena	Segou	Seguifa	–	50	100	100[a]	100[a]	350
Beleco	Koulikoro	Tiandougou	–	–	–	60[a]	60[a]	120
Kita	Kayes	Tiandougou coura	–	–	–	60[a]	60[a]	120
Sadiola	Kayes	Tiandougou	–	–	50	na	na	50
		Total	334.5	606	442	340	850	2572.5

Source: Adapted from Diourte, 2012. More recent data available from the authors.
n.a.: Not available as either the village had graduated or for some reason as accessibility the project did not revisit.
[a] Same technology and marketing program but managed by IER.
[b] After promising to sell Grinkan to the neighboring villages one village sold the taller and inferior Nachtichama.
[c] This is the number of farmers in the region receiving the program we were concentrating on when AIID took over the technology and marketing activity. The modifications AIID made are discussed in the text.

Table A2 Millet area in the program, 2008–12

Communes (Mopti region)	Villages	Number of storage units supported	Areas (ha)					Total
			2008	2009	2010	2011	2012	
Petaka	Oualo	2		60	120	180	180	540
Kanibonzon	Kanikombolé	1		–	60	80	20	160
	Sadia	1		–		60	120	180
Koro	Téré	1		–	60	80	120	260
	Pomorododiou-Bégné	1		–		60	120	180
Koporo-Nah	Kountogoro	2		–	60	120	180	360
Pel	Témégolo	1		–	–	60	120	180
Petaka	Pissa	0		60	60	na	na	120
	Mougui	0		–	60	na	na	60
Segou	Tingoni[b]	1	120	150	120	120	120	630

Communes (Mopti region)	Villages	Number of storage units supported	Areas (ha)					
			2008	2009	2010	2011	2012	Total
Segou-Global 2000[a]					494[b]	500[b]	na	999
Total		10	120	270	1034	1260	980	3664

Source: Adapted from Diourte (2012).

n.a.: Not available as either the village had graduated or for some reason as accessibility the project did not return.

[a] Includes the program accelerating our program introduction in the same region and implemented by the AID funded IICEM (Economic Growth unit within USAID program). Note the guerrilla invasion of the north in 2012 and the comments in the text on how USAID continued program activities in that year with loans to the farmers for fertilizer to be repaid to the farmers' associations at harvest in grain to set up a rotating fund for purchasing fertilizer in subsequent crop years.

[b] Tingoni is the name of the millet variety used in all the villages. This was used here to indicate the regional effect as IICEM reported 494 and 500 ha in this cultivar in 2010 and 2011 before the guerrilla attack in 2012. The column above was the project activity, which continued with the AID support.

11 References

Abdoulaye, T. and J. H. Sanders, 2006. 'New technologies, marketing strategies and public food policies for traditional food crops: Millet in Niger', *Agricultural Systems*, 90: 279–92.

Abdoulaye, T., J. Sanders and B. Ouendeba, 2006. Quelle Cereale par les Aliments de Volaille en Afrique de l'Oest: Sorgho ou Mais? INTSORMIL Bulletin No. 4, Projet Marketing-Processing, Department of Agricultural Economics, Purdue University, West Lafayette, IN.

Abdoulaye, T., J. Sanders and F. Botorou Ouendedeba 2007. *Revenues des Producteurs Effects de Technologies et Strategies de Marketing, Projet de Marketing – Transformation: Campagne agricole 2005–2006*. INTSORMIL – Bulletin No. 5, Department of Agricultural Economics, Purdue University, West Lafayette, IN. Program results 2005–06.

Abdoulaye, T., J. H. Sanders and B. Ouendeba, 2008. *Evaluation of Millet and Sorghum Technology and Market Strategy Introduction: 2006–07 Crop Year*. INTSORMIL No. 8, Lincoln, NE, 22p. Program results 2006–07.

ADA (Directorate of National Agriculture), 2015. Government of Mali, unpublished data. Bamako, Mali.

Baffles, J., 2011. *Cotton, Biotechnology and Economic Development*. Policy Research Working Paper 5896, Development Prospects Group, The World Bank, Washington, DC, 28p. mimeo.

Baquedano, F., M. Diarra and A. Ahmoudou, 2009. *Economic Analysis of the 2007–08 Crop Year*. INTSORMIL Bulletin No. 9, Purdue University, Department of Agricultural Economics, 34p. Program results 2007–08.

Baquedano, F., J. H. Sanders and J. Vitale, 2010. Increasing incomes of Malian farmers: Is elimination of U.S. subsidies the only solution? *Agricultural Economics*, 103: 418–32.

Carmencita, D. M. and F. C. Nelia, 2006. 'Feeding and economic evaluation of corn, wheat and sorghum based diets in broilers', *Philippine Journal of Science*, 135(1): 49–58.

Clement, I. M., I. D. Kwari, J. Igwebuike, I. Nkama, I. D. Mohammad and B. Hamaker, 2010. 'Performance and Economics of Production of Broiler Chickens fed Sorghum or Millet as Replacement for Maize in the semi-arid zone of Nigeria', *Agricultural Biology Journal of North America*, 1(3): 321–5.

Coulibaly, J., 2010. *Evaluation des Technologie de Production et de Commercialisation de Sorgho et du Mil dans le Cadre du Projet IER-INTSORMIL/Mali, Campagne Agricola 2008–09*. Bulletin No. 10, Purdue University, Department of Agricultural Economics, 36p. Program results 2008–09.

Coulibaly, J., G. Kamaswarmy, and J. H. Sanders, 2013. *Economic Impact of Sorghum and Millet Technologies in Mali, Agricultural Campaign 2010–11*, Bulletin IER-INTSORMIL No. 11, Purdue University, Department of Agricultural Economics, Mali, 31p. Program results 2010–11.

Coulibaly, J., J. H. Sanders, P. V. Preckel and T. G. Baker, 2015. 'Will Cotton Make a Comeback in Mali?',*Agricultural Economics*, 46: 53–67.

Département d'Agriculture, 2014. Government of Burkina Faso, *Annuaire des Statistiques Agricoles*, Ouagadougou, Burkina Faso.

Diourte, M., 2012. *Transfer of Sorghum, Millet Production, Processing and Marketing Technologies in Mali*. Report for INTSORMIL, IER, Bamako, Mali, 10p.

Dowling, L. F., C. Arndt and B. R. Hamaker, 2002. 'Economic viability of high digestibility sorghum as feed for market broilers',*Agronomy Journal*, 94: 1050–8.

Duvick, D. N., 1987 (reprinted of 1984 edition).'Genetic contributions to yield gains of US hybrid maize, 1930–1980'. In: D. N. Duvick (Ed.), *Genetic Contributions to Yield Gains of Five Major Crop Plants*. CSSA Special Publication No. 7, Crop Science Society of America and American Society of Agronomy, Madison, Wisconsin, pp. 15–47.

Hancock, J. D., 2000. 'Value of sorghum and sorghum co-products in diets for livestock'. In: W. Smith and R. A. Frederickson (Eds), *Sorghum, Origin, History, Technology and Production, Wiley Series in Crop Science*, New Jersey, pp. 731–51.

Ibrahim, A., 2012. *The Economic Impact of New Sorghum and Millet Technology Adoption in Niger: Performance and Challenges, 2010–2011 Crop Year*, INTSORMIL Bullet No. 12, Department of Agricultural Economics, Purdue University, West Lafayette, IN. Program results 2010–11. Niger.

Ibrahim, A., 2014. *The Economics of an Extension Program in Maradi, Niger*. unpublished PhD thesis, Department of Agricultural Economics, Purdue University, West Lafayette, IN Program results in 2012–13. Niger.

Jazayeri, A., 2014. *Value Chain Analysis (VCA) in Mali: Millet, Sorghum and Rice*. USAID-Mali, Bamako, Mali, 36p.

Lynam, J. K. and R. W. Herdt, 1989, 'Sense and sustainability: Sustainability as an objectvie in international agricultural research', *Agricultural Economics* (3):381–98.

Mali Best Project, 2015. *Bellman Estimation Studies for Title II, USAID-Best Analysis, Mali*, Fintrac INC., Washington, DC.

Miller, F. R. and Y. Kebede, 1987. 'Genetic Contributions to Yield Gains in Sorghum, 1950–1980', In: D. N. Duvick (Ed.), *Genetic Contributions to Yield Gains of Five Major Crop Plants*. CSSA Special Publication No. 7, Crop Science Society of America and American Society of Agronomy, Madison, Wisconsin, pp. 1–14.

Ndoye, A., B. Ouendeba and J. H. Sanders, 2016. 'Demand for millet processed products in Mali and Burkina Faso'. In: J. H. Sanders , B. Ouendeba, A. Ndoye, S. Traore and N. Teme. 2016. *Development of Sorghum and Millet Production Systems in Mali and Burkina Faso*, pp. 92–109.

OMA (*Observatoire du Marché Agricole*), 2016. Unpublished data. Bamako, Mali.

Ouendeba, B., T. Niaba and J. H. Sanders, 2013. *Fiche de Production et de Commercialisation de Sorgho Grinkan, Production-Marketing Project of INTSORMIL*. Department of Agricultural Economics, Purdue University, West Lafayette, IN 47907, 8p.

Sanders, J. H., B. Ouendeba, A. Ndoye, S. Traore and N. Teme, 2016. *Development of Sorghum and Millet Production Systems in Mali and Burkina Faso*. Report to the Gates Foundation, Department of Agricultural Economics, Purdue University, West Lafayette, IN 47907, 138p. Program results 2015–16. Mali and Burkina Faso.

Tandiang, D. M., M. T. Diop, A. Dieng, G. M.L. Yoda, N. Cisse and M. Nassim, 2014. 'Effect of corn substitution by sorghum grain with low tannin content on broiler production: Animal performance, nutrient digestibility, and carcass characteristics',*International Journal of Poultry Science*, 13(10): 568–74.

Toure, A., T. Abdoulaye, J. H. Sanders and B. Ouendeba, 2007. *Transformation Commerciale du Mil et du Sorgho au Mali*. INTSORMIL Bulletin No. 7, Department of Agricultural Economics, Purdue University, West Lafayette, IN 38p.

Chapter 7

Improving sorghum cultivation in South America

Rafael Augusto da Costa Parrella, Robert Eugene Schaffert, Cicero Bezerra de Menezes, José Avelino Santos Rodrigues, Jurandir Vieira Magalhães, Cynthia Maria Borges Damasceno, Dagma Dionísia da Silva and Simone Martins Mendes, Embrapa Milho e Sorgo, Brazil

1 Introduction

Agribusiness is a strategic sector for the Brazilian economy, representing 23% of the GDP (CEPEA 2015). Beef cattle, poultry and swine production are predominant Brazilian agribusiness activities on the world stage. Brazil has the second largest cattle herd in the world with about 211 million head. In addition, since 2004, Brazil has taken the lead in meat exports, supplying 20% of internationally traded meat, and sales in over 180 countries. The Brazilian poultry industry has become the world's third largest producer and leading exporter. The swine industry in Brazil is ranked fourth in world production and exportation of pork (MAPA 2015).

In this scenario, sorghum has experienced a significant expansion in recent years, reaching in 2014/2015 an area planted close to a million and a half hectares, most of

http://dx.doi.org/10.19103/AS.2017.0015.26

which consists of grain sorghum, followed by forage production. The sorghum crop has been an excellent option to produce grain to feed cattle, poultry and swine, as well as forage for cattle, especially in situations of drought and conditions of low soil fertility, which offer higher risks for other crops, especially maize. From the point of view of the market, grain sorghum cultivation in succession to summer crops has contributed to the sustainable supply of good-quality feed grains, and low cost, both for farmers and for the ration industry. Currently, grain sorghum grown in Central Brazil has liquidity for the farmer and great comparative advantage for the industry, which increasingly seeks alternatives to compose their rations with quality and lower cost.

The agro-ecological biome of Central Brazil, principal region of sorghum production in Brazil, is principally Cerrado. The soils of the Cerrado region are typically associated with low pH, aluminium (Al) toxicity, low levels of available phosphorus (P) and high capacity of P fixation. The high levels of toxic Al in the subsoils act as a barrier to root growth and development and can reduce crop production (Schaffert et al. 2001). The use of lime in the surface layer (0–20 cm) solves the issue of low pH, but in the subsurface layer (20–40 cm) the problem continues and limits root growth and nutrient absorption, leaving the plants more susceptible to drought stress. The availability of phosphorus (P) in the soil is another limiting factor for agricultural productivity in these tropical soils, since this nutrient has the lowest efficiency of use by plants (Baligar et al. 1997). Phosphorus fertilization is frequently carried out by applying a high dose of soluble phosphate, and most of the fertilizer added becomes unavailable due to the high phosphorus-fixing capacity of the soil clay particles (Vance and Chiou 2011), significantly increasing the costs of production. In addition, in the coming decades, there are predictions that climate change will be so intense as to change the geography of world agricultural production.

The theme 'Global Warming' is in the main forums of global debate. There is increasing evidence that if temperature increases occur and that these, however small, may bring drastic consequences on the distribution of crops around the world. Areas that are currently major grain producers may no longer be productive before the end of the century (Hansen et al. 2012). Even though grain sorghum is grown primarily as a second crop after soybeans, with post-flowering water stress, the development of more adapted cultivars will be necessary. Thus, sorghum improvement programmes focusing on tropical soil regions will focus on developing cultivars that are tolerant to aluminium toxicity, efficient in the acquisition of P, as well as more water use efficient and drought tolerant to ensure the sustainability and competitiveness of Brazilian agribusiness.

Another trait of great importance in tropical agriculture is the resistance to most sorghum diseases. The main diseases are anthracnose (*Colletotrichum sublineolum*), rust (*Puccinia purpurea*), helmintosporium (*Exserohilum turcicum*) and downy mildew (*Peronosclerospora sorghi*). Among the pathogens of major importance for the culture of sorghum in Brazil is the fungus *C. sublineolum*, causal agent of anthracnose. Reductions of over 80% in grain production have been observed in susceptible cultivars, in years and places favourable to the development and spread of the disease (Panizzi and Fernandes 1997).

The yield potential of sorghum is also reduced by damage caused by several species of pests (Waquil et al. 1986). Thus, the knowledge of these and their management is paramount to obtain high yields with quality. Among the insects of economic importance are the lesser cornstalk borer (*Elasmopalpus lignosellus*) causing the death of seedlings, the fall armyworm, the lesser cornstock bore (*Spodoptera frugiperda*), greenbug (*Schizaphis graminum*) damaging the leaves, and the sugarcane borer (*Diatraea saccharalis*) attacking the interior of the stem.

2 History of sorghum in Brazil

Modern sorghum (*Sorghum bicolor* L. Moench) is a product of human intervention, which changed the plant for various uses. Sorghum is an extraordinary protein, energy and biomass factory, and is widely used in regions where there are no good grain yields or forage production from other species, such as maize. Sorghum arrived in Brazil at the same time as it was introduced to North and Central America, through the African slaves. Names like 'Corn d'Angola' or 'Guinea Corn', found in the literature and vocabulary of the Brazilian Northeast today, indicate that the first seeds brought to Brazil probably arrived in the Northeast.

From the second decade of the twentieth century until the late 1960s, sorghum was reintroduced to Brazil in an orderly manner by public universities and research institutes. Records of research on sorghum from that period can be found at the Agronomic Institute of Campinas, the Pernambuco Agricultural Research Institute, the Agricultural Research Institute of Rio Grande do Sul; at some Agronomy Schools like ESALQ Piracicaba, College of Agriculture of Lavras, Agronomy School of Viçosa, Agronomy School of Pernambuco; and the Federal University of Ceará. Collections were introduced from Africa and the United States and resulted in forage cultivars whose names today are still remembered by the producers, especially the forage varieties Santa Elisa, Lavrense, Fartura, Atlas and Sart.

In São Paulo, the sorghum crop for animal feed received great impulse from the Agronomic Institute of Campinas in the 1940s, with the implementation of the 'Plan of pilot farms for improving milk production', by the Department of Animal Production of the Secretary of Agriculture. The institute released and distributed Santa Elisa variety seeds, a photosensitive cultivar with a long cycle (160–70 days), high productivity, juicy stem and low grain production. The greater similarity of this variety with elephant grass than with maize, the presence of dhurrin in early stages of growth and the popularity of farmers restricting sorghum to marginal areas contributed to its low acceptance and use by farmers.

In the Northeast of Brazil research on sorghum began in 1957, comparing it with maize at various locations in the state of Pernambuco. In Bahia, it started in 1963; in Ceará, in 1964. In 1966, studies on sorghum began in Alagoas, and in 1968, in Rio Grande do Norte, according to Lira and Faris (1977).

The production and distribution system of improved cultivars began in the early 1970s when the Brazilian Agricultural Research Corporation (Embrapa) was initiated, specifically the National Research Center for Maize and Sorghum, in Sete Lagoas, Minas Gerais (currently Embrapa Milho e Sorgo), and when the private sector entered into the agribusiness of sorghum cultivar development and seed production. During this period grain sorghum hybrids with short height, and forage hybrids, adapted to produce quality silage, arrived in Brazil across the southern border of the state of Rio Grande do Sul from other South American countries, and with the initiation of sorghum improvement by several Brazilian companies.

The sorghum crop was established in Brazil, initially, based on information from results of research programmes developed abroad, and more recently during the last two decades knowledge was adapted to our conditions. The initial introduction of hybrids in the early 1970s contributed significantly to the expansion of sorghum in the South, Southeast and Central regions of Brazil, but there was very little prospect of increasing production potential with imported hybrids. The production and supply of sorghum seed in Brazil was incipient before the 1980s. The research actions with the crop, mainly for the development of cultivars adapted to the conditions of Brazil, began in 1972 at the IPEACO/DNPEA of the Ministry of Agriculture, and continued at Embrapa Milho e Sorgo in 1975. The initial research objectives were to identify production constraints and the use and marketing of sorghum grain, with the close collaboration of the public and private sectors, especially the seed industry. In the 1980s, there was a redirection of sorghum research strategies, intending to change the Brazilian agricultural boundaries.

This phase had important partnerships, especially with INTSORMIL (USA), with the involvement of American universities and international institutes. This international collaboration accelerated the advancement of both public (Embrapa) and private sector sorghum improvement programmes in Brazil. They allowed greater scope of actions and the transfer of results obtained in the different institutions involved, with direct effects on the Brazilian seed market, with a tendency for demand for specific cultivars or cultivars targeted to specific agro-ecological conditions. This interaction has contributed significantly to the increase of grain and forage production and thus competitiveness and income generation in Brazilian agribusiness.

The contributions from the public and private sectors in providing cultivars with high production potential adapted to the conditions of Brazil enabled significant increases in national average productivity, reflected in increasing feed supply (grain and forage), competitiveness and income generation, being a successful case of public x private partnerships. Sorghum acreage, both grain and forage, has had an extraordinary jump from the early 1990s. Currently, the Central West is the main grain sorghum-growing region, while Minas Gerais, Rio Grande do Sul, São Paulo and Goiás are the states that are leading the area of forage sorghum cultivation for silage. The cultivation system of forage sorghum is exclusive in the summer-autumn period and most of the planted area is for producing silage. Grazing sorghum has its main market in the South of Brazil, mainly in the state of Rio Grande do Sul. In recent years, the sorghum area for grazing and/or green chop has significantly increased in the Southeast region and in the Midwest of Brazil. There has been a significant increase in the planting of this type of sorghum for stover production for no-till planting. Modern grain, forage and grazing cultivars have adapted very well to integrated agriculture and livestock systems. Sorghum has assumed an increasingly strategic role in the consolidation of a maize export policy, either in direct or aggregate form as eggs, meat and milk.

The demand for forage sorghum continues to expand for both dairy and beef production. The confinement system of beef production implanted in Brazil in recent decades and the technical level and expansion of dairy production have created a demand for roughage that exceeds the potential for fresh forage, and the production of silage is essential. The adoption and use of modern forage sorghum hybrids has contributed significantly to reduction of the problem caused by seasonality of forage production.

The advancement of modern agriculture in the Cerrado and its various production systems continue expanding the possibilities for different types of sorghum. Soybean, the main partner in the succession system of crops, continues to advance in the states of the Central West, North and Northeast regions of Brazil, with sorghum following as an alternative of the second crop. The no-till planting system and the increasing use of the integrated agriculture livestock production system are creating a demand for both biomass and grain because of its excellent production potential. The expansion of meat agribusiness also increases the search for lower cost raw materials for feeding poultry, swine and cattle. Milk and beef cattle businesses are increasingly professionalizing as consumer markets demand more quality and competitive price. Maize, the main component in animal feed in Brazil is an important component in 'packaged' exports in the form of meat (poultry, pork and beef). Sorghum has been emerging as the main alternative to maize in the confection of rations.

The national sorghum improvement programme of Embrapa in Brazil began in 1972 with the purpose of developing cultivars (mostly hybrids) for grain and forage production. A large number of grain sorghum hybrids, '300' series (BRS300, BRS301, BRS302, BRS303, BRS304, BRS305, BRS307, BRS308, BRS309, BRS310, BRS330, BRS332, BRS373 and BRS380); and forage sorghum hybrids, '600' series (BRS601, BRS610, BRS655 and BRS658), and '800' series for green chop and grazing (BRS800, BRS801, BRS802 and BRS810) have been released by this programme of sorghum improvement. In the 2013/2014 agriculture year, sorghum cultivars developed by Embrapa accounted for approximately 18% of the total grain sorghum market, 41% of the forage sorghum market and 13% of the green chop and grazing market in Brazil, indicating the importance of this programme for the development of sorghum cultivars. In 2014, Embrapa also began the licensing of sorghum lines for use by the private sector. There is currently more incentive for Embrapa to collaborate with the private seed sector in the development of new cultivars.

More recently, there has been an increasing demand for sorghum as a feedstock for bioenergy production. Sweet sorghum is in a growing demand as a feedstock for first-generation ethanol production technology and biomass sorghum as a feedstock for second-generation ethanol production and co-generation of electricity. There is a long history of ethanol production from sugarcane in Brazil and an increasing demand to include sweet sorghum to complement sugarcane feedstock. None the less, Brazil still has difficulty to supply its domestic demand for ethanol, especially in the off-season of sugarcane. There has been a reduction in the productivity of sugarcane, from the 2008 harvest when the average productivity was 85 $t.ha^{-1}$, and during the past year average productivity was 76 $t.ha^{-1}$ (CONAB 2016b). Lack of investments in the sugar and alcohol sector of Brazil, climatic factors (lack of rain), part of the sugarcane feedstock destined for the production of sugar, and seasonality of ethanol production due to the sugarcane off-season (about 5 months from December to April) are factors contributing to insufficient supply of ethanol in recent harvests in Brazil. One proposed solution to this impasse is the development of new alternative raw materials for diversification and expansion of renewable energy sources to increase the operational time of existing sugar mills.

Sorghum has emerged as a promising crop for ethanol production in Brazil, both from the agronomic as well as an industrial point of view, by producing juicy stems with sugars directly fermentable, coupled with the possibility of using the same harvesting and industrial milling infrastructure used for processing sugarcane for ethanol production. Sweet sorghum is recommended as a spring/summer crop, with sowing at the beginning of the rainy season, which runs from October to December and harvest in the months of February to April in the sugarcane-growing region of Central Brazil. This harvest season coincides with the off-season of sugarcane, when the industrial plants are idle due to the absence of raw material for milling. Consequently, the sweet sorghum harvest is just prior to the start-up and processing of sugarcane in mid-April. This allows for an expansion of the number of days of milling and processing feedstock for ethanol production during the months of March and April, and consequently optimizing the operational fixed costs. In addition to ethanol production from sweet sorghum, the bagasse can be used as a source of energy for co-generation of electricity, second-generation ethanol production or fodder for animals, contributing to a favourable energy balance.

Sweet sorghum has recently been proposed for planting in areas of sugarcane renovation. Normally after the fifth or sixth sugarcane harvest, the area is renovated and sugarcane is planted again. In this case, after the last harvest of sugarcane, there can be one crop of sweet sorghum with an October to December planting of sweet sorghum for a February to April harvest followed by planting sugarcane. Approximately 15% of the total area of sugarcane production in Brazil is renewed annually. The area planted with sugarcane in Brazil is around 9 million hectares (CONAB 2016b). Consequently about 1.3 million hectares of sugarcane is renewed annually and part of this area can be used for the cultivation of sweet sorghum. Thus, sugar mills can use their own land without the need to expand boundaries. However, special attention should be given to implementation of sweet sorghum in these environments because the excess residual organic matter on the soil surface can hamper the planting and germination of the sweet sorghum crop where a uniform plant population is essential for high biomass production.

Considering a growing demand for bioenergy feedstock, the Embrapa sorghum improvement programme has developed sweet sorghum cultivars for ethanol and sorghum biomass cultivars for co-generation and second-generation ethanol production from sorghum biomass. Several sweet sorghum varieties have been developed and released, '500' series (BRS506, BRS508, BRS509 and BRS511). In 2014, the programme launched its first sorghum biomass hybrid, BRS716. This programme is the first to launch a biomass cultivar developed in Brazil, dedicated to bioenergy production for co-generation and second-generation ethanol production technology.

3 Improvement of grain and forage sorghum

3.1 Grain sorghum improvement

Sorghum [*Sorghum bicolor* (L.) Moench] is the fifth most planted cereal in the world after wheat, rice, corn and barley, and it is grown in tropical and semi-arid regions (FAO 2014). Sorghum has more adaptive characteristics for producing grain under water stress conditions than other food and feed grains, which explains its cultivation on millions of hectares in countries in Africa, Asia, Oceania and the Americas.

In Brazil, the acreage of grain sorghum was 722 000 hectares with grain production of 2.06 million tons in the 2014/2015 season, indicating a slight decline of area in recent years (Fig. 1). The national productivity increased slightly in recent years and is around 2.8 $t.ha^{-1}$. Despite the observed increases in productivity, the national average is far below the potential of this crop. Experiments show that the productivity of hybrids can exceed 7.0 $t.ha^{-1}$ in favourable conditions of second season production (Santos et al. 2005). Producers who use a level of technology and follow correct guidelines for planting, considering appropriate planting time, soil fertilization, fungicides and adapted cultivars, have achieved this high level of production.

The Midwest and Southeast regions in Brazil account for over 88% of national production of grain sorghum, yet the South and Northeast of Brazil have areas of grain sorghum production. Goiás is the main producing state, with 41% of national production, followed by Minas Gerais (29%) and Mato Grosso (14%) (CONAB, 2016a). In Central Brazil, sorghum is planted in succession to summer crops, mainly soybeans, as a second crop, in February and March. In the South, sorghum is planted in the spring, in October/November, and is harvested in the fall. In the Northeast, the crop is planted in the rainy season, from March

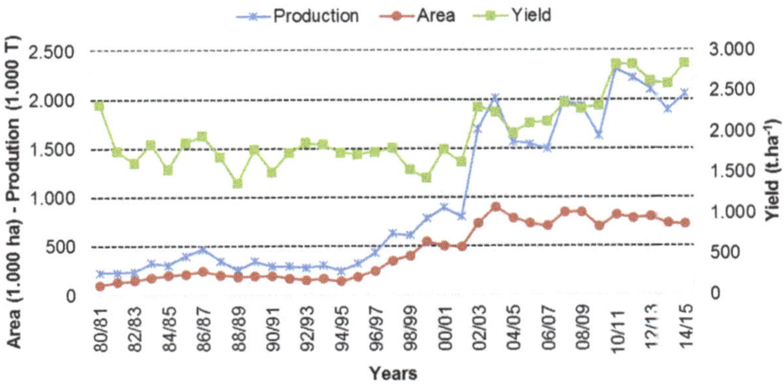

Figure 1 Historical production, area and grain sorghum productivity in Brazil, 2016. Source: Conab.

Figure 2 A commercial field with the grain sorghum hybrid (BRS 310), grown in Central Brazil as second crop.

to April. A common feature of all these areas is rainfed cultivation (Menezes et al. 2015a). In these producing areas, over 90% use hybrid seed, and only a small region in the South and Northeast use varieties. Figure 2 shows a commercial field with the grain sorghum hybrid (BRS310) grown in Central Brazil as a second crop following soybeans.

Considering the data shown above, seed companies have dedicated efforts to develop hybrids for planting as a second crop in succession to soybeans. This sowing is very conducive to long dry periods, especially at the end of the crop cycle, so it is important to plant shorter season varieties that can be harvested at the end of the rainy season. Long season cultivars are only recommended for cultivation at the very beginning of second season in February. Development of more drought-resistant cultivars with tolerance to anthracnose and aluminium toxicity, combined with a shorter cycle, are among the main objectives of the grain sorghum improvement programme. The main companies of sorghum seed production in Brazil are Embrapa, Dow AgroSciences (DowAgroSciences and Morgan Seeds), Monsanto (Agroceres, Agroeste and Dekalb) Agromen, Atlântica Seeds, Semeali, Nidera, Helix, IPA, Pioneer and CATI.

Embrapa Maize and Sorghum has two phenotyping sites for drought tolerance, located in the municipalities of Nova Porteirinha in Minas Gerais and in Teresina, the capital city of Piauí. Both locations have a well-defined rainy season, allowing for controlled irrigation to provide the desired water stress. The tests in Nova Porteirinha are carried out from May to September, and in Teresina tests are from September to December. Between the years 2010 and 2012, 243 sorghum lines from the association panel of the Institute for Genomic Diversity (Cornell University) were evaluated. The lines were evaluated in conditions with and without water stress. In Nova Porteirinha, the trials were conducted in 2010 and 2012, and in Teresina the tests were carried out in 2010 and 2011. Table 1 lists 21 lines that stood out in both moisture stress and irrigated conditions.

These lines were crossed with male sterile plants (ms_3ms_3), from a broad-based random mating population BRP13R for recurrent selection for drought tolerance. Currently, the population is in the process of recombination to obtain S2 families for continued recurrent selection. Parallel studies have also been conducted to evaluate breeding lines used by the Embrapa sorghum breeding programme. Menezes et al. (2015b) evaluated 25 sorghum lines for three consecutive years (2006, 2007 and 2008) in conditions with and without water stress. The average productivity of the lines was 39% lower in the environment with water stress when compared to the environment without water stress. Several lines including 9929020, CMSXS230B and N95B were identified as being drought tolerant, stable and responsive to environmental improvement. Santana (2013) evaluated 12 commercial hybrids in conditions with and without water stress, observing a drop of 33% in grain yield caused by stress. The hybrids BRS305 and 1G282 demonstrated tolerance to water stress and response to irrigation in these trials.

Menezes et al. (2015c) evaluated 49 hybrids, 43 experimental and 6 commercial. The experimental hybrids 1170090, 1170092, 1170064, 1167026, 1167064, 1170093, 1167008, 1167029, 0009061, 1167092, 1105647 and 1170019 were the most tolerant to drought and the most responsive to environmental improvement. Among the commercial hybrids, DKB 550 was the most productive in the condition without stress and BRS310 was the most productive in stress conditions. Considering the two environments, the hybrids 1G 282 and BRS332 were close to the overall average in both environments. Likewise, BRS304 and BRS310 were hybrids with productivities below average in both environments.

Batista (2016) evaluated 29 commercial hybrids of grain sorghum in environments with and without water stress in 2014 in Teresina-PI and in the years 2014 and 2015 in Nova Porteirinha-MG. The most productive hybrids in all environments were 1G282, DKB540, A9721 R, 1G100 and AG1090. The characteristics thousand seed weight, SPAD index at 75 and 90 days after planting, and leaf area index showed high correlations with grain yield. Some of these drought tolerance studies have shown that the interaction between genotypes and years are higher than among water regimes, indicating the need for evaluation of genotypes in several years (Menezes et al. 2015a; Santana, 2016).

3.2 Forage sorghum improvement

The sorghum crop is being increasingly used for silage because of its great versatility compared to other crops, especially maize, which is the main crop used for silage production in Brazil. Forage sorghum stands out for its wide adaptation, and can be grown throughout the country, as well as its high productivity of dry matter combined with high nutritional quality of forage. Forage conservation in the form of silage is one of the most widespread and growing practices in Brazil. As livestock production achieves a high level

Table 1 Average grain production ($t.ha^{-1}$) of the best 21 of 48 sorghum lines evaluated in conditions with and without water stress, 2016

LINES	NP10SE	NP10CE	NP12SE	NP12CE	TE10SE	TE10CE	TE11SE	TE11CE	Mean
P898012	6.55	3.65	5.31	4.19	5.58	4.28	4.00	4.05	4.70
SC84	6.50	3.51	4.87	3.88	5.20	3.94	4.04	4.38	4.54
Kuyuma	6.23	3.43	5.49	4.34	4.98	3.72	3.99	4.11	4.54
SC103	6.30	3.50	5.14	4.08	5.63	4.43	3.73	3.81	4.58
Pinolero1	6.44	3.42	4.82	3.92	5.24	3.87	4.03	4.91	4.58
White Kafer	6.74	4.00	5.07	4.22	4.87	3.78	3.32	3.89	4.49
Dorado	6.76	3.43	5.06	4.13	5.03	3.71	3.62	3.85	4.45
Tx7000	6.89	4.08	4.75	3.99	4.83	3.65	3.57	4.02	4.47
SC334	6.49	3.71	5.14	4.04	4.82	3.61	3.45	3.94	4.40
SC323	6.27	3.73	4.90	3.93	4.83	3.70	3.67	4.31	4.42
SC702	6.52	3.70	4.59	3.81	4.72	3.57	4.47	5.00	4.55
Tx7078	6.36	3.38	4.76	3.93	5.28	4.12	3.65	4.05	4.44
SC110	6.34	3.44	5.13	4.09	5.00	3.78	3.41	3.80	4.37
RTx433	6.61	3.63	4.97	3.96	5.11	3.87	3.37	3.48	4.37
SC465	6.29	3.84	5.14	4.09	4.95	3.73	3.21	3.72	4.37
SC1337	6.69	3.78	4.16	3.46	5.38	4.35	3.81	3.97	4.45
Tx7005	6.29	3.48	5.14	4.18	4.53	3.34	3.55	3.92	4.30
LianTangAi	6.34	3.45	4.97	3.69	5.14	3.77	3.66	3.81	4.35
SC704	5.94	3.40	4.94	3.93	4.97	3.84	3.67	4.08	4.34
SC720	6.48	3.47	4.92	3.82	4.85	3.71	3.42	3.83	4.31
Ajabsido	6.41	3.42	4.93	3.94	5.92	4.70	3.26	3.51	4.51

NP10SE – Nova Porteirinha 2010 without water stress, NP10CE – Nova Porterinha 2010 with water stress, NP12SE – Nova Porteirinha 2012 without water stress, NP12CE – Nova Porteirinha 2012 with water stress, TE10SE – Teresina 2010 without water stress, TE10CE – Teresina 2010 with water stress, TE11SE – Teresina 2011 without water stress, TE11CE – Teresina 2011 with water stress.

Figure 3 A commercial forage sorghum hybrid BRS 658.

of technology, the demand for better zoo-technical indexes and economic profitability has led milk and beef cattle producers to adopt this practice systematically (Teixeira et al.2014; Neves et al. 2015).

It is estimated that forage sorghum occupies approximately 600 000 to 800 000 hectares for silage production in Brazil. The number of hybrids and varieties adapted for silage, sold in the domestic market, has also increased. Sorghum is a culture that produces silage with good fermentation characteristics, ranging around 90 to 95% of the nutritional value of maize when provided as a single ingredient, or 85 to 90%, when part of a ration (Machado et al. 2014).

In Brazil, forage sorghums can be subdivided into a) forages with high biomass productivity but low grain production and low content of soluble sugars, used for conservation as silage. These cultivars have a plant height >3 m and the silage produced from this biomass often has low nutritional value, low grain content and low digestibility; b) forages with low grain production, high percentage of soluble sugars and plant height near 3 m called sweet forages, which have stems with sugar levels between 10 and 14% that can be industrialized for production of ethanol. These cultivars can also be used for preparation of silage; and c) specialized forages for grazing or producing hay, called sudangrass (*Sorghum bicolor* cv. *sudanese*) or their hybrids with *Sorghum bicolor* cv. *bicolor*. Compared with the other types, they have a thin stem, high rate of tillering and regrowth capacity. Their silage is of low quality since the moisture content of the stems is high and with a very low percentage of grain in the forage. d) Foragers with good grain production and height ranging between 2.20 and 3.0 m, called high-quality silage producers, for producing silage of high nutritional value because of the high percentage of grain in the forage, highly digestible fibre, lodging resistance and great capacity for regrowth, among other features (Magalhães et al. 2010; Cardoso et al. 2014).

In the Brazilian market, there are several forage cultivars adapted for silage of high quality, developed and marketed by both public and private institutions. They are short season hybrids characterized by high productivity (over 50 t.ha^{-1}), high animal performance, lodging resistance, high capacity for regrowth, high proportion of grain in forage (30 to 40% of grains), tolerance to abiotic stresses (Al toxicity, drought), plant nutrient efficiency (N, P), tolerance to biotic stresses, highly digestible fibre (60% IVDMD), high standard of silage fermentation, high seed productivity of parental lines, high capacity of fertility restoration in the hybrid and competitive seed prices (Rodrigues et al. 2010; Costa et al. 2016).

The criteria for selection of sorghum hybrids for silage have been mainly plant height, biomass productivity, grain yield, disease and pest resistance, and tolerance to drought. However, the identification of agronomic characteristics related to the proper fermentation process that provides low losses of dry matter and nutrients during ensiling, and high rates of digestibility and consumption are of great importance in the selection of the most appropriate cultivars for silage (Faria Jr et al. 2010; Rodrigues et al. 2010). Figure 3 shows a commercial forage sorghum hybrid BRS 658.

4 Improvement of sorghum for bioenergy

4.1 Bioenergy sorghum improvement for first-generation technology

Economic studies have shown that sweet sorghum is economically viable in Brazil with ethanol productivity between 2500 (with use of bagasse) and 3000 (without use of bagasse) $L.ha^{-1}$. This productivity can be obtained by producing 50 $t.ha^{-1}$ stalk yield and 60 L of ethanol per ton of stalks. These yields are influenced by cultivar, production system, environment and industrial management. Ethanol productivity of sweet sorghum cultivars is associated with these agro-industrial characteristics: stalk yield per hectare, moisture content and juice extraction, soluble solids in the juice (Brix), sucrose, and total sugars, which will reflect in litres of ethanol per ton of stalks (Parrella 2011). In this context, the stem productivity of sweet sorghum is highly correlated with ethanol productivity (Murray et al. 2008; Ritter et al. 2008; Lombardi et al. 2015).

Total soluble solids in the juice or Brix correlates directly with the total sugar content in the juice (Murray et al. 2008a; Ritter et al. 2008; Lombardi et al. 2015), which are consumed by yeast for ethanol production. The correlation between Brix and total fermentable sugars is greater for higher Brix values (>0.9) and less for lower Brix (<0.8). Our breeding objectives for sweet sorghum hybrids are high levels of both total sugars and sucrose (high purity) in the extracted juice, high juice extraction, small panicles and non-tillering. The selection index of tons of extracted Brix per hectare, the product of stalk production, extraction and Brix, has been a useful tool for selection of genotypes with higher levels of Brix. We have found a high correlation with stalk production, Brix and ethanol yield (Lombardi et al. 2015).

Sweet sorghum productivity and consequently ethanol productivity is a function of genotype, production system and industrial management; and can be expressed as follows:

Sweet sorghum productivity = f(Genotype + Production system + Industrial management)

We have developed a protocol of 'Industrial Planning' to deliver high-quality feedstock 24/7 (24 hours per day and seven days a week) to the Industrial Park for producing ethanol. We have also developed a PUI (Period of Industrial Utilization) protocol with minimum values of 80 $k.t^{-1}$ extracted sugar based on the hydraulic press protocol, minimum of 12.5% total sugars in the extracted juice for maximum fermentation efficiency, and a minimum period of 30 days maintaining these values (Schaffert and Parrella 2012). The PUI is necessary for developing a system of planting dates and selection of cultivars for planning harvest and processing of the feedstock.

Given the above, we have established minimum goals of productivity and quality of sweet sorghum for economic sustainable profitability. The current minimum biomass productivity is 50 $t.ha^{-1}$, minimum total sugar extraction is 80 $kg.t^{-1}$ biomass, minimum total sugar content in the juice is 12.5%, minimum alcohol production is 60 $L.t^{-1}$ biomass and the period of industrial utilization is at least 30 days.

Our R&D strategy is to develop sweet sorghum hybrids with the desired characteristics, juicy stem, and the presence of sugar in the juice with high purity and predominance of sucrose. The presence of juice in the stalk is controlled by the *D/d* gene, and the recessive alleles *dd* are responsible for the presence of juice in the stem (Rooney 2000). Thus, both parental lines of the hybrids need to have recessive alleles for juicy stem to produce high-yielding hybrids. Lines with juicy stem (*dd*) have an opaque green midrib, and dry stem (*DD* and *Dd*) lines have a white midrib.

The inheritance of sugars in the juice in sweet sorghum is complex and the presence of sugars is influenced not only by the genotype, but also by the environment and genotype x environment interaction (Murray et al. 2009). Additionally, the genetic control of sugar content is predominately additive gene action (Murray et al. 2008a). Some QTLs related to sugar production in sorghum have been identified (Murray et al. 2008a; Ritter et al. 2008; Murray et al. 2009; Souza 2016). Some studies show the presence of heterosis for biomass production, sugar production and sugar concentration in the juice (Pfeiffer et al. 2010; Durães, 2014). Our results indicate that the development of sweet sorghum hybrids is a viable strategy. However, it is noteworthy that both parental lines of the hybrid must have a high concentration of sucrose in the juice so that the hybrids have high production potential of sugar per hectare.

Embrapa Milho e Sorgo started a programme to develop sweet sorghum cultivars, in the 1970s for small distilleries, anticipating an energy shortage in Brazil and a shortage of liquid fuel for agriculture expansion in the Central West of Brazil. Sweet sorghum varieties were introduced from the US Department of Agriculture, Africa and India. These varieties were characterized for desirable agronomic traits and a small number were identified that are adapted to our conditions. In the early 1980s six varieties (BR500, BR501, BR502, BR503, BR504 and BR505) derived from the Rio, Brandes, Roma, Theis, Dale and Wray were selected as potential varieties for ethanol production in Brazil. Average biomass yields exceeded 40 $t.ha^{-1}$ and average soluble solids content ranged between 18 and 20 Brix (Easy juice). Several segregating families were developed by making crosses between these selected lines. The first variety released from these crosses was BRS506 in 1987.

The sweet sorghum breeding programme was put on hold in the mid-1980s due to a policy shift for research and development for large distilleries (250 000 to 1 million litres of ethanol per day). The sweet sorghum improvement was resumed in 2008 due to a demand of alternate feedstocks for ethanol production. With a continuation in the sweet sorghum breeding programme, three new varieties, BRS508, BRS509 and BRS511, were released in 2012. In several trials conducted by Souza et al. (2013) and Figueiro et al. (2015), these varieties showed wide adaptability and greater stability, especially BRS511 and CMSXS634 an experimental variety, compared to commercial hybrids on the market.

Figure 4 A commercial sweet sorghum variety BRS 511.

These commercial hybrids were adapted to Central Brazil with acceptable production but much lower total soluble solids content in the extracted juice resulting in low phenotypic stability for ethanol production. In a study carried out by Fernandes et al. (2014), the varieties BRS506, BRS508, BRS509 and BRS511 had a high level of sugar content in the juice with values exceeding 20 Brix and excellent fermentation efficiency in the production of ethanol, equivalent to sugarcane standards. Figure 4 shows a commercial sweet sorghum variety BRS 511.

Our current objectives are to develop and release sweet sorghum hybrids using three dwarf sweet sorghum females and tall sweet sorghum R-lines. Our early studies indicated a complete lack of historical sweet sorghum varieties that are B-lines and a complete lack of sweet sorghum A- and B-lines. Many traditional sorghum A- and B-lines are juicy but lack the genes for synthesizing sugar in the stem. Our first attempt to develop three dwarf sweet sorghum and A- and B-lines was to cross adapted B-lines to tall sweet sorghum R-lines and select for short-statured progenies with high sugar content. The first A and B sweet sorghum lines were finalized in 2014 and we are currently testing them in hybrid combinations (Rocha 2016). The Brix levels using the hydraulic press method of juice extraction and plant height of eight sweet sorghum A- and B-lines of an off-season seed increase nursery are shown in Table 2.

4.2 Development of biomass sorghum hybrids for second-generation technology and co-generation

The growing demand for biomass for renewable energy generation in Brazil is opening opportunities for the exploration of alternative feedstocks. Sorghum biomass has aroused interest in the bioenergy industry to be a fast-growing grass with high biomass production in a short period of time (May et al. 2014; Parrella et al. 2014). This type of sorghum (biomass) is photosensitive, and with flower induction when the day length is less than 12 hours and 20 minutes (Rooney and Aydin 1999), corresponding to the period between

Table 2 Mean values for flowering in days after planting, plant height in metres and soluble solids in the juice (SST) in Brix of eight pairs of A and B sweet sorghum lines in a breeding nursery in Sete Lagoas, Minas Gerais, in 2015

Line	Days to flower	Plant height (m)	SST (°Brix)
CMSXS5501A/B	85	1.20	15.80
CMSXS5502A/B	80	1.70	17.40
CMSXS5503A/B	80	1.76	16.60
CMSXS5504A/B	84	1.65	18.20
CMSXS5505A/B	83	1.95	16.00
CMSXS5506A/B	83	1.87	18.20
CMSXS5507A/B	81	1.50	16.30
CMSXS5508A/B	89	1.78	17.40

September 21 and March 21 (no flower induction between September 21 and March 21) in most of the Southern two-thirds of Brazil.

Photosensitive biomass sorghum hybrids planted during long days when day length is greater than 12 hours and 20 minutes (October through December) remain vegetative through April and produce significantly more biomass than normal photo-insensitive hybrids (Parrella et al. 2010).

These hybrids are taller, up to 5 m tall, and with a production potential exceeding 30 $t.ha^{-1}$ of dry biomass on a commercial scale (Parrella et al. 2010; Castro et al. 2015). Experimental hybrids have surpassed 60 $t.ha^{-1}$ dry biomass.

The Embrapa breeding programme develops photosensitive biomass sorghum hybrids by crossing a photo-insensitive A-line (ma_1ma_1) with a photosensitive R-line (Ma_1Ma_1). The resulting hybrids are heterozygous (Ma_1ma_1) for the Ma_1 maturity gene and are photosensitive (Parrella et al. 2010). These results are in accordance with those reported by Miller et al. (1968), Lin et al. (1995) and Rooney and Aydin (1999), reporting that the allele Ma_1 is dominant. There is a window of opportunity for the production of photosensitive hybrids in Central Brazil by planting the A- and R-lines simultaneously during the months from March to May, when the days are short enough to induce flowering of the photosensitive R-lines to coincide with the flowering of the A and B-lines (see Fig. 5).

These resulting photosensitive sorghum biomass hybrids, planted at the beginning of the rainy season in October or November, remain vegetative for 160 to 180 days before flowering in late April (Parrella et al. 2010). Castro et al. (2015) evaluated sorghum biomass hybrids with average days to flower ranging between 108 and 123 days planted at the end of November. This demonstrates the existence of genetic variability for this character and enables the selection of different maturity groups to meet the market demand. In the assessment of biomass sorghum hybrids in Lavras, Minas Gerais, there was a variation in production between 24 and 47 $t.ha^{-1}$ of dry matter, confirming the great productive potential of this type of hybrids (Castro et al. 2015). Figure 6 shows a commercial biomass sorghum hybrid BRS716 near harvest.

The moisture content of the biomass and dry matter content at harvest are another important characteristic for this type of sorghum. In this case, the photosensitive R-lines

Figure 5 Sorghum crossing nursery with simultaneous flowering of A (photo-insensitive) and R (photosensitive) lines to produce biomass sorghum hybrid seed. Images of the use of a sorghum crop nursery (a and b). A photo-insensitive male sterile A-line (short) is crossed with a photosensitive restorer R-line (tall) to produce biomass sorghum hybrid seed.

Figure 6 A commercial biomass sorghum hybrid BRS 716.

and/or the female A-line need to have the dominant gene for dry stem (*DD*). The photosensitive hybrids will be dry stem (*DD*) or (*Dd*). Castro et al. (2015) observed a variation of 30–42% in the dry matter content of different biomass hybrids evaluated in Minas Gerais. The ideal dry matter content for burning for co-generation is 50% or more.

Castro et al. (2015) observed an average fibre content of 36% for these experimental biomass sorghum hybrids. These high values are of great interest to the industry, which already uses sugarcane with fibre content around 13% and biomass with upper calorific value and lower calorific value biomass, on a dry weight basis, around 4300 and 3800 Kcal kg^1, respectively. With 50% moisture content in the biomass the lower calorific value is around 1800 Kcal kg^{-1} dry matter. This demonstrates the great potential of this raw material for generation of electricity.

Considering the high productivity levels, the phytotechnical quality of biomass, mechanized planting and harvesting, biomass sorghum is a promising crop to generate renewable energy in Brazil.

4.3 Future trends in sorghum improvement for bioenergy

Another focus of the sorghum programme is the development of photosensitive sweet sorghum hybrids to increase the green biomass productivity with high sugar content. Two strategies have been used to obtain the photosensitive hybrid. The first aims to develop photosensitive sweet sorghum R-lines (Ma_1Ma_1) with high juice extraction and high levels of sucrose. Progenies derived from crosses with photosensitive R-lines and non-photosensitive sweet sorghum R-lines are being selected during long days. These advanced photosensitive R-lines were crossed with photo-insensitive male sterile A-lines (Table 1) and the hybrids were evaluated in the 2015/2016 crop season. These new experimental photosensitive sweet sorghum hybrids are showing very promising results for ethanol production.

We are also developing sweet sorghum photosensitive hybrids using the $ma_5ma_5Ma_6Ma_6$ genotype derived from backcrosses with the sorghum line EBA-3. We are currently evaluating sweet sorghum R-lines with the genotype $ma_5ma_5Ma_6Ma_6$ developed by backcrossing and selection with molecular markers. These photo-insensitive R-lines have been crossed with new sweet sorghum A-lines (Table 1). The resulting hybrids, double heterozygotes ($Ma_5ma_5Ma_6ma_6$), are expected to be photosensitive (Rooney and Aydin 1999). Thus, all the genetic improvements from our normal R-line development programme will be maintained in these new photosensitive sweet sorghum hybrids.

Another aspect of our sorghum biomass programme is the development of photosensitive hybrids with the brown midrib gene, bmr_6, to meet the demand for biomass quality to produce second-generation ethanol or cellulosic ethanol. Thus, the bmr_6 allele was introduced through a backcrossing programme assisted by molecular markers in a set of A, B and R-lines of the best conventional sorghum biomass hybrids. The biomass sorghum hybrids with brown midrib had a high dry matter production, around 30 t ha^{-1}, but with higher efficiency in the conversion to cellulosic ethanol. In preliminary laboratory results, these bmr_6 hybrids showed around 70% efficiency in enzymatic hydrolysis compared to 40% of conventional hybrids. Considering the productivity levels achieved and the greater efficiency of enzymatic hydrolysis, which is a significant cost in obtaining cellulosic ethanol, the bmr_6 photosensitive biomass hybrid has proven to be an excellent alternative for the second-generation fuel technology.

5 Sorghum tolerance for aluminium and low levels of phosphorus

5.1 Sorghum aluminium tolerance

Aluminium (Al) toxicity is a serious constraint for sorghum production on acid soils, which are common in South America. Nevertheless, low-pH soils are widespread: it has been estimated that over 50% of the world's potentially arable lands are acidic (Von Uexküll and Mutert 1995; Wood et al. 2000). In fact, the greatest extent of Oxisols, which occupy about 18% of the total acid soil area, is found in South America and Africa (von Uexküll and Mutert 1995). In Brazil, the Cerrado biome is an area of over 2 million km^2 where toxic levels of Al are frequently found in agricultural areas.

At pH values below pH 5.0, the rhizotoxic species, Al^{3+}, is released into the soil solution causing root damage, thereby restricting root growth and ultimately causing dramatic yield losses. Genetic variation for Al tolerance has enabled the development of crops that can still produce significant yield even in the presence of high Al saturation in the soil. The mechanisms that allow plants to withstand Al act either by preventing Al^{3+} from entering cells in the root apex, which is particularly sensitive to Al toxicity, or by conferring the ability to tolerate Al^{3+} once it has crossed the plasma membrane into the symplasm.

Plasma membrane transporters belonging to the Al-activated malate transporter (*ALMT1*, Sasaki et al. 2004) family and the multidrug and toxic compound extrusion family (MATE, Magalhaes et al. 2007; Furukawa et al. 2007) are major players in exclusion mechanisms that have been agronomically relevant for crop cultivation under Al toxicity. Sorghum aluminium tolerance has been shown to be controlled by the *Al tolerance in sorghum bicolor* (Alt_{SB}) locus on chromosome 3 (Magalhaes et al. 2004). The molecular and physiological nature of Al tolerance conferred by Alt_{SB} has been shown to be due to the action of the MATE member, *SbMATE* (Magalhaes et al. 2007).

This gene is responsible for an Al exclusion mechanism based on Al-activated citrate release, whereby non-toxic, Al–citrate complexes are formed in the rhizosphere thus providing Al tolerance. *SbMATE* is induced by Al at the transcriptional level and also activated by Al as significant citrate exudation in Al-tolerant lines is only observed in the presence of Al^{3+} in the external media. Differences in Al tolerance and citrate release could be largely explained by variation in *SbMATE* transcript levels. The coding region of *SbMATE* was highly monomorphic across sorghum lines and regulatory polymorphisms are likely to play a pivotal role in controlling *SbMATE* expression and thus Al tolerance conferred by Al-activated citrate release into the rhizosphere.

Sorghum is in general a very Al-sensitive crop and Al^{3+} will completely obliterate root function in the absence of tolerance mechanisms (Magalhaes et al. 2004). Noting <80% of the association panel studied by Caniato et al. (2011) was composed by highly Al-sensitive accessions, which are expected to produce very little or no grain yield on acid soils with high Al saturation. Alt_{SB} has been shown to cause a major effect in Al tolerance, explaining a large fraction of the phenotypic variation in a sorghum $F_{2:3}$ population (Magalhaes et al. 2004). Therefore, the Embrapa molecular breeding programme has been focusing on developing methods to efficiently characterize and explore intra-locus variation for Al tolerance based on Alt_{SB}. Novel, gene-based technologies are being pursued to speed up cultivar development and to assess the impact of *SbMATE* in field conditions.

There is clear potential in exploring allelic variation of *SbMATE* for breeding purposes. Using near-isogenic lines (NILs) developed with different Alt_{SB} donor lines, Caniato et al. (2007) reported an allelic series Alt_{SB} conditioning a wide range of tolerance levels, although there is evidence pointing towards distinct Al tolerance genes or accessory genes acting in the *SbMATE* pathway.

A closer functional analysis of *SbMATE* expression with an expanded set of parents and NILs harbouring Alt_{SB} showed a consistent trend of reduction in Al tolerance when NILs are compared to the respective donor lines (Melo et al. 2013). Interestingly, this trend was paralleled by a reduction in *SbMATE* expression in NILs, indicating the importance of accessory loci regulating *SbMATE* expression in *trans*. However, large differences in *SbMATE* expression and Al tolerance across NILs emphasized that *cis* effects are very important to explain allelic variation at Alt_{SB}. Therefore, the effectivity of *SbMATE* in conferring Al tolerance is ultimately due to the joint effect of *cis* and *trans*-acting factors regulating *SbMATE* expression. Selection at both types of regulatory loci seems to be needed to fully exploit the *SbMATE* potential in enhancing sorghum Al tolerance. However, two sorghum lines with strong *cis* control of *SbMATE* expression were identified and are thus important sources of marker-assisted backcross breeding approaches based solely on Alt_{SB}.

Embrapa Milho e Sorgo has been directing efforts to characterize diverse sorghum germplasm for Al tolerance, which is the first step towards the development and use of gene-specific markers based on *SbMATE* for allele mining applications. Al tolerance is not randomly distributed in the sorghum genetic diversity, as it is more prevalent primarily in the Guinea race and to a lesser extent in caudatum accessions (Caniato et al. 2011). Therefore, careful germplasm characterization, which can be efficiently achieved with marker-based population structure analysis, can provide important insights into the odds of finding Al-tolerant sources directly in the breeder's germplasm.

An association mapping study based on the Alt_{SB} locus allowed for the identification of SNP loci highly associated with Al tolerance (Caniato et al. 2014). These SNPs have been converted into the KASP genotyping system (http://www.lgcgroup.com/) and are available for high-throughput sorghum genotyping within the Integrated Breeding Platform (https://www.integratedbreeding.net/). These markers can be used for marker-assisted backcrossing or for allele mining to identify useful Al tolerance sources directly in the target germplasm. In that regard, the SNP with the strongest association with Al tolerance was found in the second intron of *SbMATE*. Selection for the allele associated with Al tolerance led to the recovery of most of the highly Al-tolerant accessions in the association panel. Allele mining using this SNP marker should require confirmatory phenotyping of only a few accessions. This contrasts to a non-marker-based selection where, due to the low frequency of Al tolerance, extensive phenotyping is required for the identification of Al-tolerant sources.

Gene-specific markers for *SbMATE* (Caniato et al. 2014) and a variety of flanking markers tightly linked to Alt_{SB} (Caniato et al. 2007, 2011; Sabadin et al. 2012) can be used for marker-assisted introgression of Alt_{SB} simultaneously reducing linkage drag across the Al tolerance locus. Historically, aluminium tolerance has been assessed by Al inhibition of root growth in nutrient solution as sustained root growth in the presence of Al^{3+} is a hallmark of plant Al tolerance (Fig. 7). Particularly for breeding applications, it was thus necessary to assess the Alt_{SB} effect on sorghum grain yield under Al toxicity in the soil.

Embrapa maize and sorghum has been directing particular efforts to the development of field phenotyping sites that as much as possible isolate the effect of Al toxicity from other abiotic stress factors that typically occur on acid soils. An Al tolerance field phenotyping site was developed by carefully mapping spatial variation in soil Al saturation at different soil depths. Soil Al saturation was low, near 2% in the superficial layer (0–20 cm) in the control site, whereas much higher Al saturation was observed in the Al toxicity site, reaching 56% on average (Carvalho et al. 2016). The same large differences in Al saturation were observed in the subsoil (20–40 cm), with mean Al saturation values of approximately 15% and 65% for the control and Al toxicity sites, respectively. The critical level beyond which Al negatively impacts sorghum yield is around 20% Al saturation for sorghum growing in Oxisols/Ultisols (Gourley 1987). Indeed, Al toxicity reduced grain yield of our recombinant inbred line (RIL) population by approximately 30%.

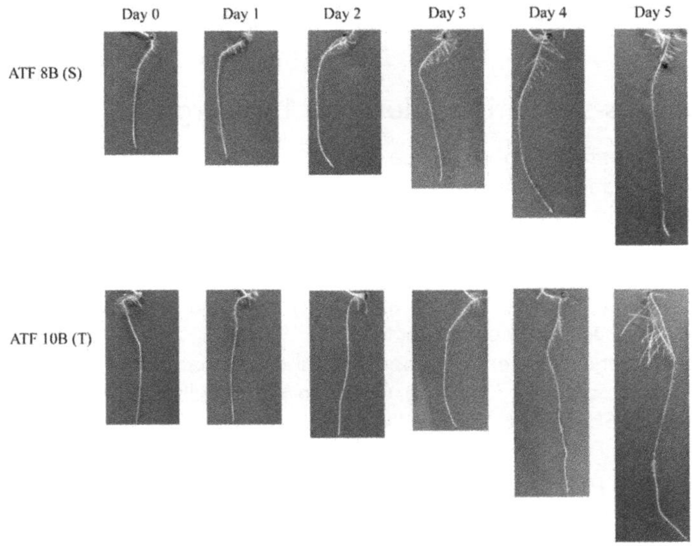

Figure 7 Hydroponics-based assessment of Al tolerance in sorghum. ATF 8B (S) and ATF 10B (T) are two near-isogenic lines sensitive (S) and tolerant (T) to Al and are derived from the parents, SC283 (T) and BR007 (S). A free Al^{3+} activity of {27} μM Al^{3+} in nutrient solution at pH 4.0 causes stronger inhibition of root growth in ATF 8B (S), which also shows root damage elicited by Al^{3+}. The symptoms of Al toxicity in ATF 8B become perceptible after 3 days of Al^{3+} exposure. The methods used to assess Al tolerance have been described (Magalhaes et al. 2004).

Using an RIL population phenotyped for Al tolerance both in nutrient solution and in the field, multi-trait QTL mapping revealed a major QTL co-localized with *SbMATE* controlling both Al tolerance assessed by Al inhibition of root growth and grain yield under high Al saturation (Carvalho et al. 2016). This validated the effect of *SbMATE* in field conditions and allowed us to estimate the gene effect as 0.6 $t.ha^{-1}$, an impressive 26% yield increase based on the RIL mean. We also developed a series of near-isogenic sorghum hybrids where the effect of Alt_{SB} could be compared across an otherwise uniform genetic background. A female effect of 0.5 ton/ha was detected arising from one copy of the tolerant allele at Alt_{SB}. There was no yield penalty related to SbMATE activity in the absence of Al toxicity. A second Al tolerance QTL was detected on chromosome 9 and the nature of this QTL is now under investigation. These results show that Alt_{SB} is an indispensable asset for sorghum yield in areas with high Al saturation in the soil. In those areas, in the absence of strong Al tolerance genes such as *SbMATE*, the yield potential of sorghum elite lines, such as one of the parents of our RIL population, will be strongly overshadowed by Al toxicity.

Several commercial male sterile sorghum lines tolerant to aluminium toxicity have been developed from crosses of elite sorghum B lines and SC283, an excellent non-restoring source of aluminium toxicity. The male sterile sorghum line ATF14 A/B is the female parent of Embrapa grain sorghum hybrids BRS330 and BRS373 are examples. The tolerance of SC283 is controlled by a single locus with partial dominance. SSR molecular markers flanking this gene have been developed and are currently used in the sorghum breeding programme at Embrapa. Restorer lines with resistance to Al toxicity were also developed from Al-tolerant sources CMSXS225, SC549 and SC566, which are in advanced stages of evaluation.

5.2 Phosphorus-starvation tolerance 1 in sorghum (SbPSTOL1)

In rice, *phosphorus-starvation tolerance 1* (*OsPSTOL1*) is similar to serine/threonine receptor-like kinases and has been shown to enhance early root development, phosphorus uptake and grain yield under low-P conditions (Gamuyao et al. 2012). We investigated a possible role of *PSTOL1* homologs in sorghum, *SbPSTOL1*, in sorghum root morphology and architecture, in addition to P uptake and grain yield performance under low P availability in the soil. On acid soils, P availability is typically low due to P fixation by aluminium and iron oxides in the clay fraction.

Using association and bi-parental mapping we showed that *SbPSTOL1* alleles reducing root diameter also increased P uptake under low P conditions in hydroponics. Most importantly, for two *SbPSTOL1* genes, alleles increasing root surface area also increased grain yield under low P availability in the soil (Hufnagel et al. 2014). Therefore, our results indicate that there is potential for exploring *SbPSTOL1* for increasing root surface area, thereby enhancing P uptake and grain yield under low P availability in the soil. This work was again possible via the development and use of a field phenotyping site to isolate the P stress effect from that of other abiotic stress factors on acid soils.

Together, Al toxicity and low P availability are two of the most important abiotic stress factors leading to yield instability for sorghum cultivated on acid soils. Using gene-specific markers for both *SbMATE* and also *SbPSTOL1*, we are using a recombination-based approach on multi-parent populations developed with the male sterility gene, *ms3*, to combine positive alleles of these genes. Our final goal is to generate cultivars with broad adaptation to acid soils containing both Al toxicity and low P availability, which are expected to show higher yield stability on acid soils.

6 Genetic and molecular approaches to studying biomass production and sorghum quality

Due to growing demand for renewable energy sources, interest in crops dedicated to biofuel production and co-generation is expanding. Sorghum has several traits that make it a promising feedstock for bioenergy production, ranging from the presence of fermentable sugars in the stalk for straightforward ethanol production, to high biomass with differential compositions for biomass combustion in furnaces or for cellulosic ethanol production. Sorghum also is an excellent model for genetic studies in grasses, due to its compact sequenced genome and diploid status.

At Embrapa Milho e Sorgo, we are interested in understanding the genetic control of traits related to the production and quality of biomass, as well as sugar content and juiciness. In order to identify potential targets that could be used in the breeding programme we are applying several approaches such as QTL mapping, GWAS studies, genome-wide selection and candidate gene approach.

A sweet sorghum RIL population, consisting of 223 lines derived from a cross between the genitors Wray and Brandes, both sweet and contrasting regarding the quality and quantity of sugars present in the stem juice, was used for genetic mapping studies.

The experiment was carried out in Sete Lagoas, Minas Gerais, Brazil, for three harvests, and biomass production and sugar-related traits were evaluated. The population was genotyped by sequencing (GBS) (Elshire et al. 2011) and applying the multiple trait multiple interval mapping approach; sixty-five QTLs were identified and all analysed variables had significant epistatic effects between QTLs (Souza 2016; Souza et al. 2017a). Though several of the identified QTLs have already been previously reported (Ritter et al. 2008; Murray et al. 2008a, 2009; Shiringani et al. 2010; Guan et al. 2011; Felderhoff et al. 2012; Mocoeur et al. 2015; Zang et al. 2015), novel sugar-related QTLs on several chromosomes were identified in this study, and validation has been carried on so future implementation of marker-assisted selection in our sweet sorghum breeding programmes can be achieved.

Lignin is a major component of plant cell walls and it can interfere with the process of biomass conversion to cellulosic ethanol. Nevertheless, lignin increases the calorific value of biomass, which is desirable for co-generation of electricity. Sorghum genotypes have a wide variation regarding lignin content and they can provide a quality feedstock for both end purposes. Therefore, studying the genetic architecture of lignin content is important for genotype selection and development for co-generation or cellulosic ethanol production. Two different approaches, GWAS on a lignin contrasting sorghum panel and expression analysis of genes involved in lignin biosynthesis, have been used to identify potential gene targets for sorghum breeding related to biomass production and quality, including lignin content.

Embrapa has put together a genetic diverse sorghum association panel composed of approximately 200 high biomass materials using accessions from Embrapa's germplasm collection and breeding programme, as well as accessions from the CIRAD core collection. A biochemical characterization of the panel using near-infrared spectroscopy and chemical analysis was performed for biomass composition traits and it shows significant variability of cell wall components, especially lignin content, among the genotypes. This diverse sorghum panel was evaluated at harvest during two crop seasons in Sete Lagoas, Brazil, for flowering date, plant height, fresh and dry mass production, and lignin content. For GWAS mapping, over 260K SNP markers were obtained using GBS analysis. Using a multi-locus mixed model approach, five SNPs were mapped for plant height; three for flowering; two for lignin and FMP; and one for DMP. The SNPs mapped for flowering and plant height corroborate previous reports (Murray et al. 2008b, 2011; Higgins et al. 2014) and one SNP was also simultaneously significant for flowering, FMP (Fresh Biomass Production) and DMP (Dry Biomass Production) (Souza et al. 2016, 2017b). The most significant SNP for lignin is on chromosome 6 within a region containing a gene belonging to the Small Auxin Upregulated RNA (SAUR) gene family, which are proteins responsive to auxin. These results and future studies will validate and contribute to developing new breeding strategies for biomass quality and composition in sorghum.

Lignin gene expression analysis was performed in young plants of the diverse panel to identify potential breeding targets and also better understand lignin synthesis in sorghum. Gene expression was evaluated by RT-qPCR using SybrGreen in a set of over 20 sorghum genes homologous to previously identified lignin genes in *Arabidopsis*. From the several genes tested, only one gene belonging to hydroxycinnamoyl transferase (HCT) family showed significant correlation ($P < 0.01$, $r = 0.58$) with lignin content (Souza et al. 2016). Functional studies and allelic variation of the HCT candidate gene within the diverse panel are being pursued so this gene may be used for future studies on lignin biosynthesis in sorghum, in addition to becoming a target for marker-assisted selection.

Overall, these findings contribute to a better understanding of the genetic architecture of traits related to bioenergy and forage production, and to the future implementation of marker-assisted selection in sorghum breeding programmes.

7 Improving disease resistance

The tropical conditions in Brazil favour the high incidence of several sorghum diseases, and various pathogens affect grain and biomass yield. The five major pathogens that occur in Brazil are *Colletotrichum sublineolum* Henn (anthracnose), *Exserohilum turcicum* (Pass.) K. J. Leonard and E. G. Suggs (sorghum leaf blight), *Puccinia purpurea* Cooke (rust), *Peronosclerospora sorghi* (W. Weston and Uppal) C. G. Shaw (downy mildew) and *Sphacelia sorghi* McRae (sugary disease or ergot). The diseases caused by these pathogens can result in severe losses in productivity and as there are not yet registered fungicides for foliar diseases of sorghum in Brazil, genetic resistance should be focused as a primary disease management strategy in breeding programmes.

First, in the quest to develop resistant hybrids, it is necessary to determine if there is variability in the pathogenic populations. This is because, with some fungi such as *C. sublineolum*, which can be highly aggressive, the relationship between the susceptibility and severity of anthracnose may reflect the high variability of the pathogen. Variability assessments have been conducted in the greenhouse using a differential series, consisting of hybrids or lines that are inoculated with single spore isolates of the pathogen, collected in different regions of the country. In the field, lines and hybrids, both experimental and pre-commercial, are grown in different regions of the country, and the severity of the disease is evaluated. The evaluation of disease severity in the greenhouse and in field environments allows for the classification of hybrids and lines as resistant, moderately resistant or susceptible, and allows the physiological differentiation of pathogen races.

The physiological identification of pathogenic races does not provide the genotypic distinction between them, which is important for developing gene introduction strategies, such as pyramiding. However, based on the population structure and the existence of physiological races it is possible to trace management strategies of resistance, such as the rotation of cultivars. As the identification of races, using molecular techniques, has not been efficient in genotypic distinction, the identification based on the reaction of resistance of lines or hybrids is still the most used.

The information obtained for lines is used to define the existence of association/dissociation of virulence in fungal population. Having virulence association, there is an assumption that the combination of two lines may not be effective to generate hybrids with resistance, since most of the physiological races are capable of infecting both strains. On the other hand, when the virulence is dissociated in the pathogen population, the derived resistance can be more stable.

The control of anthracnose is obtained using genetically resistant cultivars. Pereira et al. (2011) inoculated ten sorghum hybrids and their inbred lines, with twenty single spore isolates from different sorghum-producing regions of Brazil. The authors found no line or hybrid resistant to all isolates used, which shows the variability of existing isolates in Brazil. The lines CMSXS657, ATF14B, ATF08B and CMSXS210 had higher levels of resistance in the study, and the hybrid BRS305 was resistant to most isolates. BR001B (derived from Wheatland) and CMSXS222 were susceptible to 90% of the isolates. The hybrid BRS304 when released in the 1980s was one of the most planted in Brazil. It was in the market for a long time until becoming susceptible to the major existing races. This hybrid was subsequently replaced by the hybrids BRS308 and BRS310, with greater resistance to anthracnose.

In recent years, these hybrids also became susceptible and have been replaced by the newer hybrids BRS330 and BRS332, earlier this decade. Two new hybrids, BRS373 and BRS380, have been launched recently and are expected to replace the prior materials. The use of genotype rotations and the use of several hybrids reduce the severity of anthracnose and other diseases of sorghum resulting in higher productivity compared to continuous cultivation. The severity of the disease in the susceptible hybrid BR304 decreased by 42% and 37%, respectively, in rotations with the resistant hybrids IG150 and DAS740 (Silva et al. 2014b).

The results of several experiments conducted by Embrapa have generated knowledge about the virulence structure in the population of *C. sublineolum* that was used in selecting sorghum hybrids grown in cultivar rotation systems. Three commercial sorghum hybrids, one of them susceptible (BR304) and two resistant (1G150 and DAS740), were grown in a rotation system for seven seasons. Each hybrid was also grown in continuous plantations in the same trial, during all seasons (Table 3). There was a reduction in the severity of anthracnose on BR304 in all seasons when grown after one of the resistant hybrids,

Table 3 Sequence of sorghum hybrids in rotation system for seven planting seasons

	Crop season						
Cultivation system	2006*	2007	2007/08	2008/09**	2009/10	2010	2011
BR 304 continuous	BR 304	BR 304	BR 304	BR 304	BR 304	BR 304	BR 304
BR 304 after 1G150	1G150	1G150	1G150	1G150	1G150	1G150	1G150
BR 304 after DAS740	DAS740	DAS740	DAS740	DAS740	DAS740	DAS740	DAS740
1G150 continuous	BR 304	1G150	BR 304	1G150	1G150	BR 304	BR 304
DAS740 continuous	BR 304	DAS740	BR 304	DAS740	BR 304	DAS740	BR 304
1G150 after DAS740	DAS740	1G150	DAS740	1G150	DAS740	1G150	DAS740

2006* multiplication of inoculum; 2008/09** only the final disease severity was assessed. Table based on the article of Silva et al. (2015).

Table 4 Area under the disease progress curve for three sorghum genotypes, grown in continuous planting and rotation in five seasons

Cultivation system	2007	2007/08	2008/09	2009/10	2010	2011
BR 304 continuous	652.7 b	1743.2 c	*	2236.7 c	1507.9 c	2244.9 d
BR 304 after 1G150	29.0 a	1027.3 b	*	1G150**	865.6b	1929.6 c
BR 304 after DAS740	29.0 a	1093.8 b	*	654.6b	62.5 a	1223.3 b
1G150 continuous	39.7 a	51.7 a	*	62.1 a	67.2 a	42.7 a
DAS740 continuous	29.0 a	50.5 a	*	47.5 a	28.1 a	45.4 a
1G150 after DAS740	29.0 a	50.5 a	*	60.4 a	16.5 a	59.6 a

Table based on the article of Silva et al. (2015).

Table 5 Reduction in anthracnose severity in BR 304 (susceptible genotype) rotated with 1G150 and DAS740 (resistant genotypes)

	Crop season			
Cultivation system	2007/08	2009/10	2010	2011
BR 304 rotated with 1G150	41.1%	NE	42.6%	14.0%
BR 304 rotated with DAS 740	37.3%	70.7%	NE	45.5%

NE = Not evaluated. Percentages of disease reduction calculated by comparison with the total disease severity of BR 304, at each season and continuous planting. Table based on the article of Silva et al. (2015).

reaching up to 70% (Tables 4 and 5) (Silva et al. 2015). Also in this study, a reduction of sorghum leaf blight was observed in the hybrids 1G150 and DAS740 cultivated after BR 304 that is more resistant to the disease.

The management of genetic resistance was also evaluated by Costa et al. (2012), Valério et al. (2004) and Guimarães et al. (1998 a,b) who studied the use of dynamic multiline and mixtures of cultivars in order to reduce losses and preserve resistance genes to anthracnose for a longer period of time. Other studies refer to the integrated use of genetic resistance associated with the application of fungicides to reduce the use of pesticides based on genetic resistance levels to diseases (Costa et al. 2010).

Sorghum downy mildew, caused by *Peronosclerospora sorghi* [Weston & Uppal (Shaw)], was a disease restricted, until recently, to the Southern states of Brazil, but has acquired in recent years a greater importance in the Southeast and Midwest regions. It has caused problems primarily in seed production, particularly in the northwest of the state of Minas Gerais and in certain areas of São Paulo.

Considering that plants infected with *P. sorghi* in the early stages of development are sterile, it is easy to imagine the losses that may occur in susceptible or partially susceptible female lines when the conditions are favourable for disease development (Silva et al. 2014a). In a study to identify sources of resistance to sorghum downy mildew, Barbosa et al. (2004) evaluated 42 genotypes in natural infection conditions in field conditions. The genotypes CMSXS156 CMSXS157, CMSXS243, ARG-1, 8902, 9902054, 9910032, 9910296, Tx430, QL-3, SC170-6-17, CMSXS762 and BR304 were classified as highly resistant. Among these lines, SC170-6-17 and 9910296 presented 0% of systemic infection.

Freitas et al. (2008) evaluated four restorer lines (BR012R, CMSXS180R, CMSXS182R and 9910032R), four male sterile lines (ATF8A, ATF14A, ATF54A and CMSXS206A) and their hybrids, with inoculation of six different races of the fungus in the greenhouse. The authors observed that the restorer line 9910032R was resistant to six races of *Peronosclerospora sorghi*, while BR012R was resistant to one race, and the other strains were susceptible to all races. The line ATF14 A/B has excellent combining ability, and is the parent of several commercial BRS hybrids, but is highly susceptible to downy mildew. Resistant lines are in advanced stages of development.

Sorghum leaf blight, *H. turcicum*, is a disease that is increasing in importance, especially in sorghum in succession to soybeans in Central Brazil, because its causal agent is favoured by warmer temperatures prevailing in this region. The pathogen persists in the soil or glumes of seeds as mycelium and as conidia on infected crop debris. In the absence of a resistance reaction, the hyphae penetrate the vascular bundles and initiate the absorption of nutrients, and proliferate. It is assumed that the damage is caused by obstruction of the vascular bundles. Most restorer lines used in the Embrapa sorghum improvement programme are resistant, but there are interesting cases like {BR012R*(BR012R*SC566)}, which has good combining ability and is tolerant to aluminium toxicity, but its hybrids are highly susceptible to sorghum leaf blight. Incorporating resistance to this disease in new lines is an important component in the Embrapa sorghum improvement programme.

The 'sugary disease' of sorghum, known also as ergot, was diagnosed for the first time in Brazil in 1995. Currently, this disease has been severe and widespread in all regions of Brazil, becoming a serious problem for the seed industry. As the pathogen infects only the unfertilized ovaries, during anthesis all the climatic and biological factors that affect the production and vigour of pollen and/or prevent the normal opening of the anthers will favour the pathogen to induce and develop the sugary disease. The causal agent of this disease is the fungus *Sphacelia sorghi*, an imperfect form of *Claviceps africana*. There is no genetic resistance recorded for ergot control in sorghum. The use of fungicides to protect the flowers is an important and indispensable tool in the control of ergot in hybrid seed production fields. In the case of forage sorghum production, care should be used to select well-adapted cultivars more tolerant to low temperatures.

Integrated management systems of pests for sorghum using crop rotation, reduced tillage, rational use of fertilizers and resistant varieties are used to improve performance, reduce production costs and decrease environmental impact.

8 Sorghum pests

The occurrence of insect pests is one of the loss factors in sorghum production, whatever its purpose, and may, depending on infestation and kinds of pest species, derail production (Mendes et al. 2014). Thus, integrated pest management (IPM) is a fundamental part of the sorghum production system. Within the philosophy of IPM, in which the use of various control strategies is recommended, the natural resistance of plants plays an important role, especially in the context of the limited number of chemical insecticides registered for sorghum by the Ministry of Agriculture of Brazil.

Among the phytophagous insects present in the vegetative phase of sorghum, only a few species cause damage to the plant. The most harmful species for the sorghum crop are *Spodoptera frugiperda* (Smith) (Lepidoptera: Noctuidae), known as fall armyworm, and *Diatraea saccharalis* (Fabricius) (Lepidoptera: Crambidae), the sugarcane borer or stem borer. These insect pests are defoliating and borers, respectively, and can cause damage and production losses if they are not managed effectively. In addition, aphids, which carry viral diseases, considered secondary pests, are also present in sorghum-production areas in Brazil.

The fall armyworm emerges and scrapes the leaf blade. Then, it heads to the whorl of the plant, where it continues feeding and causing severe damage. The fall armyworm is brownish with three light stripes, having an inverted 'Y' on the head, and feeds mainly on new leaves, and later on the sorghum leaf whorl (Fig. 8).

The sugarcane borer is an insect that opens galleries in the plant stem, at an early stage of plant development, causing the death of the apical bud, a symptom known as 'dead heart'. In more developed plants, it causes symptoms of red rot in the stem, plant lodging, and reduction of plant weight and white panicle, among other symptoms (Fig. 9). The

Figure 8 Forage sorghum plants with symptoms of *Spodoptera frugiperda* (fall armyworm) infestation (a) and *S. frugiperda* (fall armyworm) (b).

Figure 9 Symptoms of damage caused by *Diatraea saccharalis* (a) and pupal *D. saccharalis* (b) in sorghum stem.

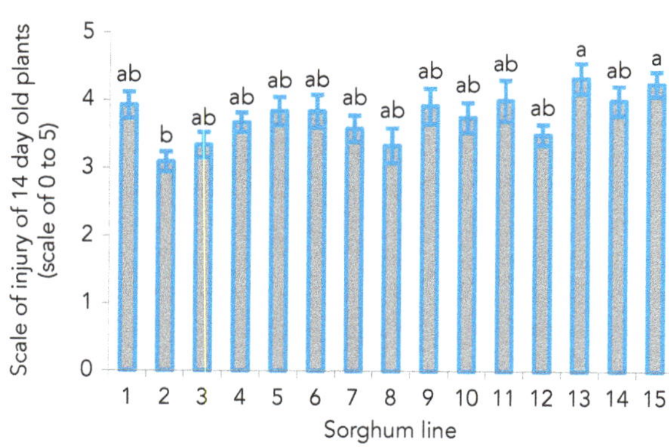

Figure 10 Average grade of injuries (± ep) fourteen days after infestation with *Spodoptera frugiperda* in sorghum lines. Means followed by the same letter do not differ by *t*-test.

symptom of sugarcane borer attack in sorghum is often little noticed in the development of the crop, as the attack takes place within the stem. In grain sorghum, especially, a typical symptom is white panicle, in which there is no filling of grains due to the retention of sap flow caused by the fall armyworm. The red rot disease appears because of the action of anthocyanins present in sorghum. These substances help the plant defend against pests and pathogens, and because of their accumulation in the affected areas, red pigmentation may occur.

Plant resistance is one of the main strategies for IPM. According to Vendramim and Guzzo (2009), a resistant plant can be defined as one that, because of their genotypic constitution, is less damaged than others that are on equal terms to the attack of a kind of insect. Based on these variations, the resistance can be classified into three types: non-preference or antixenosis, antibiosis, and tolerance.

There are, in sorghum genotypes, varying levels of natural resistance to *S. frugiperda*, as verified by Carvalho et al. (2014) for sweet sorghum genotypes that had different injury notes (Fig. 10).

Antibiosis aspects have been detected in different genotypes of sweet sorghum, indicating that this is a likely mechanism of resistance to be exploited in breeding programmes. According to Cortez and Waquil (1997), there are sorghum hybrids with high resistance to *S. frugiperda* and hybrids as susceptible as maize, whose losses, as for the grain sorghum hybrid BR300, can reach 25% of grain production. AF-28, EA 261 and SC 599-6-3 have been reported to be resistant to *S. frugiperda* (Freed and Honey, McMillian and Starks 1978) (Lucena 1978). Viana et al. (1998) observed mortality rates for grain sorghum commercial genotypes ranging from 45 to 60%, showing that, even for these commercial genotypes, the levels are variable and confirm the hypothesis that antibiosis is a common resistance mechanism.

Varying levels of resistance in sorghum genotypes were also detected for *D. saccharalis*. For grain sorghum, Waquil et al. (2001) reported infestation rates ranging between 15 and 95% for 50 different sorghum genotypes. In this case, the authors related resistance

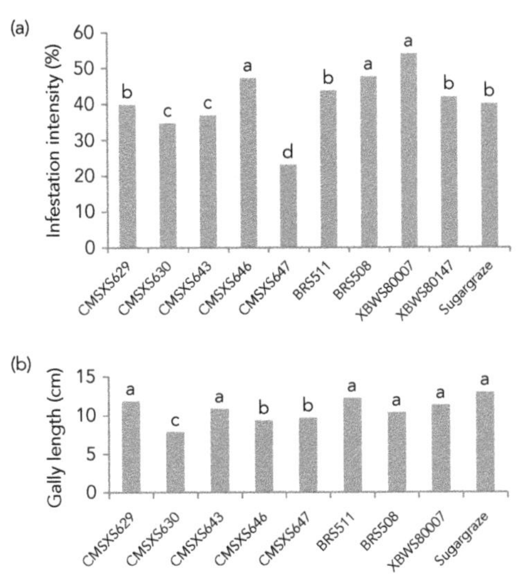

Figure 11 Infestation intensity (a) and gallery size (b) of *Diatraea saccharalis* in different genotypes of sweet sorghum. Extracted from Araújo 2014.

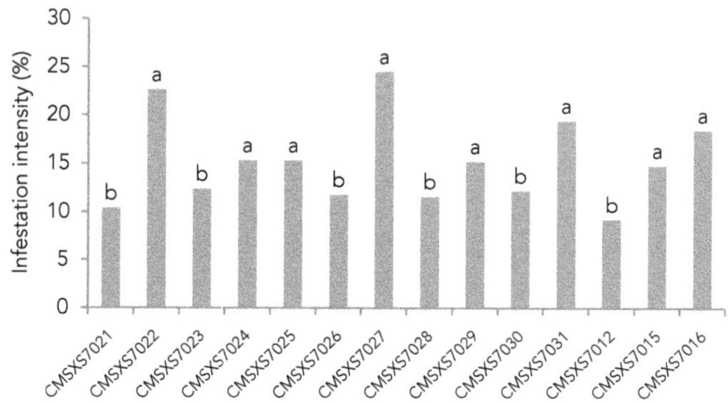

Figure 12 Infestation intensity of *Diatraea saccharalis* in different genotypes of biomass sorghum. Extracted from Araújo 2014.

with non-preference and antibiosis. Among the types of sorghum, sweet sorghum has greater damage under the same intensity of infestation. Pereira et al. (1987) showed that the sweet sorghum genotype BRS505 (Wray) is a probable source of resistance to this pest species.

The difference among the genotypes of sweet sorghum for resistance to *D. saccharalis* can be seen in the intensity of infestation, which reflects the number of galleries caused by the pest on each stem, as well as the gallery size. In this sense, it has been suggested that the plant has less resistance to insect pest attack (Fig. 11). These authors also noted that a high intensity of infestation did not necessarily result in larger major galleries, suggesting that different resistance mechanisms are involved.

Even for biomass sorghum genotypes, which present lower infestation indices than those presented by sweet sorghum, some genotypes may be important in the breeding programme because they have lower infestation rates (Fig. 12).

9 Conclusion

Over the past four decades, both the public and private sectors have collaborated in developing and improving sorghum cultivars for the tropical savannas of Central South America. Currently, there is a movement to increase the use of molecular breeding tools to improve productivity and quality of grain, forage and energy sorghum cultivars. Strides have been made to overcome both biotic and abiotic stresses not frequently occurring in temperate environments. Continuous gains are expected in the future with continued collaboration between the public and private sectors.

10 Where to look for further information

Future advances and breakthroughs in sorghum molecular breeding can be accessed at: https://www.embrapa.br/milho-e-sorgo. Questions and comments can be directed to: www.cnpms.embrapa.br/sac/.

11 References

Baligar, V. C. and Fageria, N. K. 1997. Nutrient use efficiency in acid soils: nutrient management and plant use efficiency. In Moniz, A. C., Furlani, A. M. C., Fageria, N. K. and et al. (Eds), *Plant-soil Interactions at Low pH: Sustainable Agriculture and Forestry Production*. Campinas: Sociedade Brasileira de Ciência do Solo, pp. 75–95.

Barbosa, F. C. R., Casela, C. R., Pfenning, L. H. and Santos, F. G. 2004. Identification of sources of resistance in sorghum to Peronosclerospora sorghi. *Fitopatologia Brasileira*, 30: 522–4.

Batista, P. S. C. 2016. Fenotipagem de genótipos de sorgo granífero para tolerância à seca. Dissertação de mestrado. Universidade Estadual de Montes Claros, 129p.

Caniato, F. F., Guimarães, C. T., Hamblin, M., Billot, C., Rami, J.-F., Hufnagel, B., Kochian, L. V., Liu, J., Garcia, A. A. F., Hash, C. T., Ramu, P., Mitchell, S., Kresovich, S., Oliveira, A. C., Avellar, G., Borém, A., Glaszmann, J.-C., Schaffert, R. E. and Magalhaes, J. V. 2011. The relationship between population structure and aluminum tolerance in cultivated sorghum. *PLoS ONE*, 6: e20830.

Caniato, F. F., Guimarães, C. T., Schaffert, R. E., Alves, V. M. C., Kochian, L. V. Borém, A., Klein, P. E. and Magalhaes, J. V. 2007. Genetic diversity for aluminum tolerance in sorghum. *Theoretical and Applied Genetics*, 114: 863–76.

Caniato, F. F., Hamblin, M. T., Guimaraes, C. T., Zhang, Z., Schaffert, R. E., Kochian, L. V. and Magalhaes, J. V. 2014. Association mapping provides insights into the origin and the fine structure of the sorghum aluminum tolerance locus, *Alt_{SB}*., *PLoS ONE*, 9: e87438.

Cardoso, R. M., Pires, D. A. A., Rocha Junior, V. R., Reis, S. T., Sales, E. C. J., Sales, D. D., Gerassev, L. C., Rodrigues, J. A. S. and Lima, L. O. B. 2012. Avaliação de híbridos de sorgo para silagem por meio da degradabilidade in situ. *Revista Brasileira de Milho e Sorgo*, 11: 106–14.

Carvalho, Jr. G., Schaffert, R. E., Malosetti, M., Viana, J. H. M., Menezes, C. B., Silva, L. A., Guimaraes, C. T., Coelho, A. M., Kochian, L. V., van Eeuwijk, F. A. L. and Magalhaes, J. V. 2016. Back to acid soil fields: The citrate transporter SbMATE is a major asset for sustainable grain yield for sorghum cultivated on acid soils. *G3: Genes|Genomes| Genetics*, 6: 476–84. https://doi.org/10.1534/g3.115.025791

Carvalho, S. S. de S., Carvalho, E. A. R. de, Santos, A. E. dos, Silva, L. O., Mendes, S. M. and Menezes, C. B. de. 2014. Resistência de linhagens de sorgo à Spodoptera frugiperda (Smith, 1797) (Lepidoptera: Noctuidae). In *CONGRESSO NACIONAL DE MILHO E SORGO, 30.; SIMPÓSIO SOBRE LEPDÓPTEROS COMUNS A MILHO, SOJA E ALGODÃO, 1., 2014, Salvador*. Eficiência nas cadeias produtivas e o abastecimento global: resumos expandidos. Sete Lagoas: Associação Brasileira de Milho e Sorgo.

Castro, F. M. R., Bruzi, A. T., Nunes, J. A. R., Parrella, R. A. C., Lombardi, G. M. R., albuquerque, C. J. A. and Lopes, M. July 2015. Agronomic and energetic potential of biomass sogrhum genotypes. *American Journal of Plant Sciences*, 6: 1862–73.

CEPEA. 2015. Site: www.cepea.esalq.usp.br/.../Cepea_Perspectivas%20Agroneg2015_relator... Acesso em 14 agosto de 2015.

CONAB. 2016a. Acompanhamento de safra brasileira: grãos. Brasília, DF: Companhia Nacional de Abastecimento, vol. 7, pp. 1–158.

CONAB. 2016b. Companhia Nacional de Abastecimento. 1° Levantamento da cana-de-açúcar: maio/2016. Brasília. Acesso em: 31 mai.2016.

Cortez, M. G. R. and Waquil, J. M. 1997. Influência de cultivar e nível de infestação de Spodoptera frugiperda (JE Smith) (Lepidoptera: Noctuidae) no rendimento do sorgo. *An. Soc. Entomol. Brasil*, 26(n. 2): 407–10.

Costa, R. F., Pires, D. A. de A., Moura, M. M. A., Rodrigues, J. A. S., Rocha Junior, V. R. and Tolentino, D. C. Apr./June 2016. In situ degradability of dry matter and fibrous fraction of sorghum silage. Acta Scientiarum. *Animal Sciences, Maringá*, 38. (n 2): 171–6. https://doi.org/10.4025/actascianimsci.v38i2.29576

Costa, R. V., Cota, L. V., Silva, D. D. and Parreira, D. F. 2010. *Uso Integrado da Resistência Genética e Aplicação de Fungicidas para o Manejo da Antracnose do Sorgo*. Sete Lagoas: Embrapa Milho e Sorgo, 7p. (Embrapa Milho e Sorgo. Circular técnica, 143).

Costa, R. V., Zambolim, L., Silva, D. D., Cota, L. V. and Casela, C. R. fev. 2012. Utilização de multilinhas dinâmicas para o manejo da antracnose do sorgo. *Pesq. Agropec. Bras., Brasília*, 47(2): 173–80. https://doi.org/10.1590/s0100-204x2012000200004.

Durães, N. N. L. 2014. Heterose em sorgo sacarino, 96p. Dissertação (Mestrado em Genética e Melhoramento de Plantas) - Universidade Federal de Lavras, Lavras.

Elshire, R. J., Glaubitz, J. C., Sun, Q., Poland, J. A., Kawamoto, K., Buckler, E. S. and Mitchell, S. E. 2011. A robust, simple genotyping-by-sequencing (GBS) approach for high diversity species. *PLoS ONE*, 6(5): e19379.

Faria Junior, W. G., Gonçalves, L. C., Mauricio, R. M., Rodrigues, N. M., Colodo, J. C. N., Faria, W. G. and Souza, L. F. 2010. Avaliação das silagens de sorgo BRS-610 em sete estádios de maturação pela técnica in vitro semiautomática de produção de gases. *Arquivo brasileiro de medicina veterinária e zootecnia*, 62: 898–905.https://doi.org/10.1590/s0102-09352010000400020.

Felderhoff, T. J., Murray, S. C., Klein, P. E., Sharma, A., Hamblin, M. T., Kresovich, S., Vermerris, W. and Rooney, W. L. 2012. QTLs for Energy-related Traits in a Sweet × Grain Sorghum [*Sorghum bicolor* (L.) Moench] Mapping Population. *Crop Science*, 52: 2040–9.

Fernandes, G., Braga, T. G., Fischer, J., Parrella, R. A. C., Resende, M. M. de and Cardoso, V. L. 2014. Evaluation of potential ethanol production and nutrients for four varieties of sweet sorghum during maturation. *Renewable Energy*, Oxford, 71: 518–24.

Figueiredo, U. J. de, Nunes, J. A. R., Parrella, R. A. da C., Souza, E. D., Silva, A. R. da, Emygdio, B. M., Machado, J. R. A. and Tardin, F. D. 2015. Adaptability and stability of genotypes of sweet sorghum by GGEBiplot and Toler methods. *Genetics and Molecular Research, Ribeirão Preto*, 14(n. 3): 11211–21. https://doi.org/10.4238/2015.september.22.15.

Freitas, M. E., Casela, C. R., Silva, D. D., Maciel, C. T. and Costa, R. V. da. 2008. Resistência de linhagens de sorgo e seus híbridos a Peronosclerospora sorghi. *Tropical Plant Pathology*, Brasília, 33: S200. Edição dos Resumos do XLI Congresso Brasileiro de Fitopatologia, Belo Horizonte, MG, ago.

Furukawa, J., Yamaji, N., Wang, H., Mitani, N., Murata, Y., Sato, K., Katsuhara, M., Takeda, K. and Ma, J. F. 2007. An aluminum-activated citrate transporter in barley. *Plant and Cell Physiology*, 48: 1081–91.

Gourley, L. M. 1987. Finding and utilizing exotic Al-tolerant sorghum germplasm. In Gourley, L. and Salinas, J. (Eds), *Sorghum for Acid Soils*. Proceedings of a workshop on evaluating soghum for tolerance to Al-toxic tropical soils in Latin America, CA, Colombia, pp. 293 –309.

Guan, Y. A., Wang, H. L., Qin, L., Zhang, H. W., Yang, Y. B., Gao, F. J., Li, R. Y. and Wang, H. G. 2011. QTL mapping of bio-energy related traits in Sorghum. *Euphytica*, 182(3): 431–40.

Guimarães, F. B.Casela, C. R., do Vale, R., Zambolim, F. X., Santos, L. and dos, F. G. 1998a. Resistência dilatória de genótipos de sorgo a diferentes raças de *Colletotrichum graminicola. Summa Phytopathologica*, 24(n.2): 136 –40.

Guimarães, F. B., Casela, C. R., Santos, F. G. dos and Ferreira, A. S. 1998b. Controle da antracnose do sorgo através da utilização de misturas de cultivares. *Summa Phytopathologica*, 24: 131–5.

Gamuyao, R., Chin, J. H., Pariasca-Tanaka, J., Pesaresi, P., Catausan, S., Dalid, C., Slamet-Loedin, I., Tecson-Mendoza, E. M., Wissuwa, M. and Heuer, S. 2012. The protein kinase Pstol1 from traditional rice confers tolerance of phosphorus deficiency. *Nature*, 488: 535–39.

Hansen, J., Sato, M. and Ruedy , R. September 2012. Perception of climate change. *PNAS*, 109: 3.

Higgins, R. H., Thurber, C. S., Assaranurak, I. and Brown, P. J. 2014. Multiparental mapping of plant height and flowering time QTL in partially isogenic sorghum families. *G3: Genes|Genomes|Genetics*, 4(9): 1593–602.

Hufnagel, B., Sousa, S. M., Assis, L., Guimaraes, C. T., Leiser, W., Corradi, G., Negri, B., Larson, B. G., Shaff, J. E., Pastina, M. M., Barros, B. A., Weltzien, E., Rattunde, H. F. W., Viana, J. H., Clark, R. T., Falcão, A., Gazaffi, R., Garcia, A. A. F., Schaffert, R. E., Kochian, L. V. and Magalhaes, J. V. 2014. Duplicate and conquer: multiple homologs of phosphorus-starvation tolerance 1 enhance phosphorus acquisition and sorghum performance on low-P soils. *Plant Physiology*, 166: 659–77.

Lin, Y. R., Schertz, K. F. and Paterson, A. H. 1995. Comparative analysis of QTLs affecting plant height and maturity across the Poaceae, in reference to na interespecific sorghum population. *Genetics*, 141: 391–411.

Lira, M. A. and Faris, M. A. 1979. Aspectos gerais da cultura do sorgo no Brasil-Região Nordeste. In *IMPOSIO BRASILEIRO DE SORGO, 1., 1977, Brasília, DF*. Anais... Sete Lagoas: EMBRAPA-CNPMS, pp. 31–8.

Lombardi, G. M. R., Nunes, J. A. R., Parrella, R. A. C., Teixeira, D. H. L., Bruzi, A. T., Durães, N. N. L., Fagundes, T. G. 2015. Path analysis of agro-industrial traits in sweet sorghum. *Genetics and Molecular Research, Ribeirão Preto*, 14. (n 4): 16392–402. https://doi.org/10.4238/2015.december.9.8.

Lucena, A. I. T. 1978. Comportamento de genotipos de sorgo granífero em relação à *Spodoptera frugiperda* (Smith, 1797) e tipos de resistência envolvidos. Tese de mestrado, ESALQ/USP, Piracicaba, SP, 105p.

Machado, F. S., Rodríguez, M. N., Gonçalves, L. C., Rodrigues, J. A. S., Ribas, M. N., Lobato, F. C. L., Veiga, I. R. F. M., Guimarães Júnior, R. and Pereira, L. G. R. 2014. Valor nutricional de híbridos de sorgo em diferentes estádios de maturação. *Arquivo Brasileiro de Medicina Veterinária e Zootecnia*, 66(n.1): 244–52. https://doi.org/10.1590/s0102-09352014000100033.

Magalhaes, J. V., Garvin, D. F., Wang, Y., Sorrells, M. E., Klein, P. E., Schaffert, R. E., Li, L. and Kochian, L.V. 2004. Comparative mapping of a major aluminum tolerance gene in sorghum and other species in the Poaceae. *Genetics*, 167: 1905–14.

Magalhaes, J. V., Liu, J., Guimarães, C. T., Lana, U. G. P., Alves, V. M. C., Wang, Y.-H., Schaffert, R. E., Hoekenga, O. A., Piñeros, M. A., Shaff, J. E., Klein, P. E., Carneiro, N. P., Coelho, C. M., Trick, H. N. and Kochian, L. V. 2007. A gene in the multidrug and toxic compound extrusion (MATE) family confers aluminum tolerance in sorghum. *Nature Genetics* 39: 1156–61.

Magalhães, R. T., Gonçalves, L. C., Borges, I., Rodrigues, J. A. S. and Fonseca, J. F. 2010. Produção e composição bromatológica de vinte e cinco genótipos de sorgo (*Sorghum bicolor* (L.) Moench). *Arquivo Brasileiro de Medicina Veterinária e Zootecnia*, 62(n.3).https://doi.org/10.1590/s0102-09352010000300034.

MAPA. 2015. Site: www.agricultura.gov.br/.../Dados%20de%20rebanho%20bovino%20e%... Acesso em 14 agosto de 2015.

McMillan, W. W. and Starks, K. J. 1978. Greenhouse and laboratory screening of sorghum lines for resistance to fall armyworm laruae. *Journal of Economics and Entomology*, 59: 1462–3. https://doi.org/10.1093/jee/60.5.1462

Melo, J. O., Lana, U. G. P., Piñeros, M. A., Alves, V. M. C., Guimaraes, C. T., Liu, J., Zheng, Y., Zhong, S., Fei, Z., Maron, L., Schaffert, R. E., Kochian, L. V. and Magalhaes, J. V. 2013. Incomplete transfer of accessory loci influencing SbMATE expression underlies genetic background effects for aluminum tolerance in sorghum. *Plant Journal*, 73: 276–88.

Mendes, S. M., Waquil, J. M., Rodrigues, J. A. S., Sampaio, M. V. and Viana, P. A. jan./fev. 2014. Manejo de pragas na cultura do sorgo. *Informe Agropecuário, Belo Horizonte*, 35(n. 278): 89–99.

Menezes, C. B. de (Ed.). 2015a. Sorgo granífero: estenda sua safrinha com segurança. Sete Lagoas: Embrapa Milho e Sorgo, 65p. (Embrapa Milho e Sorgo. Documentos, 176). Autores: Alexandre Ferreira da Silva, Álvaro Vilela de Resende, André May, Antônio Marcos Coelho, Cicero Beserra de Menezes, Dagma Dionísia da Silva, Emerson Borghi, Evandro Chartuni Mantovani, Flávio Dessaune Tardin.

Menezes, C. B. de, Ribeiro, A. da S., Tardin, F. D., Carvalho, A. J. de, Bastos, E. A., Cardoso, M. J., Portugal, A. F., Silva, K. J. da, Santos, C. V. dos and Almeida, F. H. L. de. 2015b. Adaptabilidade e estabilidade de linhagens de sorgo em ambientes com e sem restrição hídrica. *Revista Brasileira de Milho e Sorgo,* Sete Lagoas, 14(n. 1): 101–15.

Menezes, C. B., Saldanha, D. C., Santos, C. V., Andrade, L. C., Julio, M. F. M., portugal, A. F. and Tardin, F. D. 2015c. Evaluation of grain yield in sorghum hybrids under water stress. *Genetics and Molecular Research*, Ribeirão Preto, 14(n. 4): 12675–83. Lagoas, v 14, n. 1, pp. 101–15.

Miller, F. R., Barnes, D. K. and Cruzado, H. J. 1968. Effect of tropical photoperiods on the growth of sorghum when grown in 12 monthly plantings. *Crop Science* 8: 499–502.

Mocoeur, A., Zhang, Y., Liu, Z., Shen, X., Zhang, L., Rasmussen, S. K. and Jing, H. 2015. Stability and genetic control of morphological, biomass and biofuel traits under temperate maritime and continental conditions in sweet sorghum (*Sorghum bicolour*). *Theoretical Applied Genetics*, 128, 1685–701.

Murphy, R. L., Klein, R. R., Morishige, D. T., Brady, J. A., Rooney, W. L., Miller, F. R., Dugas, D. V., Klein, P. E. and Mullet, J. E. 2011. Coincident light and clock regulation of pseudoresponse regulator protein 37 (PRR37) controls photoperiodicflowering in sorghum. *PNAS*, 108(39): 16469–74.

Murray, S. C., Rooney, W. L., Hamblin, M. T., MitchelL, S. E. and Kresovich, S. 2009. Sweet sorghum genetic diversity and association mapping for brix and height. *The Plant Genome*, 2(1): 48.

Murray, S. C., Rooney, W. L., Mitchell, S. E., Sharma, A., Klein, P. E., Mullet, J. E. and Kresovich, S. 2008b. Genetic improvement of sorghum as a biofuel feedstock: II. QTL for stem and leaf structural carbohydrates. *Crop Science*, 48(6): 2180–93.

Murray, S. C., Sharm, A., Rooney, W. L., Klein, P. E., Mullet, J. E., Mitchell, S. E. and Kresovich, S. 2008a. Genetic improvement of sorghum as a biofuel feedstock: I. QTL for stem sugar and grain nonstructural carbohydrates. *Crop Science*, 48: 2165–79. doi: 10.2135/cropsci2008.01.0016.

Neves, A. L. A., Santos, R. D., Pereira, L. G. R., Oliveira, G. F., Schierer, C. B., Verneque, R. S. and Mcallister, T. 2015. Agronomic characteristics, silage quality, Intake and digestibility of five new Brazilian sorghum cultivars. *Journal of Agricultural Science*, 153: 371–80.

Panizzi, R. C. and Fernandes, N. G. 1977. Doenças do sorgo (*Sorghum bicolor* (L.) Moench). In Camargo, E. A. and Rezende, J. A. M. (Eds), *Manual de Fitopatologia – Doenças das Plantas Cultivadas*, vol. 2. São Paulo: editora Agronômica Ceres, pp. 676–89.

Parrella, R. A. da C. 2011. Melhoramento genético do sorgo sacarino. *Agroenergia em Revista, Brasília*, 2. (no. 3): 8–9.

Parrella, R. A. C., Rodrigues, J. A. S., Tardin, F. D., Damasceno, C. M. B. and Schaffert, R. E. 2010. *Desenvolvimento de híbridos de sorgo sensíveis ao fotoperíodo visando alta produtividade de biomassa*. Sete Lagoas: Embrapa Milho e Sorgo, 25. (Boletim de Pesquisa e Desenvolvimento, 28).

Pereira, I. S., Silva, D. D. DA, Casela. C. R., Tardin, F. D. and Abreu, M. S. de. abr./jun. 2011. Resistência de linhagens genitoras e híbridos simples de sorgo a Colletotrichum sublineolum, agente causal da antracnose. *Revista Caatinga, Mossoró*, 24(n. 2): 46–51.

Pereira, R. P., da Costa, R. A. and Ferreira, J. M. 1987. Incidência de Diatraea em genótipos de sorgo sacarino em diferentes épocas de plantio. *Pesquisa Agropecuária Brasileira*, 22. (n 1): 23–5.

Pfeiffer, T. W. J. Bitzer, M. J., Toy, J. J. and Pedersen, J. F. September-October 2010. Heterosis in sweet sorghum and selection of a new sweet sorghum hybrid for use in syrup production in Appalachia. *Crop Science*, 50: 1788.

Ritter, K. B., Jordan, D. R., Chapman, S. C., Godwin, I. D., Mace, E. S. and Lynne, C. 2008. Identification of QTL for sugar-related traits in a sweet 3 grain sorghum (*Sorghum bicolor* L. Moench) recombinant inbred population. *McIntyre Molecular Breeding*, 22: 367–84, DOI: 10.1007/s11032-008-9182-6.

Rocha, M. J. da. 2016. Capacidade combinatória de linhagens e seleção de híbridos de sorgo sacarino. 2016, 77p. Dissertação (Mestrado em Genética e Melhoramento de Plantas) - Universidade Federal de Lavras, Lavras.

Rodrigues, J. A. S., Santos, F. G., Shaffert, R. E., Ferreira, A. S., Casela, C. S. and Pitta, G. V. E. 2004. BRS 610- hibrido de sorgo forrageiro para produção de silagem de alta qualidade. Sete Lagoas: Embrapa Milho e Sorgo. Comunicado Técnico Embrapa Milho e Sorgo.

Rooney, W. L. 2010. Genetics and Cytogenetics In Smith, C. W. and Frederiksen, R. A. (Eds), *Sorghum: Origin, History, Technology, and Production*. Wile Series in Crop Science, Series Editor, Texas A & M University, College Station, 261–308 pp.

Rooney, W. L. and Aydin, S. 1999. Genetic control of a photoperiod-sensitive response in *Sorghum bicolor* (L.) Moench. *Crop Science*, Madison, 39: 397–400 https://doi.org/10.2135/cropsci1999.0011183x0039000200016x.

Sabadin, P., Malosetti, M., Boer, M., Tardin, F., Santos, F., Guimarães, C., Gomide, R., Camilo, A., Albuquerque, P., Caniato, F., Mollinari, M., Margarido, G., Oliveira, B., Schaffert, R., Garcia, A., Eeuwijk, F. and Magalhaes, J. V. 2012. Studying the genetic basis of drought tolerance in sorghum by managed stress trials and adjustments for phenological and plant height differences. *Theoretical and Applied Genetics*, 124: 1389–402.

Santana, M. C. B. 2013. Desempenho produtivo de híbridos comerciais de sorgo granífero sob deficiência hídrica. 57 f. Dissertação (Mestrado em Agronomia) - Universidade Federal do Piauí, Teresina.

Santos, F. G., Casela, C. R. and WAquil, J. M. 2005. Melhoramento de sorgo. In Borém, A. (Ed.), *Melhoramento de espécies cultivadas* , 2ª Ed. Viçosa: UFV

Sasaki, T., Yamamoto, Y., Ezaki, B., Katsuhara, M., Ahn, S. J., Ryan, P. R., Delhaize, E. and Matsumoto, H. 2004. A wheat gene encoding an aluminum-activated malate transporter. *The Plant Journal*, 37: 645–53.

Schaffert, R. E. and Parrella, R. A. da C. 2012. Planejamento Industrial. In May, A., Duraes, F. O. M., Pereira Filho, I. A., Schaffert, R. E. and Parrella, R. A. da C. (Eds), *Sistema Embrapa de produção agroindustrial de sorgo sacarino para bioetanol: Sistema BRS1GTecnologia Qualidade Embrapa.* Sete Lagoas: Embrapa Milho e Sorgo, 118p. (Embrapa Milho e Sorgo. Documentos, 139).

Schaffert, R. E., Alves, V. M. C., Pitta, G., Bahia, A. and Santos, F. 2001. Genetic variability in sorghum for efficiency and responsiveness. In *Developments in Plant and Soil, Symposium. Plant Nutrition: Food Security and Sustainability of Agro-ecosystems.* Netherlands: Academics, pp. 72–3.

Shiringani, A. L., Frisch, M. and Friedt, W. 2010. Genetic mapping of QTLs for sugar-related traits in a RIL population of *Sorghum bicolor* L. Moench. *Theoretical and Applied Genetics*, 121(2): 323–36.

Silva, D. D. da, Cota, L. V., Costa, R. V. da, Parreira, D. F. 2014a. Manejo de doenças. In Borém, A., Pimentel, L. D. and Parrella, R. A. da C. (Ed.). *Sorgo: do plantio à colheita.* Viçosa, MG: UFV. cap. 10, pp. 242–65.

Silva, D. D., Costa, R. V., Cota, L. V., Figueiredo, J. E. F., Casela, C. R. and Lanza, F. E. 2015. Genotype rotation for leaf anthracnose disease management in sorghum. *Crop Protection*, Guildford, 67: 145–50. Publicado online em 3 nov. 2014b.

Silva, D. D., Costa, R. V., Cota, L. V., Figueiredo, J. E. F., Casela, C. R. and Lanza, F. E. 2015. Genotype rotation for leaf anthracnose disease management in sorghum. *Crop Protection*, 67: 145–50.

Souza, V. F. de, Parrella, R. A. da C., Tardin, F. D., Costa, M. R., Carvalho Junior, G. A. de and Schaffert, R. E. 2013. Adaptability and stability of sweet sorghum cultivars. *Crop Breeding and Applied Biotechnology*, Londrina, 13: 144–51.

Souza, V. F. 2016. Genetic mapping and functional analysis of agroindustrial traits in sorghum for bioenergy production. Thesis, D.Sc., Universidade Federal de São João del-Rei, UFSJ, Brazil.

Souza, V. F., Pastina, M. M., Parrella, R. A. C., Noda, R. W., Simeone, M. L. F., Magalhães, J. V., Schaffert, R. E. and Damasceno, C. M. B. 2017b. 'Genome-Wide Association Studies for Lignin Content and Biomass Production in a Diverse Sorghum Panel Using Genotyping-by-Sequencing' (Manuscript in preparation).

Teixeira, A. De M., Ribeiro Junior, G. De O., Velasco, F. O., Faria Júnior, W. G.De, Rodriguez, N. M., Rodrigues, J.A.S., Mcallister, T. I. M.; Gonçalves, L. C. 2014. Intake and digestibility of sorghum (*Sorghum bicolor* L. Moench) silages with different tannin contents in sheep. *Revista Brasileira de Zootecnia*, 43: 14–19.

Valério, H. M., Casela, C. R., Resende, M. A. and Santos, F. G. 2004. Variability of the anthracnose fungus *Colletotrichum graminicola* in sorghum genotype mixtures. *Fitopatologia Brasileira*, 29: 567–9. https://doi.org/10.1590/s0100-41582004000500019.

Vance, C. P. and Chiou, T. J. 2011. Phosphorus focus editorial. *Plant Physiology*, 156: 987–8.

Vendramim, J. D. and Guzzo, E. C. 2009. Resistência de plantas Informação Tecnológica; Londrina: Embrapa Soja, 2009. e a bioecologia e nutrição dos insetos. In: Panizzi, A. R. and Parra, J. R. P. (Eds), *Bioecologia e nutrição de insetos: base para o manejo integrado de pragas.* Brasília: Embrapa Informação Tecnológica; Londrina: Embrapa Soja, pp.1055–105.

Viana, P. A., Waquil, J. M., Cruz, I., Santos, J. P. and Rodrigues, J. A. S. 1998. Avaliação das cultivares de sorgo granífero incluídas no Ensaio Nacional (97/98) para resistência a Elasmopalpus lignosellus (Zeller) e Spodoptera frugiperda (J. E. Smith). In *CONGRESSO NACIONAL DE MILHO E SORGO, 22., 1998, Recife*. Globalização e segurança alimentar: anais. Recife: IPA.

Von Uexküll, H. R. and Mutert, E. 1995. Global extent, development and economic impact of acid soils. *Plant Soil*, 171: 1–15.

Waquil, J. M., Cruz, I. and Viana, P. A. 1986. Pragas do sorgo. In: Sorgo: Uma opção agrícola. *Informe Agropecuário*, 12(144): 46–51.

Waquil, J. M., Rodrigues, J. A. S., Santos, F. G., Ferreira, A. S., Vilella, F. M. F. and Foster, J. E. 2001. Resistance of commercial hybrids and lines of sorghum, *Sorghum bicolor* (L.) Moench., to Diatraea saccharalis (Fabr.) (Lepidoptera: Pyralidae). *Neotropical Entomology*, Londrina, 30(4): 661–8.

Chapter 8

Improving production and utilization of sorghum in Asia

Aruna C., B. Dayakar Rao, Vilas A. Tonapi and T. G. Nageshwar Rao, Indian Institute of Millets Research, India

1 Introduction

Sorghum (*Sorghum bicolor* (L.) Moench) is the fifth major cereal of the world after maize, paddy, wheat and barley as per Food and Agriculture Organization (FAO) production data of 2014. It is crucial to the world food economy and contributes to household food security in many of the world's poorest regions of both Africa and Asia. Sorghum is an important crop in the semi-arid tropics of the world because of its tolerance to various stresses, including drought, heat and salinity as well as its potential to produce substantial yields even under low input conditions (Ashok Kumar et al. 2011). Sorghum which is a C_4 species has a higher photosynthetic ability and an ability to utilize nitrogen and water more efficiently. Hence, it is genetically suited to hot and dry agroecological conditions with low rainfall and drought which are mostly unsuitable for growing other food crops. Both grain and stover are highly valued products from sorghum which is grown widely for both food and feed. In the developing countries of Asia and Africa, it is predominantly grown for food purposes, while in the United States, Australia and China, it is grown for livestock feed and fodder purposes. In India, sorghum is grown in both the rainy and post-rainy seasons. While rainy-season sorghum grain is used for both human consumption and livestock feed, post-rainy-season produce is used primarily for human consumption.

http://dx.doi.org/10.19103/AS.2017.0015.28

Production is generally more intensive in Asia, where fertilizer and improved seed are used far more widely compared to Africa. Currently, among the Asian countries, India, China and Thailand are playing a major role in the development of sorghum cultivars in the public and private sectors. In many countries, the predominant industry is agriculture, and sorghum is typically grown by resource-poor farmers with small landholdings (House et al. 2000).

The origin and early domestication of sorghum are hypothesized to have taken place in northeastern Africa or at the Egyptian–Sudanese border around 5000–8000 years ago (Mann et al. 1983). The Indian subcontinent is considered to be the secondary origin of sorghum, with evidence of early cereal cultivation discovered at an archaeological site in western parts of Rojdi (Saurashtra) dating back about 4500 years (Vavilov 1992; Damania 2002). Sorghum has a long history of cultivation in China. As in Africa and India, China has a significant diversity in climate and vegetation but unlike them, it extends far north (over 55 °N) (Kimber 2000). The monsoon is deeper in China, because the interior cools further in winter and heats up higher in summer, than in India. There are many theories regarding the introduction of sorghum into China; one of them was that sorghum from India could have been transported to China as soon as it arrived in India. Kimber et al. (2000) discussed at length different ideas about the introduction of sorghum into China.

Over the years, the region of sorghum cultivation has reduced drastically worldwide and, particularly, in Asia. There are many factors contributing to this, including competing crops, the low economic viability of sorghum and the changing economic scenario of agriculture. This chapter attempts to analyse the sorghum situation in Asia, its production constraints, the improvement that has been made so far and further probable measures to improve its cultivation.

2 The sorghum production scenario in Asia

Sorghum has offered considerable opportunities for poverty reduction and food security in the developing countries. Production and consumption patterns of sorghum in Asia have changed significantly in the last decade, and this has important implications for cereal improvement programmes such as the National Food Security Mission in India, particularly those seeking to support the livelihoods of rural households. While food use is declining, non-food or industrial uses of sorghum are on the rise. This has implications for crop production and marketing with the introduction of the private industry in coarse cereal economies (Parthasarathy Rao et al. 2004). India and China are among the top ten producers of sorghum across the world (FAOSTAT 2016). So, the major focus has been on these two countries when presenting data and discussions. In India, sorghum is predominantly grown in southern and southwestern dryland parts of the country. In China, it is grown primarily in the northern part of the country, especially in the region north of the Yellow River and the Qinling Mountains (about 65%) where it is sown in spring. Another important area is between the Yellow River and the Yangtze River (about 29% of the area in the country) where it is sown in summer (House et al. 2000).

2.1 Sorghum production in Asia vs. global production

The global sorghum economy can be broadly categorized into production and utilization systems. Intensive, commercialized production systems, mainly for livestock feed, characterize the developed world and parts of Latin America and the Caribbean. Here,

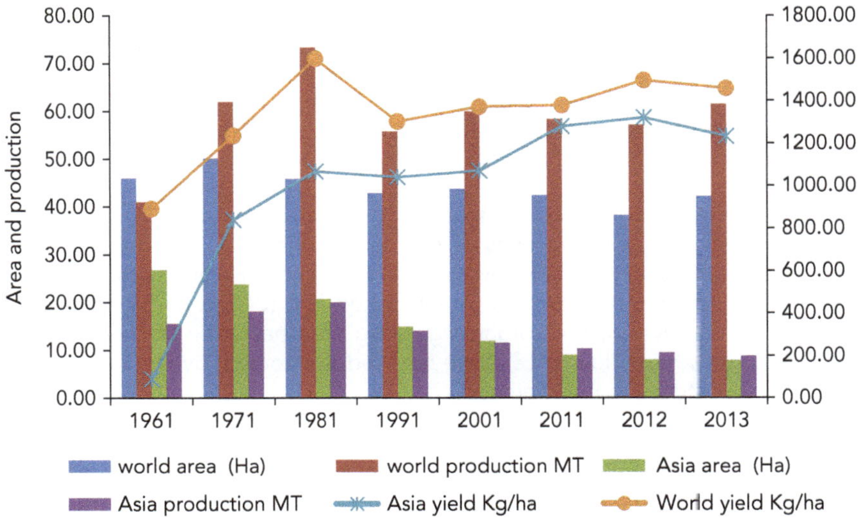

Figure 1 Area, production and yield of sorghum in Asia and world (source: FAO 2016).

improved hybrid seed and crop-management technologies are widely used and average yields hover around 3–5 t/ha. Such commercialized production systems cover less than 15% of the world's sorghum area, but account for over 40% of global output. In sharp contrast are the low-input, extensive production systems in most of the developing world, where sorghum is grown mainly for food. While improved varieties are being adopted, particularly in Asia, management practices such as fertilizer application and moisture conservation technologies generally remain less intensive in such systems than in the commercialized systems. As a result, average yields remain between 0.5 and 1.0 t/ha in many areas (Parthasarathy Rao et al. 2004).

The areas of sorghum cultivation have decreased over the years both globally and in Asia (Fig. 1); however, the Asian sorghum area reduced at a compound annual growth rate (CAGR) of −2.36%. Further, global sorghum production has increased over the years, while in Asia, annual sorghum production has reduced at a CAGR of −1.37%. Productivity has increased both in Asia and globally; however, the increase was greater at the global level. About 17% and 14% of global sorghum areas and production are made up by the Asian countries. The decline in sorghum area, especially in India, as documented in the National Academy of Agricultural Sciences (NAAS, 2013) was due to low remuneration for sorghum compared to other competing crops, lack of input subsidies and price incentives, the subsidized supply of fine cereals through a public distribution system (PDS) and changes in consumer preferences. These factors had led to a shift from production of sorghum to other competing crops such as soybean, maize, cotton, sugarcane and sunflower.

2.2 Spatial distribution of sorghum in Asian countries

In Asia, sorghum was grown in around 27 countries during 2014. In Central Asia, sorghum is grown in Uzbekistan, Kazakhstan, Kyrgyzstan and Tajikistan, while in Eastern Asia, it is grown in mainland China, the Democratic People's Republic of Korea, Taiwan Province of

China and the Republic of Korea. Similarly, sorghum is grown in Southern Asian countries such as India, Pakistan, Sri Lanka, Bangladesh and the Maldives. In Southeastern Asia, sorghum is grown in Myanmar, Thailand and the Philippines. Further, the countries which cultivate sorghum in Western Asia are Yemen, Saudi Arabia, Iraq, Israel, the United Arab Emirates, Oman, Syrian Arab Republic, Jordan, the occupied Palestinian territory and Lebanon. Sorghum cultivation is highly concentrated in southern and eastern Asian countries (Fig. 2).

Within Asia, India ranked first both in area (78%) and production (57%) with 5.55 M T from 6.08 M ha, showing a productivity of 912 kg/ha during TE 2014 (Table 1). China occupied the second position followed by Yemen, Myanmar, Pakistan, Saudi Arabia and so on. However, in terms of productivity, Israel ranked first (24 476 kg/ha) followed by Oman (21 780 Kg/ha). Overall, the sorghum-growing area has been declining in Asia. Slow productivity growth and low prices have reduced the competitiveness of sorghum, resulting in crop substitution in many areas.

In some cases, sorghum has shifted into more marginal lands which can be a major hindrance to its cultivation. There is a vast yield gap between the productivity of sorghum in demonstration fields vs. farmer fields (NAAS, 2013).

Yield gaps: In a study by the International Crops Research Institute for Semi-arid Tropics (ICRISAT), using the Crop Estimation through Resource and Environment Synthesis (CERES)–sorghum crop growth models and historical weather data, potential rain-fed yields of rainy and post-rainy sorghums were estimated in different production zones for selected locations in India. The total yield gap (simulated potential rain-fed yield – farmer's yield) in production zones ranged from 2130 to 2560 kg/ha for rainy (*kharif*) sorghum and

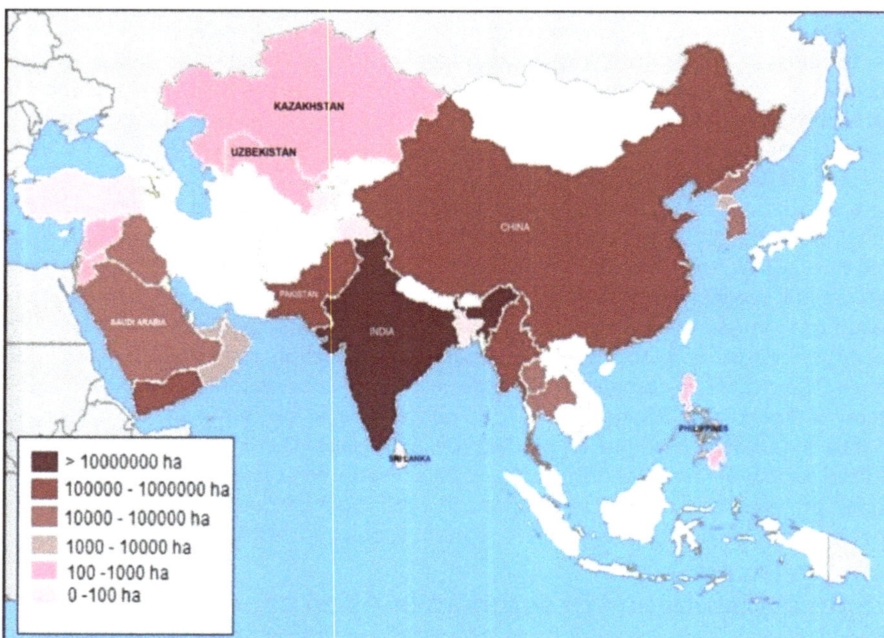

Figure 2 Distribution of sorghum area in Asia, 2014 (source: FAO 2016).

Table 1 Area, production and yield of sorghum in different sorghum-producing Asian countries, TE 2014

S. no.	Country	Area (M Ha)	Production (M T)	Productivity (kg/ha)	Rank based on Area	Rank based on Production	Rank based on Yield
1	India	6.08	5.55	912.33	1 (78.39)	1 (57.57)	24
2	China	0.60	2.80	4682.26	2 (7.71)	2 (29.07)	6
3	Yemen	0.47	0.41	872.69	3 (6.10)	3 (4.29)	25
4	Myanmar	0.23	0.22	957.30	4 (2.91)	5 (2.24)	23
5	Pakistan	0.19	0.11	607.55	5 (2.44)	6 (1.19)	27
6	Saudi Arabia	0.08	0.27	3167.33	6 (1.08)	4 (2.75)	9
	Total Asia	7.76	9.64	1242.98			

Note: Figure in parentheses indicates percentage to total Asia. Source: FAO 2016.

from 280 to 830 kg/ha for post-rainy (*rabi*) sorghum. This indicates that the productivity of *rabi* sorghum can be increased 1.4 to 2.7 times and that of *kharif* sorghum can be increased 3.0 to 4.0 times from current levels of productivity (Murty et al. 2007).

The Farm Level Demonstration (FLD) programme organized by agricultural scientists of the Indian Council of Agricultural Research (ICAR)/SAUs of India covering an area of more than 5000 ha over the last 5 years in various states indicates a larger yield gap of 6–236% over local practices and provides substantial scope for productivity improvement (Singh and Dayakar 2010). An average yield of 6500 kg/ha of sorghum has been recorded in the rice fallows of the Guntur district of Andhra Pradesh, India, over an area of 24 000 ha during *rabi* (2011–12) with the highest FLD yield being recorded as 8500 kg/ha (Chapke et al. 2011).

To arrest the decrease in the cultivation of sorghum and millets and increase the demand for their products, the Indian government has made an allocation of Rs. 300 crores in the budget of 2011–12 under the Initiative for Nutritional Security through Intensive Millet Promotion (INSIMP) of *Rashtriya Krishi Vikas Yojna* to help improve integrated production and post-harvest technologies. The following steps may be taken to increase the demand for sorghum and other millets as suggested by NAAS (2013):

- creating an awareness of their environmental sustainability, and nutritional and other health benefits
- making sorghum available through PDS
- value addition.

2.3 International trade in sorghum

The international trade in sorghum has largely been dominated by the developed countries such as the United States as the major exporting country. On the contrary, the developing countries of Asia and Africa are the major importing countries of sorghum. The export potential of the Asian countries is limited. The quantity of sorghum exports from Asia was only 2.79% of the total global export volume, while Asia was the major sorghum-importing continent (32%) during TE 2014 (FAO 2016). The trend of exports and imports of sorghum in Asia is depicted in Fig. 3. The quantity of sorghum imports is greater than the

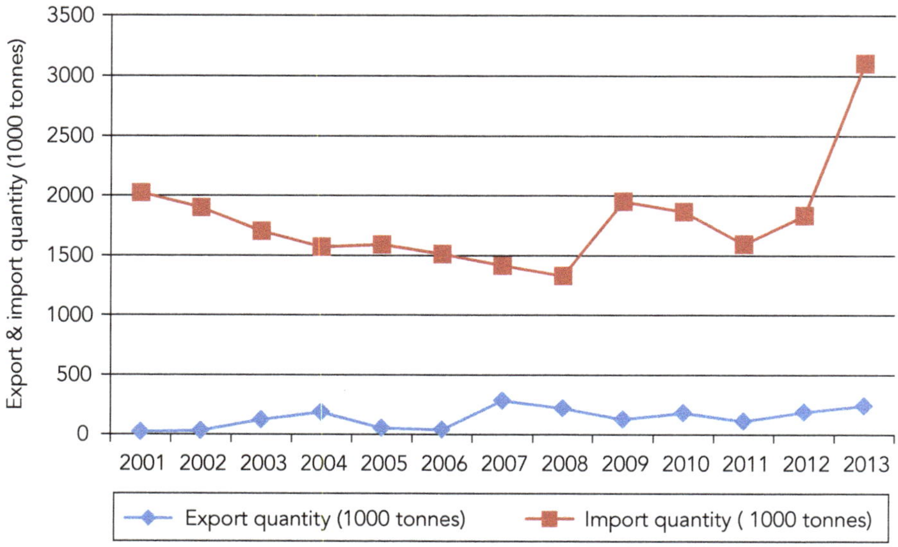

Figure 3 Trends in export and import of sorghum in Asia.

export quantities throughout. This means the gap between the production and demand of sorghum in Asia is bridged by imports from other parts of the world. The quantity of sorghum exports increased over the years from 19 thousand tonnes during 2001 to 239 thousand tonnes during 2013 with a CAGR of 15%. Similarly, the quantity of sorghum imports in Asia has increased over the years from 2023 thousand tonnes during 2001 to 3103 thousand tonnes in 2013 with a CAGR of 1.62%. Among the Asian countries, India contributes 91% of total sorghum exports followed by China (7%). The major sorghum-importing countries in Asia were Japan (57%), China (34%), Taiwan (3%) and Israel (2%) in the year 2013 (FAO 2016).

Asian countries contribute 32% of world sorghum imports from various regions to bridge the gap between production and demand. The major sorghum-exporting countries to Asia are Argentina (38%) followed by the United States (25%) and Australia (25%). India contributes around 4% of total sorghum imports to Asian countries and the major importers of sorghum from India are Israel, Taiwan Province of China, Pakistan, Yemen and Saudi Arabia (FAO 2016).

World trade in sorghum is strongly linked to demand for livestock products, dominated by feed requirements and prices. Only 6% of world sorghum trade (about 500000 tons per year) is for food use (FAO and ICRISAT, 1996). Since trade is primarily for animal feed, volumes are very sensitive to sorghum/maize price differentials and can fluctuate considerably. Competition between different grains for animal feed depends on relative feed values and prices. Sorghum has total digestible nutrients, which are 95% of those in maize. Sorghum, therefore, becomes attractive as a feed only when its price drops to below 95% of the maize price. Consequently, international sorghum prices mirror very closely those of maize, the world's most important feed grain, but are usually slightly lower.

World prices are perceived as an efficiency benchmark in determining the optimal use of domestic resources. The price of sorghum does fluctuate due to the non-food market and there are frequent shifts between sorghum and maize prices depending on the price ratio of the two commodities. One important trend is that sorghum imports from developed countries have fallen sharply, while those from developing countries have increased considerably in response to a growth in livestock production. The price tariff support on the import of edible oils since the late 1980s has artificially increased the competitiveness of soybean and sunflower at the expense of sorghum in many Asian countries, particularly, in India. As a result, the share of the developing countries in world sorghum imports has increased substantially, from 3% to 4% in the early 1960s to about 55% in the recent years.

The sorghum export competitiveness in most of the Asian countries has deteriorated in the global market compared to that in the developed countries. To tap the potential of the international export market, it is of the utmost importance that the Asian countries, especially India, should revive the factors that enhance sorghum's export competitiveness. The trend in international market prices significantly influences the competitiveness of a particular product. The trend in domestic and international prices of sorghum has been depicted in Fig. 4.

It can be seen from this figure that the domestic prices of sorghum have remained well above international market prices. Various measures that could be taken to improve the export competitiveness of Indian as well as Asian sorghum are as follows:

i Effective measures should be taken to improve grain quality through the release of improved and nutrient-enriched varieties of sorghum.
ii The cost of production should be reduced which will significantly enhance the export competitiveness of both Indian and Asian sorghum. This could be done by providing low-cost cultivation practices to the farmers.
iii Value addition to sorghum is the most effective way to improve its export competitiveness. Value-added products produced by following international food safety standards will increase its demand.
iv Creation of value addition chains across the country through public-private partnership would enhance the effectiveness of the value addition network
v Emerging business opportunities should be exploited, increasing diversification by developing niche export markets such as gluten-free products, sweet sorghum syrup, alcoholic beer, stalk-based ethanol/grain-based potable spirit, value-added ready-to-eat (RTE)/ready-to-cook (RTC) foods, and so on.

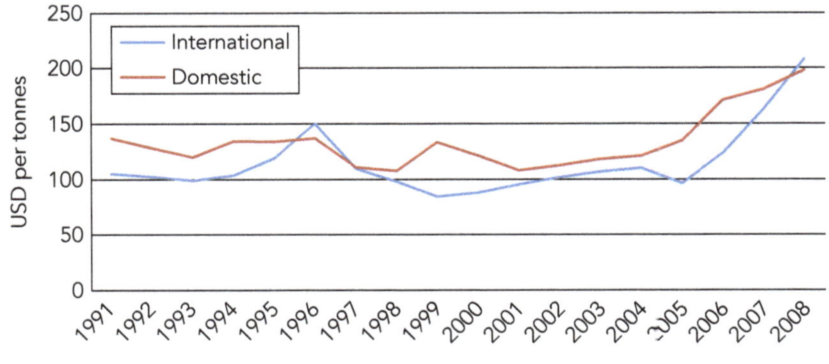

Figure 4 Domestic and international pricing trends for sorghum.

vi It should be made competitive in the feed industry for the export markets with respect to maize, which is now diverted for ethanol production.

3 Sorghum utilization in Asia

Over the last two decades, several factors have contributed to a change in the traditional role of sorghum as a food security crop. The factor from the supply side is the reduced sorghum-planting area, due to slow productivity growth compared to competing crops and low prices. On the demand side, food use for sorghum is reduced due to income growth, urbanization and changes in tastes and preferences, further lowering the prices. However, utilization patterns in Asia are in a dynamic phase, moving from traditional food use to alternative uses such as livestock and poultry feed, bakery products, potable alcohol and in the starch industry (Parthasarathy Rao et al. 2004).

In Asia, sorghum is utilized as food, fodder, feed and, recently, for making fuel. The pattern of utilization of sorghum as food in selected Asian countries is given in Table 2. During 1980, the availability of sorghum and its products was 5.72 kg/capita/year, which reduced to 1.89 kg/capita/year during 2011. In most Asian countries, the utilization of sorghum as food has reduced over the years except in Korea. The highest percentage of reduction was seen in Pakistan followed by Yemen and India. The highest per capita consumption of sorghum in Asia during 1980 and 2013 was by Yemen followed by India and Myanmar (Table 3).

3.1 Food uses of sorghum

Sorghum is a key staple in many parts of the developing world, especially in the drier and more marginal areas of semi-arid tropics. There are diverse traditional sorghum foods consumed in various countries of Asia. In India, particularly in southern regions, boiled sorghum (rice-like) called *annam* or *soru* is one of the common uses, sorghum *sankati* is a type of thick porridge and *Kanji* is a thin porridge (Ratnavathi and Patil 2013). *Upma* is a very important south Indian breakfast food and popped or parched sorghum is used as a snack in India. Aromatic or scented sorghums (basmati) are grown in India (Prasada Rao

Table 2 Utilization of sorghum in Asia

Country	Grain	Stalk
India	Food, feed and alcohol	Dry fodder, forage and alcohol
Myanmar	Food	Fodder
Pakistan	Feed	Dry fodder and forage
Thailand	Feed	Dry fodder
China	Feed, alcohol, minor alternative uses	Sweet sorghum (alcohol)
Philippines	Feed	
Indonesia	Feed	

Source: Gowda and Stenhouse 1993.

Table 3 Food use (kg/capita/year) of sorghum and its products in Asia

S. no.	Country	1980	2011	2012	2013	% reduction (1980 to 2013)
1	Myanmar	–	1.56	1.46	1.46	–
2	China	4.63	0.69	1.12	1.34	−71
3	India	13.25	5.1	4.18	3.58	−72
4	Democratic People's Republic of Korea	0.85	1.17	1.13	1.33	56
5	Pakistan	2.32	0.68	0.6	0.57	−75
6	Yemen	64.02	16.61	18.15	16.95	−73
7	Asia	5.72	1.89	–	–	−66

Source: FAO 2016.

and Murty 1982), and special vani (sugary) sorghums are harvested in the milk or dough stage, roasted and consumed as a snack called *hurda* (Rooney and Waniska 2000). Apart from traditional foods, sorghum is used in other foods such as bakery products such as bread, cakes and biscuits

A study by Li Guiying et al. (2004) documented traditional sorghum foods found in China such as *Conjee, Hele, Dumpling* and Chinese liquor. About 40 kinds of foods are made from sorghum in China (Zhao 1985), which are categorized into rice-like food, flour food and others (House et al. 2000). In some parts of China, snacks are made by popping or extruding sorghum rice, and sweets are made with waxy sorghum. In recent years, the amount of sorghum used for food has reduced drastically, the proportion for wine making has increased and the portion used for feed has remained unchanged.

3.2 Feed and forage uses

About 48% of world sorghum grain production is fed to livestock (human food use constitutes about 42%) (FAO 1988). In contrast to food utilization, which is relatively stable, utilization for feed sorghum changes significantly in response to two factors: rising incomes, which stimulate the consumption of livestock products, and the price competitiveness of sorghum *vis-à-vis* other cereals, especially maize. Demand for animal feed is concentrated in the developed countries and middle-income countries in Asia, where demand for meat is high and the livestock industry is correspondingly intensive. Three countries (United States, Mexico and Japan) together absorb nearly 70% of the world total. These countries along with Argentina, the former Soviet Union and Venezuela account for over 80% of global use of sorghum as animal feed (FAO 1995). Roughly 5–10% of the sorghum produced in India and a considerably higher proportion in China is used for livestock and poultry feed (FAO and ICRISAT 1996).

One important factor is consumer preference for meat colour. Maize contains higher carotene levels than sorghum, so meat from maize-fed animals tends to be more yellow than meat from sorghum-fed animals. In Japan for example, consumers generally prefer white-coloured meat. Therefore, sorghum is a valued ingredient in some compound feed rations. In contrast, sorghum is not preferred by producers in India because consumers there generally prefer poultry meat and egg yolks with a deeper yellow colour.

In addition, farmers in Asia have shown a growing interest in the sale and purchase of sorghum fodder. There is a large and growing market in Asia for traded sorghum residues to meet both rural feed shortfalls and urban agricultural demands, the latter largely for maintenance of dairy animals. The use of forage sorghum is common in north India, scattered primarily among commercial farmers. The use of sorghum silage is rare in Africa and Asia, essentially used only by large commercial farmers.

3.3 Other uses

Major breweries in Africa and Asia use sorghum grits as an inexpensive source of fermentable carbohydrates in brewing barley beer. In most cases, red sorghums are used, but white sorghums with tan plant colour and straw-coloured glumes are preferred due to their improved milling properties, producing high yields of excellent quality grits (Rooney 1996; Rooney and Waniska 2000). Thus, an important outlet for grain sorghum is in the preparation of alcoholic beverages. The grain is used for malt or as an adjunct in the production of beer. Although statistics on the quantities of sorghum used in beer preparation are lacking for many countries, the available data show that sorghum is traditionally a major ingredient in home-brewed beer, the growing demand for which has created a commercial industry in some countries. In China, about one-third of sorghum grain production is reported to be used to make alcoholic beverages, mainly a strong traditional liquor.

Sorghum kernels and plants are used in a variety of other products in China. Sorghum grain is used to produce starch, threshed panicles for making brooms, stalks for making paper, construction materials and fuel. The outer cortex is stripped from the stem and used to weave mats and handicraft articles.

Sweet-stemmed sorghum has been used as a source of sugar and biofuel, but this use has been limited because processing costs for this are greater than those for sugarcane. It has been used as a source of fuel alcohol, particularly in Brazil, where its use along with sugarcane extended the operating period of processing plants, as the crops are ready for harvest at different times. There is a minor use for sorghum syrup, jaggery, in India. In many places in Africa and India, sweet-stemmed sorghum is grown and the stems used/sold for chewing.

The use of sorghum in the developed world has been primarily restricted to animal feed and, hence, little work has been done to evaluate them in industrial food systems. Recent work has suggested that sorghum possesses unique characteristics that have both nutritional and functional properties that lead to the development of healthy, nutritious foods. It is gluten-free and has unique phenolic compounds, which are identified as having medicinal properties and contain protein and starch characteristics that lend themselves to functional food uses that may impact health. In Africa and Asia, sorghum has been used in traditional food products, but now their use in industrial settings is being explored (Dahlberg 2000). Recently, interest in sorghum in high-end food products, especially in health food markets, has begun to develop in regions of the world that traditionally have not used these grains in human food systems. Japan has taken the lead in developing snack food products using sorghum and interest in sorghum's gluten-free properties has developed a demand for high-quality food sorghums within the United States (Dahlberg 2000).

4 Constraints on sorghum production in Asia

Sorghum is becoming less important in the economies of the SAT countries, and the demand for these grains as human food has been declining (Ryan and Spencer 2001). This is particularly evident in Asia (especially in South Asia). Analysis of data in the top 10 sorghum-producing countries of the world from 1970 to 2009 showed that three countries, China, India and the United States, recorded a drastic reduction in harvested area (Rakshit et al. 2014). The biggest reduction compared to the 1970 baseline was in China (~89%) followed by the United States (~59%) and India (~56%). The reasons behind the loss of area in China and India are different to those in the United States. Over recent years, both China and India experienced a drop in per capita consumption of sorghum as food (House et al. 2000; Parthasarathy Rao et al. 2010). An overall reduction in importance of sorghum as a food grain coincided with a rapid loss in area in two major Asian countries (FAOSTAT 2012). Several factors are responsible for the decline in sorghum cultivation (Chand 2007), which are listed below:

4.1 Production constraints

- The majority of smallholder farmers see sorghum as a semi-subsistence enterprise that offers smaller returns compared to alternative farm and non-farm enterprises, and tend not to invest in fertilizers or pesticides.
- Rising labour costs have affected most farm operations, from land preparation, weeding and bird scaring to harvesting and grain processing.
- Grain molds cause significant losses in both grain yield and quality, particularly in areas where improved early cultivars have been adopted. Other important diseases include anthracnose, charcoal rot, downy mildew, ergot and leaf blight.
- Another constraint is *Striga*, a parasitic weed, which inhibits plant growth and reduces yields. Some *Striga*-resistant sorghum varieties have been developed, but these generally offer lower yields than other improved varieties.
- Insect pests such as shoot fly, stemborer, head bugs and midge adversely affect production in many areas.

4.2 Social constraints

- Most sorghum-growing areas are characterized by high population, poverty and malnutrition, and low agricultural productivity is one of the major reasons contributing to these problems.
- Another factor, important throughout Asia, is changing food preferences. As incomes rise, consumers tend to purchase wheat, rice and in some cases maize, which are easier and faster to cook, rather than traditional coarse grains.
- This trend is accentuated by rapid urbanization and the growing availability of a range of convenience foods based on wheat and rice.
- Urbanization has also taken away land from all crops including sorghum.

4.3 Marketing and policy constraints

- In Asia, sorghum is largely a subsistence crop and, therefore, only small volumes enter the marketing chain. Transactions take place mostly in rural markets near the production areas. Marketing channels between producers and the major urban centres are poorly developed.
- A major problem is that variable rainfall leads to large fluctuations in production. Prices fall abruptly in good years, leaving traders reluctant to enter the market, especially since stockholding infrastructure is usually inadequate.
- In India, rainy-season production is predominantly used for industrial purposes, while most of the harvest of post-rainy sorghum is consumed as food (Parthasarathy Rao et al. 2010). Lower prices received for rainy-season sorghum due to grain mold damage have reduced the profitability of rainy-season sorghum compared to other competing crops such as sunflower, soybean, groundnut and corn (Basavaraja et al. 2005; Awaghad et al. 2010).
- Government policies in a number of countries have also contributed to the decline in food utilization of sorghum. For instance, imports of relatively cheap wheat and rice by many countries discouraged the consumption of locally produced cereals. In other countries (China and India) government purchases and sales of sorghum under PDSs were discontinued, reducing use in urban areas.
- In Asia, particularly in India, irrigation and fertilizer subsidies have increasingly favoured rice, wheat and cash crops at the expense of coarse grains.
- The food security strategy of the Indian government, in which subsidized rice and wheat are given to the poor through the PDS, is largely responsible for declining consumption of sorghum and other millets.
- In a number of developing countries that had long-standing price support policies for sorghum, this support has been drastically reduced or eliminated, mainly as a result of market deregulation.

To halt the declining trend in sorghum production and utilization, R&D efforts are required to intensify production and help in value addition to make it more remunerative for the farmer. Further, the nutritional benefits of this crop should be highlighted to generate greater consumer demand (Ashok kumar et al. 2011).

5 Genetic improvement of sorghum

The scientific community has paid little attention to sorghum improvement in the past, especially in Africa and Asia, because it is a coarse grain and much of its production is at subsistence level. Increased pressures of population growth on food supplies, enhanced use of animal products and depleted fossil fuel reserves have focused attention on exploiting the full potential of this crop as food, feed, fodder and fuel. The objectives of sorghum improvement programmes have been different in different parts of the world and depend on the production environment and specific utilization.

5.1 Sorghum improvement in India

The sorghum improvement in India is being carried out at the Indian Institute of Millets Research (IIMR) in collaboration with the All India Co-ordinated Sorghum Improvement Project (AICSIP) having 21 centres distributed across 10 states of India and the ICRISAT. Sorghum research in India can be identified as a dramatic example, which has significantly contributed towards the green revolution in drier areas. In India, unlike other countries, sorghum is cultivated in two seasons – the rainy season (*kharif*), which extends from June–July to September–October and the post-rainy (*rabi*) season extending from September–October to January–February. The production environments and constraints are different for these two seasons and, hence, the cultivar options are quite different (Rana et al. 1997). Besides grain sorghum genotypes, efforts are being made to develop forage sorghum and sweet sorghum cultivars with increased demand for forage crops as an alternate source of bioethanol.

Grain sorghum improvement

Systematic efforts for the improvement of rainy-season sorghum began in the 1960s by introducing American germ plasm to manipulate the genes controlling height and maturity and adopting temperate × tropical crosses (Rao and Rana 1982). Sorghum improvement has been conducted in a network model under the AICSIP since 1969, and remarkable progress has been achieved over the following five decades (Patil et al. 2015). The yield gain in rainy sorghum is at 12.4 kg/ha/year and in post-rainy sorghum, it is 8.7 kg/ha/year. This has been achieved both through genetic gains as a result of breeding efforts and improved crop management (Rakshit et al. 2015).

The *Zera-zera* landraces from Ethiopia and Sudan were sources of many yield and grain quality traits and were used extensively in the sorghum-breeding programmes (Reddy et al. 2008). The Indian *Durras* as a source of resistance to drought, shoot fly and stem borer, and *kafir* race as the basis of the nuclear cytoplasmic male sterility system were also heavily used in the Indian programmes (Audilakshmi et al. 2003; Aruna and Audilakshmi 2008). Similarly, the *guinea* race has provided resistance to grain mold and *bicolor* has helped the breeding of forage sorghum. As a result of efforts under the accelerated hybrid sorghum project initiated by the ICAR, the first commercial sorghum hybrid CSH 1 was released in 1964. Since then, 30 hybrids have been released, with some of them being adaptable to the post-rainy season. The yield potential has improved from 3.0 t/ha of CSH 1 to 4.5 t/ha in latest releases.

The post-rainy sorghum crop accounts for 45% of the total cultivated sorghum area and 32% of the total sorghum production in India (Sajjanar et al. 2011). The productivity of post-rainy sorghum is generally low (850 kg/ha) compared to rainy-season sorghum (1100 kg/ha) because of its unique situation as the crop is grown in medium to shallow soils with receding soil moisture. Post-rainy adapted sorghums are characterized by photoperiod sensitivity, thermo-insensitivity, tolerance to terminal moisture stress and resistance to stalk rot/charcoal rot, coupled with high biomass, lustrous grain and tolerance to shoot fly (Sanjana Reddy et al. 2012). All these characteristics are best exemplified in M 35-1, a variety selected from a local land race nearly 75 years ago at Maharashtra, India. In the recent past, efforts have been made to develop varieties adapted to specific soil situations (shallow, medium and deep). Focused breeding efforts on *rabi* sorghum led to the development of several *rabi* sorghum varieties and hybrids. However, the varieties have become more popular compared to hybrids, the exact opposite to *kharif* (Sanjana and Patil 2015) because though the hybrids are heterotic for grain yield, they have poor grain quality and are vulnerable to biotic and abiotic stresses.

Forage sorghum improvement

Sorghum is the major source of green and dry fodder in India and its role becomes important during the lean winter and summer months. It is estimated that sorghum fodder constitutes 20–45% of the dry weight of feed of dairy animals during normal seasons and up to 60% during the lean summer and winter seasons (Anonymous, 2015). Forage sorghum in India is grown either as a single-cut (rainy season) or as a multi-cut (summer and rainy) forage crop. Intensive cropping, short growing season, nutritional quality and the need for a continuous supply of green fodder created demand for forage sorghum in northern India. In view of its importance, research on improving sorghum for forage was initiated during the 1970s under AICSIP, with the major objectives being to develop varieties both for single-cut and multi-cut with high tonnage, better quality, good seed yield and resistance to insect pests and diseases. Many single-cut and multi-cut varieties and hybrids were released at the national level with high fodder yield potential and good quality in terms of dry matter digestibility and total soluble sugars with comparable protein content. Several private sector hybrids are also popular for multi-cut forage.

Sweet sorghum improvement

Sweet sorghum is considered to be a potential bioethanol feedstock and with the Indian Government's initiatives for the production and use of biofuels, considerable progress has been made in the development of sweet sorghum cultivars. Some sweet sorghum lines attain juice yields of about 78% of total plant biomass, containing 15% to 23% soluble fermentable sugars (Reddy et al. 2008). Concerted research efforts at the IIMR, its co-operating centres in different State Agricultural Universities and at ICRISAT have resulted in excellent sweet sorghum varieties for use in ethanol production by the sugar industries/alcohol distilleries and for use as green/dry fodder. Promising nationally released varieties/hybrids were SSV 84, CSV 19SS, CSV 24SS and the hybrid CSH 22SS. The productivity ranged between 40 and 50 tha^{-1} of fresh stalk yield.

These research efforts led to the release of 22 hybrids and 15 varieties of rainy-season adaptations, 6 hybrids and 10 varieties of post-rainy adaptations, 2 forage hybrids and 2 forage varieties, 2 sweet sorghum varieties and one hybrid (Rakshit et al. 2014) at the national level and more than 130 cultivars were released by various State government programmes. Private seed companies also participated actively in the development and release of sorghum hybrids, especially for the rainy season. The ICRISAT located in India also played a significant role providing improved germ plasm to complement the sorghum improvement programmes in both public and private sectors (Ashok Kumar et al. 2015).

Besides productivity, research is focused on resistance to important biotic stresses such as grain molds and foliar diseases shoot fly in the rainy season as well as charcoal and shoot fly in the post-rainy season. Besides conventional breeding, biotechnology tools are being employed to tackle these. Major quantitative trait loci (QTL) governing shoot fly resistance have been identified and, currently, efforts are being made to use them through marker-assisted breeding (Satish et al. 2009; Aruna et al. 2011). Transgenic approaches are employed to infuse resistance against stem borer using *Cry1b* gene (Visarada et al.

2014), and initial field evaluation of advanced events shows much promise. Drought is the major production constraint in post-rainy sorghum. Major QTL contributing towards stay-green, an important trait attributing drought tolerance, have been identified and efforts are underway to transfer them into superior cultivars (Reddy et al. 2014).

Thus, a mission-oriented approach for sorghum improvement has made significant contributions towards increased productivity in terms of food, fodder and fuel. There are still some issues to be tackled in the future for each of the product lines. For rainy-season sorghum, grain mold resistance and improving grain quality remain the research priority along with the development of genotypes for specific end uses. In post-rainy sorghum, the emphasis is on the development of varieties and hybrids for specific soil types and exploiting heterosis. In the case of forage sorghum, developing genotypes with high digestibility along with high seed yield is important, while for sweet sorghum, the emphasis is on improving the keeping quality of juice and second-generation biofuels.

5.2 Sorghum improvement in China

Before 1949, all sorghum cultivars used by Chinese farmers were local varieties (landraces) (Li and Li 1998). In 1951, an intensive selection in sorghum landraces was carried out and some elite cultivars were released. Sorghum breeding in China has progressed through three stages: (i) selecting within the best local varieties; (ii) making crosses between local varieties or local × exotic and (iii) exploiting heterosis through the development of single-cross hybrids (Zheng Yang 1997). Heterosis breeding has been the main method of breeding since 1965. The major objectives of sorghum improvement research in China include grain yield, multiple resistance to abiotic (low temperature and drought) and biotic (aphids and head smut) stresses, grain quality, grain feed and forages for livestock (Reddy et al. 2008).

There are two main groups of sorghum in China: the amber cane sorgos and the *kaoliangs*. The amber cane sorgos are tall with loose panicles and are useful for forage and syrup making (Doggett 1988). Chinese sorghum (*kaoliang*) is one of the unique sorghum types in the world and while comparing Chinese sorghums with the world collection, de Oliveira et al. (1996) discovered that the Chinese lines formed a distinct cluster. Chinese sorghum landraces have played an important role in the development of R lines, while foreign germ plasm have played an essential role in the development of A lines (Li and Li 1998). The development of sorghum hybrids has become the predominant activity in sorghum improvement after the introduction of CK 60A from the United States and has led to increased yield several times.

It is worth noting that Chinese sorghum may contribute greatly to the use of cytoplasmic male sterility systems other than A_1 (Li and Li 1998). In China, several hybrids on A_2 cytoplasm have been grown annually on about 200000 ha since the early 1980s (Chen and Shi 1995). In China, sweet sorghum is primarily used for silage preparation. With the introduction of improved sweet sorghum varieties such as Rio, Roma, Ramada and Wray from several countries, a systematic sweet sorghum breeding programme was initiated in 1985 and two sweet sorghum hybrids were developed and released during 1989 and 1995, respectively (Reddy et al. 2008). China has launched numerous research projects and pilot production activities involving sweet sorghum as feedstock for bioethanol production and several International scientific and technological co-operation projects and national key technology R&D programmes have been funded by the Ministry of Science and Technology of China (Li et al. 2013; Wang et al. 2014). Qu et al. (2014) reported that the stable nutritional value and multiple harvests of sweet sorghum offer great potential as a source of green fodder and an alternative to corn as a silage crop in the relatively humid environment of eastern China.

6 Seed systems

Sorghum seed in Asia needs focus in terms of seed and varietal replacement, as only 30% quality seed is supplied by the seed industry with the remaining 70% being farm saved. Seed is the basic and most critical input for sustainable agriculture and is the cheapest of all inputs in rain-fed agriculture. Improvements in the quality of the seed can enhance production by 20% and with improved crop production practices the yield can increase by 35–40% (Ravinder Reddy et al. 2015). The seed industry in India is well established and is supported by a strong crop improvement programme in both the public and private sectors.

6.1 Seed industry in India

India's seed programme has a strong seed production base in terms of diverse and ideal agro climates spread throughout the country for producing high-quality seeds from several crop varieties in sufficient quantities at competitive prices (http://seednet.gov.in/material/indianseedsector.htm). The developments in the seed industry in India, particularly in the last 30–40 years, are very significant. A major re-structuring of the seed industry by the Indian government through the National Seed Project Phase-I (1977–8), Phase-II (1978–9) and Phase-III (1990–1) was carried out including a crucially needed strengthening of the seed infrastructure. To facilitate efficient production and distribution of seed, the Indian government established a central production organization, the National Seed Corporation (NSC) in 1963 and the State Farm Corporation of India (SFCI) in 1969. The National Seeds Programme (NSP) was established in 1967 to coordinate the efforts of the NSC, SFCI and private companies (Ramakrishna et al. 2006). The Indian seed industry today is set to work with a 'farmer-centric' approach and is market-driven.

There are about 50 small- to medium-sized seed companies engaged in the production and marketing of improved seed through both public and private channels. The private sector focuses largely on hybrid seed and is a major player in the hybrid seed markets of vegetables, sorghum, oilseeds, maize, cotton and pearl millet. The seed industry is one of the earliest examples of contract farming in India, as the seed firms usually outsource the production of seeds to contract growers. This helps increase rural employment in the seed sector in addition to improving the economic conditions of the seed-producing farmers.

6.2 Seed production and supply systems in India

The seed production process has a sequence of stages in which the original quantity of seed of a new variety is multiplied to obtain sufficient commercial seed. The breeder and foundation seeds together are the precursors to commercial seed production and are known as source seed. Source seed production constitutes a serious bottleneck in many national seed systems, but India has taken major steps to improve access to source seed (Ramakrishna et al. 2006) and adequate safeguards to maintain the purity of the variety as it flows from breeder to farmers. Both public and private sector organizations meet the requirement of the

Indian seed industry, with the public sector contributing 41%, large private sector companies 33% and the small companies and unorganized segment 13% each across all the crops.

The seed supply system in India can be divided into formal and informal systems. The formal seed sector is an official or private control of seed, monitored through the entire process of breeding, multiplication, processing and storage leading to the final product, characterized by inclusion of only identified/notified varieties and a system of quality control. The informal seed sector is simply the farmers themselves providing each other and themselves with seed. Informal seed supply systems are traditional, operating mainly at the community level and involving limited quantities per transaction (Cromwell et al. 1992). Although the informal seed sector provides a dynamic and flexible system of seed supply, continuous use of untested seed inevitably leads to a degeneration of seed quality. Farmers depend on their own seed for sowing, not only because of inadequate access to seed from the formal seed sector, but also because the formal seed sector more often provides seeds of a limited range of cultivars and varieties, which do not always fulfil the needs of farmers (Ramakrishna et al. 2006).

6.3 Sorghum seed systems in India

The sorghum seed system in India is unique with contrasting situations. Hybrids are the cultivar choice in rainy-season sorghum and hybrid adoption by farmers is up to 95% in states like Maharashtra though there are wide variations across the states. The seed requirement is predominantly met by the vibrant formal seed system led by the private seed sector and public sector seed agencies like the NSC and state seed development corporations in different states. In the case of post-rainy season sorghum, open pollinated varieties (OPVs) are the cultivar choice because of stringent quality considerations and lack of appropriate hybrids (Ravinder Reddy et al. 2015). High yield and improved cultivar seed availability are not constraints in rainy-season sorghum, but they are important issues in post-rainy-season sorghum in India and the majority of varieties, especially M 35-1 (Maldandi) and Dagadi are still dominating most cultivated areas (Reddy et al. 2012). Of late, several improved varieties have been released both at state and national levels. However, the private sector is not forthcoming in multiplying the OPVs of sorghum since (i) the margins in production and marketing of OPVs are low compared with the hybrid market and (ii) the seed of post-rainy genotypes produced in the post-rainy season can only be marketed in the subsequent post-rainy season, which means a delay of 8 months in marketing and, hence, the returns on investment are only realized later.

The constraints regarding the rate and level of adoption of new varieties mostly relate to market conditions and associated seed production and trade policies. Hence, there is a need to identify options for strengthening seed multiplication and trade based on issues of national seed policy, commercial profitability and the opportunity to exploit alternative seed supply systems. In order to improve the seed delivery system, interventions are required to strengthen the informal seed supply systems, such as establishing village-based seed banks (Ravinder Reddy and Tonapi 2015). The alternative village-based seed delivery models ensure sustainability of community seed systems in dry land ecosystems and improve timely seed availability and access to improved varieties of seed to small and resource-poor farmers. The concept of a 'seed bank', which advocates village self-sufficiency in production and distribution of quality seeds, is being promoted to reduce the farmers' dependence on external inputs.

7 Measures to improve sorghum cultivation in Asia and future prospects

7.1 Policy support, improved varieties and practices, supply enhancements and novel products

It is evident that though sorghum yield has increased with time, it has lost ground to other competing crops, particularly in Asia. Thus, an increase in yield has not been realized as a proportional increase in total production. With favourable policy support, this crop is likely to regain lost ground, particularly in countries like India. In 2011–12, the Indian government began the INSIMP, which allocated over 3000 million Indian rupees for the production and promotion of sorghum and other millets as nutri-cereals. This initiative has given impetus to the crop through value addition and processing. Being a drought-resilient crop, sorghum is likely to play an important role with changing climate (Nedumaran et al. 2013).

The demand for millet products has been estimated to be 30 million tonnes by the end of 2025, which has to be addressed through improved varieties, good agronomic practices, effective extension services and major policy initiatives (NAAS 2013). This emphasizes the need to address issues regarding the declining trend in the area and production of sorghum and other millets. The decreasing area and crop production can be addressed through both demand creation and supply enhancement. (i) Creation of awareness about the nutritional and health benefits of sorghum products over other traditional food grains among consumers could be a major step in demand creation. This could be done through proper and regular advertisements and awareness campaigning. ii) The removal of inconveniences in preparation could be achieved by processing diversification and product development catering through RTC/RTE technologies. (iii) The awareness programmes should aim to eliminate the low social status attached to the consumption of millet products to further enhance the demand for these products. (iv) Introduction of sorghum products in the PDS and other food- and nutrition-related programmes such as the midday meal scheme will also create a greater demand.

Besides the above-mentioned demand-creating policies, the decreasing trend of area and production of millets could also be addressed through supply enhancements. This can be done by (i) Creating awareness among the cultivators about the various positive aspects of sorghum cultivation such as its suitability in dryland agriculture and greater adaptability to climate change using demonstrations and training. (ii) Developing and distributing high-yielding varieties and hybrids of sorghum among the farmers. The breeding target should be to develop genotypes with specific purposes. (iii) Expanding the area of cultivation to nontraditional areas such as rice fallows and the northeastern region in India. (iv) Creating a proper systematic channel for the timely distribution of improved agronomic practices and other technical assistances. Any difficulties faced and reported by the farmers should be addressed as a priority. (v) Strengthening seed production and supply systems to speed up the flow of improved varieties to the farmers. (vi) Assuring a remunerative price for the farm produce to the farmers through proper buy-back arrangements. (vii) Covering farmers under insurance schemes to avoid any loss due to crop failure and other natural calamities. (viii) Motivating and supporting entrepreneurs to take up sorghum in their new ventures. Subsidized loans and other equipment should be provided to assist new initiatives.

In order to create demand for sorghum and millets in India and restore food and nutritional security, especially in the arid regions of the country, several initiatives have been undertaken by the research institutes, universities and other non-profit organizations. For instance, the ICAR–IIMR under the National Agricultural Innovation Project (NAIP) has developed novel sorghum products such as sorghum-based multigrain flour (*atta*), flakes, biscuits and extruded products, which have been successfully commercialized in pilot mode.

7.2 A value chain strategy to enhance demand for sorghum in India

To augment the production and consumption of sorghum, a value chain on sorghum processed foods is needed with emphasis on value addition and development of value-added RTC/RTE products from sorghum. A value chain model developed as a pilot scheme under the leadership of IIMR in India in the private–public partnership mode was indeed successful in a southern Indian state and currently efforts are being made to upscale it and replicate it in other parts of the country where sorghum consumption is in vogue. Due to the availability of processing technologies including machinery and product development processes already in place, it would be an easier task. However, the challenge is to bring all the stakeholders involved to a common platform. Consequently, there is significant value creation at each level. The farmers (our primary stakeholder) will be assured a higher income in the long run, as value addition and commercialization of millet-based convenience products take place, drawing attention to their nutritional benefits, thus creating future demand. This could be done by involving private sectors, public research institutes, policy makers, entrepreneurs and consumers (Dayakar et al. 2014). The value chain efforts on sorghum and millets by IIMR and many partners have generated renewed interest in sorghum as the crop is now promoted as a health food and alternative crop for those suffering from gluten intolerance or other lifestyle diseases. Thus, it is relevant not only in those states where sorghum is produced and consumed, but even in non-traditional areas.

A project on Harnessing Opportunities for Productivity Enhancement (HOPE) funded by the Bill and Melinda Gates Foundation (BMGF) was undertaken by ICRISAT with the objective of discovering and developing new market opportunities for post-rainy-season sorghum in the Maharashtra state of India (Basavaraj et al. 2014). The research findings from the value chain mapping helped to understand the limiting factors of supply, trade and demand and opportunities identified along the chain have been highlighted to upgrade and enhance market demand and value chain functioning for post-rainy-season sorghum. Process upgrading and product upgrading are found to be two critical areas that need to be addressed as a priority for post-rainy sorghum value chain interventions. Choice of the product is an important determinant of consumption and its acceptance. On the post-harvest front, farmers are constrained by lack of grading and cleaning facilities, which, if properly addressed, will help achieve higher prices. Value addition in the existing value chain of sorghum is limited to physical processing, and processing as a specialized activity for product development is still a minor activity. Thus, product upgrading through production of RTE products as a way of developing the value chain will provide benefits to different stakeholders of the value chain and more importantly have the potential to benefit small and marginal farmers growing sorghum.

Thus, globally, the constraints and opportunities of value chain development of sorghum and millets are very similar except for those factors which are particular to the individual country. However, the value chain development in millets is restricted to whole sorghum grain. Most of the researchers opine on the development of processed food using sorghum and millets (RTE/RTC). So, efforts should be made to overcome the constraints and come up with a new technology, which may enhance the smooth running of value chain of sorghum in the long-term as the future of sorghum is brighter owing to its adaptability to climate change.

In addition, sorghum utilization can be enhanced by developing value chains around densified fodder blocks through its stover being value-added and supplied across fodder-deficit regions during peak shortages in summer in India. Similarly, owing to its potential to act as a raw material for brewing industries, there is renewed interest in setting up microbreweries with active government support in some Indian states. Sorghum is currently used in poultry feed production and potable alcohol production. Building value chains of individual products will generate demand for crop cultivation as a whole and ultimately improve productivity levels.

8 Conclusion

Thus, yield improvements and value addition through industrial utilization may enhance the profitability of sorghum and alleviate rural poverty in the semi-arid tropics. Marketing, contract farming and farmer-industry linkages are niches for the commercialization of sorghum. Production of feed for domestic and industrial uses as well as international trade may generate adequate demand for sorghum. In the context of climate change in Asia, sorghum cultivation is going to be revived as evidenced by its resurgence as a food crop in India as well as its multiple uses, which should arrest the decline in the short-to-medium term. With increased profitability, there will be increased private investment in food processing of sorghum and other industrial uses, stabilizing crop production and increasing its long-term area of cultivation. This may require cultivar development with use of specific grain quality attributes, high productivity to enhance cost competitiveness and other required standards. Different organizations like ICRISAT at global level, ICAR-IIMR in India and the Chinese Academy of Sciences are focusing their research in this direction. Mission-oriented R&D activity on these aspects involving client institutions, state-of-the-art know-how and appropriate governmental policies may provide solutions to sorghum cultivation problems in Asia.

9 References

Anonymous. 2015. *Vision 2050*, Indian Institute of Millets Research, Rajendranagar, Hyderabad 500 030, Telangana, India, p. 60

Aruna, C. and Audilakshmi, S. 2008. A strategy to identify potential germplasm for improving yield attributes using diversity analysis in sorghum. *Plant Genetic Resources: Characterization and Utilization*. 6(3): 187–94.

Aruna, C., Bhagwat, V., Madhusudhana, R., Sharma, V., Hussain, T., Ghorade, R., Khandalkar, H., Audilakshmi, S. and Seetharama, N. 2011. Identification and validation of genomic regions that affect shoot fly resistance in sorghum [*Sorghum bicolor* (L.) Moench]. *Theoretical and Applied Genetics*. 122: 1617–30.

Ashok Kumar, A., Reddy, B. V. S., Sharma, H. C., Hash, C. T., Rao, P. S., Ramaiah, B. and Reddy, P. S. 2011. Recent advances in sorghum genetic enhancement research at ICRISAT. *American Journal of Plant Science*. 2: 589–600.

Ashok Kumar, A., Reddy, B. V. S. and Grando, Stefania. 2015. Global millets improvement and its relevance to India and developing world. In *Millets-Promotion for food, feed, fodder, nutritional and environment security*. Vilas A. Tonapi, B. Dayakar Rao and J. V. Patil, Society for Millets Research, ICAR-Indian Institute of Millets Research, Rajendaranagar, Hyderabad- 500 030, India. pp 154–72.

Audilakshimi, S., Aruna, C., and Kiran, V. V. S. 2003. Utilization of germplasm for improvement of varieties in India. In: Report of the AICSIP Coordinating Team, 2002-03, AICSIP Annual Meeting of the Sorghum Group, 1–3 April, held at Sorghum Research Station (GAU), Athwa Farm, Surat, Gujarat, India, pp. 47–53.

Awaghad, P. R., Ganvir, B. N. and Bhopale, A. A. 2010. Growth and instability of kahrif sorghum in Western Vidarbha region. *Journal of Soils and Crops*. 20: 111–17.

Basavaraj, G., Parthasarathy Rao, P., More, Sachin and Pokharkar, V. G. 2014. Analyzing pro-poor value chains- Study of post rainy season sorghum in Maharashtra state of India. Presented at Global value chains for food and nutrition security 25–26 September 2014, Roma Tre University Sala delle Lauree, School of Economics and Management Via Silvio d'Amico 77, 00145, Rome.

Basavaraja, H., Hugar, A. Y., Mahajanshetti, S. B., Angadi, V. V. and Dayakar Rao, B. 2005. Kharif sorghum in Karnataka: An economic analysis. *Agricultural Economics Research Review* 18: 223–40.

Chand, R. 2007. Demand for food grains. *Economic and Political Weekly* 42 (52).

Chapke, R. R., Mondal, B. and Mishra, J. S. 2011. Resource-use efficiency of sorghum (*Sorghum bicolor*) production in rice (*Oryza sativa*) –fallows in Andhra Pradesh, India. *Journal of Human Ecology* 34(2): 87–90.

Chen, Y. and Shi, Y. 1995. Review, current situation and prospect of sorghum production in China. In Proc. Chinese sorghum symposium, Shenyang, China, pp. 8–13.

Cromwell, E., Friis-Hansen, E. and Turner, M. 1992. *The Seed Sector in Developing Countries: A Framework for Performance Analysis*. ODI, London.

Dahlberg, J. A. 2000. Classification and characterization of sorghum. In: C. W. Smith and R. A. Frederiksen (eds), *Sorghum Origin, History, Technology and Production. Wiley Series in Crop Science*. John Wiley & Sons, Hoboken, NJ, pp. 99–130.

Damania, A. B. 2002. The Hindustani centre of origin of important plants. *Asian Agri-History* 6: 333–41.

Dayakar, R. B., Patil, J. V., Hymavathi, T. V., Nirmal Reddy, K. and Rajendra Prasad, M. P. 2014. Creation of Demand for Millet Foods through PCS Value Chain. Final report of NAIP (ICAR). Directorate of Sorghum Research, Rajendranagar, India.

de Oliveira, A. C., Richter, T. and Bennetzen, J. L. 1996. Regional and racial specificities in sorghum germplasm assessed with DNA markers. *Genome*, 39: 579–87.

Doggett, H. 1988. *Sorghum*. 2nd ed. Longman, London; published in the United States with Wiley, New York.

FAO. 1995. *Sorghum and Millets in Human Nutrition*. (FAO Food and Nutrition Series, No. 27). Rome, Italy. ISBN 92-5-103381-1

FAO. 1998. Food and Agriculture Organization of the United Nations. FAOSTAT Statistics Database. Rome, Italy.

FAO. 2012. Food and Agriculture Organization of the United Nations. FAOSTAT Statistics Database. Rome, Italy.

FAO. 2016. Food and Agriculture Organization of the United Nations. FAOSTAT Statistics Database. Rome, Italy.

FAO and ICRISAT. 1996. *The World Sorghum and Millet Economies: Facts, Trends and Outlook*. Published by International Crops Research Institute for Semi-Arid Tropics. P 68. ISBN 92-5-103861-9.

Gowda, C. L. L. and Stenhouse, J. W., eds. 1993. Collaborative Sorghum Research in Asia Report of the Asian Sorghum Researchers' Consultative Meeting 27–29 September 1993. International Crops Research Institute for the Semi-Arid Tropics, Patancheru, Andhra Pradesh, India. ISBN 9290662794.

House, L. R., Gomez, M., Murty, D. S., Sun, Y. and Verma, B. N. 2000. Development of some agricultural industries in several African and Asian countries. In: C. W. Smith and R. A. Frederiksen (eds), *Sorghum Origin, History, Technology and Production. Wiley Series in Crop Science*. John Wiley & Sons, Hoboken, NJ, pp. 131–90.

Kimber C. T. 2000. Origins of domesticated sorghum and its early diffusion to India and China. In: W. C. Smith and R. A. Frederiksen (eds), *Sorghum: Origin, History, Technology and Production*. John Wiley & Sons, Inc., pp. 3–98.

Li, Guiying, Lu, Qingshan and Zou, Jianqiu. 2004. Alternative Uses of Sorghum- Methods and Feasibility: Chinese perspectives in Alternative uses of sorghum and pearl millet in Asia: Proceedings of the Expert Meeting held at International Crops Research Institute for the Semi-Arid Tropics.

Li, Y. and Li, C. 1998. Genetic contribution of Chinese landraces to the development of sorghum hybrids. *Euphytica*. 102: 47–55.

Li, S. Z., Li, G. M., Zhang, L., Zhou, Z. X., Han, B., Hou, W. H., Wang, J. B. and Li, T. C. 2013. A demonstration study of ethanol production from sweet sorghum stems with advanced fermentation technology. *Applied Energy*. 102: 260–5.

Mann, J. A., Kimber, C. T. and Miller, F. R. 1983. The origin and early cultivation of sorghums in Africa. Texas Agricultural Experiment Station Bulletin 1454.

Murthy, M. V. R., Singh, Piara, Wani, S. P., Khairwal, L. S. and Srinivas, K. 2007. Yield gap analysis of sorghum and pearl millet in India using simulation modellling. GTAES, Report No. 37. Patancheru 502 324. A.P., India: International Crops Research Institute for the Semi-Arid Tropics (ICRISAT). 82 pp.

NAAS, 2013. Role of millets in Nutritional Security of India. Policy Paper No. 66, National Academy of Agricultural Sciences, New Delhi: 16 p.

Nedumaran, S., Abinaya, P. and Bantilan, M. C. S. 2013. Sorghum and millets futures in Asia under changing socio-economic and climatic scenarios. Discussion paper. International Crops Research Institute for the Semi-Arid Tropics, Patancheru, Andhra Pradesh, India. http://oar.icrisat.org/6427/7/Nedumeran_etal_2013_IEDPS_2.pdf

Parthasarathy Rao, P., Basavaraj, G., Ahmad, W. and Bhagavatula, S. 2010. An analysis of availability and utilization of sorghum grain in India. *SAT e Journal* 8: 1–8.

Parthasarathy Rao, P., Hall, A. J. and Bantilan, M. C. S. 2004. Dynamics of utilization, markets, trade and coalition: sorghum and millets in Asia. In: CFC and ICRISAT. Alternate uses of sorghum and pearl millet in Asia: proceedings of the Expert meeting, ICRISAT Patancheru, India, 1–4 July 2003. CFC Technical paper No. 34, pp. 93–112.

Patil, J. V., Aruna, C., Prabhakar, P. Sanjana, Bhat, B. V. and Umakanth, A. V. 2015. Genetic improvement of sorghum in India. In: Tonapi V. A., Dayakar Rao B., Patil J. V. (eds), *Millets-Promotion for Food, Feed, Fodder, Nutritional and Environment Security*, Society for Millets Research, ICAR-Indian Institute of Millets Research, Rajendranagar, Hyderabad 500 030, India, pp. 173–89.

Prasada Rao, K. E., Murthy, D. S. 1982. Sorghum for special uses. In L. W. Rooney and D. S. Murthy (eds.), *Proc. International Symposium on Sorghum Grain Quality*. International Crops Research Institute for Semi-arid Tropics, Patancheru, AP, India, pp. 129–34.

Qu, H., Liu, X. B., Dong, C. F., Lu, X. Y., Shen, Y. X. 2014. Field performance and nutritive value of sweet sorghum in eastern China. *Field Crops Research*. 157: 84–8.

Rakshit, S., Hariprasanna, K., Gomashe, S., Ganapathy, K. N., Das, I. K., Ramana, O. V., Dhandapani, A. and Patil, J. V. 2014. Changes in area, yield gains and yield stability of sorghum in major sorghum producing countries, 1970 to 2009. *Crop Science* 54: 1571–84

Rakshit, S., Prabhakar, Aruna C., Umakanth, A. V. and Bhat, B. V. 2015. Crop improvement research in sorghum for enhancing productivity. *Indian Farming- Special issue on Small Millets*: 65 (4): 12–18.

Ramakrishna, C., Tonapi, V. A. and Dayakar Rao, B. 2006. Sorghum seed production and marketing scenario in India. In: N. Seetharama and V. A. Tonapi (eds), *Strategies for Millets Development and Utilization*. Society for Millets Research, 11-127, National Research Centre for Sorghum, Rajendranagar, Hyderabad 500 030, Andhra Pradesh, India, pp. 213–25.

Rana, B. S., Kaul, S. and Rao, M. H. 1997. Impact of genetic improvement on sorghum productivity in India. Proceedings of the International Conference on Genetic Improvement of Sorghum and Pearl millet. 22–27 September 1996, Lubbock, Texas, USA, pp. 142–65.

Rao, N. G. P. and Rana, B. S. 1982. Selection in temperate-tropical crosses of sorghum. In: Sorghum in the eighties vol. 1: Proceedings of the International symposium on sorghum, 2–7 November 81, ICRISAT, Patancheru, AP, India, pp. 403–19.

Ratnavathi, C. V. and Patil, J. V. 2013. Sorghum Utilization as Food. *Journal of Nutrition and Food Science* 4: 247.

Ravinder Reddy, Ch, Ashok Kumar, A., Gadakh, Sharad, Chavan, Uttam, Kalpande, H. K. and Tonapi, V. A. 2015. Developing post rainy sorghum seed system in India. In: V. A. Tonapi and Patil J. V. (Eds), *Millets-Ensuring Climate Resilience and Nutritional Security*. Daya Publishing house, a division of Astral International Pvt. Ltd. New Delhi 110 002, pp. 607–26.

Ravinder Reddy, Ch. and Tonapi, V. A. 2015. Sustaining viability of informal millet seed systems. In: Tonapi V. A. and J. V. Patil (eds), *Millets-Ensuring Climate Resilience and Nutritional Security*. Daya Publishing house, a division of Astral International Pvt. Ltd. New Delhi 110 002, pp. 595–606.

Reddy, B. V. S., Ashok Kumar A. and Sanjana Reddy, P. 2008. Genetic improvement of sorghum in the semi-arid tropics. In: Reddy B. V. S., Ramesh, S., Ashok Kumar, A. and Gowda, C. L. L. (eds), *Sorghum Improvement in the New Millennium*. International Crops Research Institute for the Semi-Arid Tropics. ISBN 978-92-9066-512-0, pp. 105–23.

Reddy, B. V. S., Ramesh, S., Ashok Kumar, A., Wani, S. P. and Ortiz, R. 2008. Biofuel crops research for energy security and rural development in developing countries. *Bioenergy Research* 1: 248–58.

Reddy, B. V. S., Sanjana Reddy, P., Sadananda, A. R., Dinakaran, Elango, Ashok Kumar, A., Deshpande, S. P., Srinivasa Rao, P., Sharma, H. C., Sharma, Rajan, Krishnamurthy, L. and Patil, J. V. 2012. Postrainy season sorghum: Constraints and breeding approaches. *SAT e Journal* 10: 1–12.

Reddy, N. R. R., Ragimasalawada, M., Sabbavarapu, M. M., Nadoor, S. and Patil, J. V. 2014. Detection and validation of stay-green QTL in post-rainy sorghum involving widely adapted cultivar, M35-1 and a popular stay-green genotype B35. *BMC Genomics*, 15 (1). 909.

Rooney L. W. 1996. Attributes of improved quality sorghums for value added marketing. In ASTA Proc. 51st Annual Corn and Sorghum Research Conference, Chicago, Dec, 10, pp. 112–24.

Rooney, L. W. and Waniska, R. D. 2000. Sorghum food and industrial utilization. In: C. W. Smith and R. A. Frederiksen, editors, *Sorghum origin, history, technology and production. Wiley Series in Crop Science*. John Wiley & Sons, Hoboken, NJ, pp. 689–729.

Ryan, J. G. and Spencer, D. C. 2001. Future challenges and opportunities for agricultural R&D in the Semi-Arid Tropics, Patancheru, 502 324, India. International Crops Research Institute for Semi-arid Tropics. 83pp.

Sajjanar, G. M., Biradar, B. D. and Biradar, S. S. 2011. Evaluation of crosses involving rabi land races of sorghum for productivity traits. *Karnataka Journal of Agricultural Sciences* 24: 227–9.

Sanjana Reddy, P. and Patil, J. V. 2015. *Genetic Enhancement of rabi Sorghum-Adapting to the Indian Durras*, Elsevier, p. 240.

Sanjana Reddy, P., Patil, J. V., Nirmal, S. V. and Gadakh, S. R. 2012. Improving post-rainy season sorghum productivity in medium soils: does ideotype breeding hold a clue? *Current Science*. 102: 904–8.
Satish, K., Srinivas, G., Madhusudhana, R., Padmaja, P., Reddy, R. N., Mohan, S. M. and Seetharama, N. 2009. Identification of quantitative trait loci for resistance to shoot fly in sorghum [*Sorghum bicolor* (L.) Moench]. *Theoretical and Applied Genetics* 119: 1425–39.
Singh, J. P. and Dayakar Rao, B. 2010. Status paper on millets in India, 2010 Background paper prepared for National Seminar on Millets, organized by Directorate of Millets Development, Jaipur, and Directorate of Sorghum Research, Hyderabad.
Vavilov, N. I. 1992. *Origin and Geography of Cultivated Plants*, edited by V. F. Dorofeev, Cambridge University Press, Cambridge, U.K. 332 p.
Visarada, K. B. R. S., Padmaja, P. G., Saikishore, N., Pashupatinath, E., Royer, M., Seetharama, N. and Patil, J. V. 2014. Production and evaluation of transgenic sorghum for resistance to stem borer. In *Vitro Cellular and Developmental Biology-Plant* 50: 176–89. doi:10.1007/s11627-013-9561-5.
Wang, M., Chen, Y., Xia, X., Li, J. and Liu, J. 2014. Energy efficiency and environmental performance of bioethanol production from sweet sorghum stem based on life cycle analysis. *Bioresource Technology* 163: 74–81.
Zhao S. 1985. Various sorghum foods in northern China. *Liaoning Agric. Sci.* 3.
Zhen, Yang. 1997. Sorghum breeding research in China. *International Sorghum and Millets Newsletter* 38: 15–18.

Chapter 9

Sorghum cultivation and improvement in West and Central Africa

E. Weltzien and H. F. W. Rattunde, University of Wisconsin-Madison, USA, formerly International Crop Research Institute for the Semi-Arid Tropics (ICRISAT), Mali; T. A. van Mourik, International Potato Center, Ghana; and H. A. Ajeigbe, International Crop Research Institute for the Semi-Arid Tropics (ICRISAT), Nigeria

1 Introduction

The diversity of sorghum cultivated in Africa attests to the African origin of this crop. Ten to 25 or more varieties of sorghum may be cultivated as distinct pure stands in a single village in Mali (Siart, 2008) or Burkina Faso (Barro-Kondombo et al., 2010). In Northern Cameroon, varietal mixtures are cultivated, with each mixture containing 12 varieties on average (Barnaud et al., 2006). Farmers have developed strategies for using varietal diversity to minimize risk and maximize productivity in the context of complex and diverse adaptive challenges, strategies developed over several thousand years of cultivating sorghum. The diversity of sorghum types cultivated reflects the wide and contrasting ecosystems in which it is cultivated and the range of ways it is used (Rooney, 2004).

Sorghum in West and Central Africa (WCA) is primarily produced between the 500- and 1200-mm rainfall isohyets. These rainfall isohyets, following an east-west orientation parallel to the southern edge of the Sahara (Fig. 1), as well as biotic constraints and soil

http://dx.doi.org/10.19103/AS.2017.0015.29

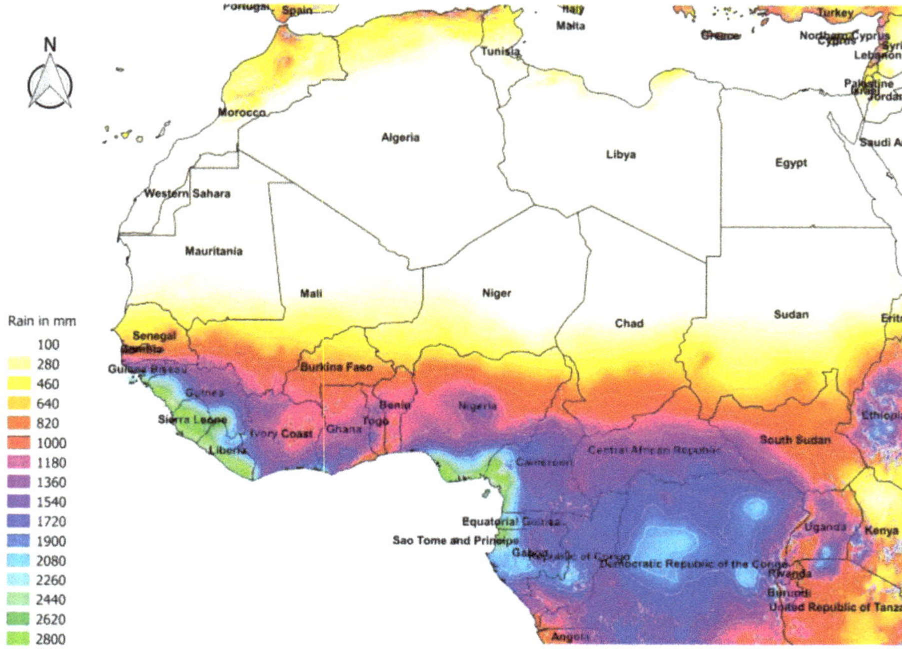

Figure 1 Rainfall isohyets of West and Central Africa (Source: www.esri.noaa.gov).

typology, determine the major varietal adaptation zones. Sorghum continues to be a basic staple food for large portions of the population in this region. The total sorghum production is larger and increasing in West Africa (1980, 5.5 m tons vs 2010, 12.9 m tons) as compared to East Africa (1980, 3.3 m tons vs 2010, 6.4 m tons) and Southern Africa (1980, 0.8 m tons vs 2010, 0.3 m tons) (FAOSTAT, 2017). This chapter aims to provide insights and entry points for understanding sorghum cultivation in the WCA region. Chapter sections describe the major sorghum production systems, options for intensification and sustainable production, genetic diversity and opportunities to breed for enhancing sorghum crop value, and progress in strengthening seed systems.

2 Overview of sorghum production systems in West and Central Africa

Sorghum is grown first and foremost for food throughout most of WCA. This crop therefore is not just a commodity but has been a determinant of survival and an integral part of life and culture of this region over millennia. For example, sorghum in WCA is typically considered a 'men's' crop (Weltzien et al., 2006), with men being responsible for sustaining the families by providing the basic staple cereals used to cook the main meals. As such it was traditionally taboo for men to sell sorghum grain and put the survival of their family in jeopardy. Thus sorghum certainly was a subsistence crop, grown for home consumption as tô (thick porridge), degué (thin porridge), couscous or other preparations, including beer in some locations. Yet, farmers are ever more integrated into a cash economy, and sorghum is increasingly grown not only for food but also for cash income and fodder, with strong demand for sorghum grain from both rural and the fast-growing urban populations and processing companies. Women often cultivate sorghum as well, but typically in smaller areas. The grain they produce is often for special needs, such as for children's meals outside of the main family meals, during hunger times when the men's grain stores are exhausted, or for sale to meet household expenses. The use of sorghum grain for malting in artisanal and/or industrial brewing can be locally very important, such as in parts of Burkina Faso, Ghana and Nigeria. And finally, the sorghum residue, the stover, can be important for feeding livestock, particularly for dry season survival in lower rainfall zones and areas with more intense land use.

2.1 Sahelian region

The Sahel zone, on the edge of the Sahara, is a zone with annual rainfall between 100 and 600 mm (Nicholson, 2013), and agricultural production between 300 and 600 mm. The rainfall occurs in a single rainy season, primarily between July and September. The Sahel is known for its high inter-annual variability and unpredictable rainfall, leading to frequent droughts and even famine. The predominant soils in this zone are aridisols with sand content often 80% or higher. This zone spans sub-Saharan Africa from Senegal in the west to Sudan in the east (Fig. 1).

Pearl millet is the dominant staple crop in this zone, with sorghum cultivated in areas where there are heavier soils such as in the Gedaref of Sudan or in lower lying areas where water accumulates or stagnates. However, sorghum is the dominant cereal in certain sandy soil areas where pests or diseases prevent pearl millet cultivation. One such area is the north-west region of Mali, where blister beetles feeding on the pollen and anthers of pearl millet frequently cause poor seed set. Sorghum is also intercropped with pearl millet, especially in deep sandy soils where water reserves at lower depths can be reached by late-flowering sorghums. In Niger this type of 'dune sorghum' also benefits from dew during November and December to complete its growth cycle. Sorghum may be cultivated in rotation with pearl millet to interrupt the multiplication of the parasitic weed *Striga hermonthica* in areas where this parasite is specific to pearl millet.

A sorghum cropping system that uses receding water following the end of the rainy season in Sahelian areas is identified by region-specific names; 'Decrue' in Mali and Mauritania, 'Fadama' in Northern Nigeria or 'Marwari' in Chad and Cameroon (Comas and Gomez MacPherson, 2002). Bird damage and stemborer attacks are strong biotic constraints in this production system. Sorghum may also be irrigated in these areas, especially where there is a high demand for animal fodder as in Mauritania (Mohamed et al., 2016).

Animal husbandry is the basis for livelihoods and resilience in the drought-prone Sahelian zone (Mohamed et al., 2016). Sorghum and pearl millet stover therefore are highly valued, with nearly all stover being consumed by livestock grazing or managed feeding during the dry season.

2.2 Savannah region

The Savannah region encompasses a broad swath of land with annual precipitation ranging from 600 to 1300 mm, with rainfall declining as one heads northwards towards the Sahel. Subzones are identified within this region based on rainfall levels; the Sahel Sudan Savannah or Northern Sudanian zone (approximately 600 to 800 mm), the Sudanian-Savannah or Southern Sudanian zone (800 to 1100 mm), and the Northern Guinea-Savannah zone (1100 to 1300 mm). These agroecological zones stretch across WCA immediately south of the Sahel zone, extending from Senegal and Ghana to Chad and parts of South Sudan (Fig. 1). The sowing period for sorghum depends on the onset and length of the rainy season, with sowing in the wetter Northern Guinea zone spanning the period from May to early August, whereas in the Sudanian zone it occurs in the months of June and July, with typically later onset in the Northern Sudanian relative to the Southern Sudanian zone.

Sorghum is the dominant traditional staple crop in the Savannah region. It is often grown in rotation or intercropped with cotton, maize, millet, groundnuts, cowpeas or soybean in varying proportions and intercrop arrangements. Sorghum typically follows cotton or maize in these rotations and benefits from residual fertility, receiving little direct application of inorganic fertilizer. Sorghum's ability to more reliably produce under poorer soil fertility conditions contributes to its important role in these farming systems and its importance for food security. Pearl millet, rice and diverse minor millets, roots and tubers and other legumes are also grown in this region for food security and sale in local markets.

Research and development interventions for intensifying sorghum production have greatest potential for positive and rapid outcomes in the Savannah region, especially in the context of higher grain prices. Recently sorghum demand in the malting and confectionery industries has increased significantly, especially in Nigeria (Office of the Honorable Minister of Agriculture, 2012) and Ghana, with sorghum being substituted for other cereals in the production of malt and composite flours. The opportunities are greatest in the Southern Sudanian zone, where the largest population and sorghum production areas are found, and in the Northern Guinea Savannah, where historically relatively less research and development efforts have been made. There are some indications of intensification (Sanogo et al., 2010), but it depends on the individual farm unit and location (Falconnier et al., 2015; Ollenburger et al., 2016).

Labour shortage, resulting from limited mechanization and low population densities, especially in higher rainfall areas, is one of the main constraints to sorghum production and intensification in these zones (Kenga et al., 2003). Weed control is a primary constraint, although the use of herbicide is rapidly increasing. Women sometimes lead in adopting herbicide use to cope with the higher weed pressures resulting from later allocation and ploughing of their fields relative to men's fields. Low soil fertility is another serious constraint to sorghum productivity in this region (Buerkert et al., 2001; Leiser et al., 2012), with sorghum typically cultivated in fields deemed to be too poor for maize production. Furthermore, what little inorganic fertilizer is applied has relatively low efficacy (Buerkert and Hiernaux, 1998) due to labour limitations and near absence of mechanization for optimal application methods (Leiser et al., 2012). Thus the value and sustainability of new sorghum production technologies will be determined to a considerable extent by their labour requirements and their contributions under specific soil fertility and weed management contexts.

The integration of livestock and crop production is progressing where land use is intensifying and where improved control of the Tsetse fly reduces the risk of Trypanosomiasis for ruminants. Intensified agroforestry and horticulture are also important emerging options. Maize has become an important staple in the higher rainfall areas, complementing or even replacing sorghum in these zones. Maize can produce higher yields with sufficient fertilization, and due to its shorter growth cycle, spreads the work load and provides food during the 'hungry period' prior to the end of the rainy season, although with serious risks of aflatoxin contamination.

Sorghum, unlike maize, normally matures after the rains end which, together with a lax panicle form, minimizes grain mould infestations of the openly exposed grains (Haussmann et al., 2012). Photoperiod sensitivity enables sorghum to flower at a relatively fixed calendar date and to mature after the rains end, despite highly variable sowing dates. This ability not only helps to minimize grain mould and aflatoxin contamination but also reduces the potentially devastating damage by bird and livestock grazing, by facilitating synchronous, optimally timed, maturation relative to the end of the rains. It also reduces risks of poor grain filling encountered by late-maturing plants that run out of water. Photoperiod sensitivity is thus a key mechanism for yield stability and grain quality in these zones (Clerget et al., 2008; Haussmann et al., 2012). Although photoperiod-insensitive varieties can be grown in niche environments, photoperiod-sensitive varieties are best for the predominant, full-season, production environments, and hybrids with photoperiod sensitivities similar to those of local varieties have recently been bred (Rattunde et al., 2013; Kante et al., 2017). Photosensitive varieties typically have adaptation ranges in east-west bands of only about 100 km width, corresponding to the narrow east-west rainfall isohyets of WCA (Fig. 1).

3 Intensification and sustainability of sorghum production systems in WCA

3.1 Soil fertility management and varietal options for farmers

WCA has some of the most geologically ancient soil parent materials and a preponderance of highly weathered soils with low soil organic matter content and low fertility (Breman, 1998). The low plant-available phosphorus (P) in these soils is a major limiting factor to sorghum production (Doumbia et al., 1998; Buerkert et al., 2002; Ouédraogo et al., 2005; Kihara et al., 2016). Whereas the critical P threshold required for sorghum growth is 11.6 ppm Bray1 P (Doumbia et al., 1993), sorghum production fields in Mali averaged only 8.3 ppm Bray1 P in the more intensified cotton-growing Dioila region and 4.5 ppm Bray1 P in the Mande region where little or no cotton is grown (unpublished data). Soil organic matter was also low (0.6% organic C) in both zones.

Sorghum's ability to reliably produce harvestable grain under low soil fertility conditions without mineral fertilization is one reason why it remains an important staple cereal in WCA. Approximately 90% of sorghum in the region is produced without inorganic fertilization, primarily due to farmers' financial constraints. The crop relies to a great extent on residual fertilizer effects from previous crops and limited applications of organic amendments. The low fertility conditions under which sorghum is grown is a major contributor to the generally low (approximately 1 ton/ha) grain yields obtained by farmers in the region (Buerkert et al., 2001; vom Brocke et al., 2010).

Agronomic studies already in the 1960s showed that sorghum yield responses to nitrogen (N) and phosphorus (P) applications were attainable, although there was no response to N without P applications in Northern Nigeria (Kaduna to Zaria) (Goldsworthy, 1967). A recent meta-analysis of sorghum response to soil fertility options in Africa reported 47–98% yield increases with N and P chemical fertilizers and 43–87% yield increases with use of organic nutrients across 29 studies in Africa (Tonitto and Ricker-Gilbert, 2016). This study estimated that net returns from N+P applications were higher than those from N alone, but the net returns with both mineral fertilizer treatments were considerably lower than those with sorghum–cowpea (*Vigna unguiculata*) rotations in Burkina Faso where in-depth economic analyses were done. Intercropping sorghum and cowpea, however, gave net returns that were considerably lower than those obtained with rotations and slightly lower than those obtained with mineral fertilizer treatments.

Organic soil amendments are vital for intensification of WCA farming systems (de Ridder and van Keulen, 1990). The integration of livestock with crop production offers important opportunities for enhancing sorghum producers' access to animal manure (McIntire et al., 1993).

The method of fertilizer application is a major determinant of fertilizer efficiency for sorghum. Broadcasting fertilizer without incorporation is expected to result in losses through run-off, N volatilization and poor plant uptake (Leiser et al., 2014). Also, farmers' practice of applying fertilizer only after the first weeding, three to five weeks after sowing, is expected to deliver P too late to benefit the current crop. Placement of small quantities of fertilizer [4 kg NPK ha^{-1} (Buerkert and Hiernaux, 1998), 3 g per planting hole (Aune et al., 2007)] near the seed at or shortly after sowing were found to increase yields at least 30% and give a threefold increase of P use efficiency relative to traditional practices (Buerkert and Hiernaux, 1998). Microdosing just 0.6 g NPK per planting hole was able to increase grain yield on a poor heavy clay soil (Abdalla et al., 2015). The development of appropriately scaled implements for mechanized application of fertilizer could greatly enhance fertilizer efficiency and yields by enabling farmers to incorporate fertilizer in a timely and effective manner despite their labour constraints.

The complexity of WCA farming systems, with trade-offs between income generation, soil conservation and community agreements, calls for focus on farm and farming system and not just individual fields for targeting technologies (Giller et al., 2011). Although context-specific information on fertilization would be useful to sorghum farmers seeking to optimize fertilization rates and profitability, such information is extremely rare. For example, information on rate of returns to fertilizer use by soil type and fertility level, rainfall zone and preceding cropping history would help farmers make informed decisions.

Sorghum fertilization and varietal options were jointly examined for their effects on productivity and profitability in a series of 81 on-farm variety trials conducted in Mali by the Institut d'Economie Rurale, ICRISAT and farmer organizations from 2009 to 2012. Three different types of varieties (farmers' local, pure-line bred and Guinea-race hybrids) and two fertilization options [without and with fertilization (100 kg di-ammonium phosphate and 50 kg urea/ha)] were tested. The trials were conducted by equal numbers of women and men farmers in the Dioila, Koutiala and Mande zones of Mali. Fertilization improved grain yields on average by 55% to 70% depending of the type of variety (Table 1). The net returns, estimated by subtracting the expenses for fertilizer, seed and seed treatment and field labour from the value of the grain produced, were higher with fertilization than without for all types of varieties. The research-bred pure-line and hybrid varieties, however, showed approximately double the returns to fertilizer use relative to the farmers' local variety. The use of these improved sorghum varieties, with considerably lower cost for seed relative to that of fertilizer, appears to offer farmers a type of insurance for obtaining returns on their investment in fertilizer purchase.

Table 1 Mean yields and net returns under unfertilized and fertilized conditions of farmers' local varieties, research-bred pure-line and hybrid varieties over 81 on-farm trials in Mali, 2009–12

	Unfertilized control		Fertilized		Gain fertilized over control	
Variety type	Yield (kg/ha)	Net return (Fcfa*)	Yield (kg/ha)	Net return (Fcfa)	Yield (%)	Net return (%)
Local	685	117 436	1062	146 164	55	24
Bred pure-line	785	134 083	1332	192 083	70	43
Hybrid	893	150 843	1516	222 360	70	47

* Monetary unit in West Africa with fixed exchange rate to the Euro (656 Fcfa:1 Euro).

3.2 Soil and water conservation

An array of techniques have been developed for soil and water conservation (SWC) that are pertinent for the predominant sorghum production areas of WCA where water run-off can cause serious soil, nutrient and water losses (Zougmoré et al., 2009).

Contour bunds, stone or earthen mounds with permanent grass cover installed on the contour are the most widely followed and economically important practices. Contour earthen bunds are inexpensive and profitable (Gigou et al., 2006). The installation of contour bunds with oxen-drawn plough requires external help for determining the contour lines. A local Malian NGO has assisted with installation of contour bunds, and farmers' demand has exceeded the NGO's capacity (Gigou et al., 2006). Grass strips with *Andropogon gayanus* is another option (Zougmoré et al., 2009).

Contour bunds have substantially reduced losses of organic carbon, N and P by reducing soil losses by 84% with stone lines and 71% with grass strips (Zougmoré et al., 2009). Application of compost even without installation of barriers was also found to substantially reduce soil loss (Zougmoré et al., 2009).

The installation of erosion barriers alone is not necessarily sufficient for obtaining good sorghum performance on nutrient-deficient soils without application of soil fertility measures (Zougmoré et al., 2003; Fatondji et al., 2009). Further, by combining water conservation and soil fertility measures, even greater reductions of run-off can be obtained. Whereas SWC measures alone reduced run-off by 59%, these measures plus application of mineral N or organic N reduced run-off by 67% and 84%, respectively (Zougmoré et al., 2010). This synergy of soil fertility and SWC measures was also observed with compost application with SWC barriers reducing soil loss by 79%, whereas compost alone gave only a 52% reduction (Zougmoré et al., 2010).

The Zai technique offers possibilities for enhancing sorghum yields on degraded, highly crusted soils as found in parts of Burkina Faso. This technique involves digging planting holes that not only break hard surface layers but enable capture of windblown sediment and effective placement of scarce organic amendments (Fatondji et al., 2009). The half-moon basin technique, introduced in Northern Burkina Faso (Zougmoré et al., 2003), also offers the possibility of point applications of soil amendments and water capture in highly crusting degraded soils.

The management of excess water is also critical for successful sorghum production as waterlogging can severely impede early vegetative growth. Farmers often sow sorghum on ridges, typically made with animal traction, to favour good crop growth despite the presence of standing water.

3.3 Pest, disease and weed management

Sorghum productivity in WCA can be constrained by a wide range of insect pests, diseases and competition from weeds. Each agroecology and dominant production system has specific complexes of biological threats to both yield and quality of grain for food and stover for ruminant fodder. Options for pest, disease and weed management for WCA sorghum production are presented below, considered by stage of growth.

4 Biological constraints encountered in specific growth phases

4.1 Seed storage and seed vigour

Most sorghum farmers in WCA store their own seed. Panicles harvested for seed are typically tied together by braiding the peduncles, and are thus identifiable as seed and kept separate from grain. These bundles are dried more carefully than grain and often stored in specific places, such as near the smoke of cooking fires, to minimize insect attacks. Some farmers and seed cooperatives have successfully started to use triple layered bags for storage of seed. The germination percentage and seedling vigour of farmer-produced sorghum seed was found to be excellent, equal to or better than researcher-produced seeds in Mali (M. Diallo, 2009, unpublished PhD thesis).

Important varietal differences in storability of grain and seeds exist in sorghum. Softer endosperm grain types, common among Durra and Caudatum-race varieties, are more susceptible to storage weevils than harder endosperm, Guinea-race, grain types.

4.2 Crop establishment and early growth

Bird feeding on sown grain or young seedlings can lead to huge losses during sorghum establishment. Sorghum emergence is rapid, occurring in three to five days, and bird scaring during the brief period of establishment is feasible and often practised. Some forms of seed treatment and fertilizer placement in or next to planting holes can enhance seedling growth through this critical period. Seed priming can also reduce the time taken from sowing to emergence and enhance early vigour (Harris, 2006), minimizing the time the sorghum crop is exposed to bird damage or damping off diseases.

Sorghum shoot fly (*Atherigona soccata*) and spittle bugs [*Poophilus costalis* (Walker)] although relatively rare, can be problematic where sowing dates differ widely and pest population buildups on early sown fields can devastate those that are sown later. Timely placement of fertilizer, and particularly P (Leiser et al., 2014), can help sorghum crops grow out of this susceptible stage more quickly. Stemborers in Northern Nigeria, with *Busseola fusca* (Fuller) being the most important species (Ajayi, 1989), can cause serious damage to sorghum starting in the establishment phase and continuing throughout the growth duration.

Weed control is a major concern for sorghum producers in the early growth stages. Farmers prepare land for sowing sorghum by removing weeds, bushes and residues of previous crops. Some farmers sow sorghum with the very first rains, possibly into zai holes, without any further land preparation so as to get a head start over weed growth and benefit from the early season nitrogen flush. The use of pre-emergent herbicide, particularly for later sown sorghums, has been more frequently observed in recent years.

4.3 Vegetative growth phase

Anthracnose (*Colletotrichum sublineolum*) is the most economically important disease (Cuevas et al., 2016), with host resistance specific for stem, rachis and grain (Hess et al., 2002). This pathogen is genetically variable and goes through several cycles of sexual reproduction per season (Gale, 2003). Thus, certain sorghum varieties that were resistant to Anthracnose have shown susceptibility after multiple years of cultivation on the same research station fields. Crop rotation therefore offers farmers a good control option. Removal of sorghum residues from the field can also reduce inoculum levels, but the economic value of the practice needs to be examined. As the microclimate in a sorghum canopy may be altered by plant architecture (e.g. stem internode lengths, leaf size and angle), plant spacing and density, these factors may influence incidence and development of Anthracnose and other diseases.

A wide range of foliar diseases, mostly fungal but also bacterial and viral, occur as well. Although these diseases rarely cause significant grain yield losses, they can cause serious reductions in green leaf area and reduce the fodder quality of stover. Monitoring and incorporating foliar disease resistance are therefore important for dual-purpose grain/fodder varieties.

Various insect infestations can occur but are typically local in their distribution, with high inter-annual variation (Chantereau and Fliedel, 2013). Stemborers (*Busseola fusca, Chilo partellus* and *Sesamia spp.*) may cause serious damage in areas with continuous cultivation of sorghum in some systems such as in Northern Nigeria Savannah region or in residual moisture areas. *Busseola fusca* is the most damaging species (Muturi et al., 2012). Many species of aphids occur but rarely justify chemical treatment, which can disturb natural regulation of insect populations (Chantereau and Fliedel, 2013).

4.4 Reproductive phase

Sorghum midge (*Stenodiplosis sorghicola*) lay eggs in sorghum flower spikelets, and the larvae devour the ovary. The incidence can be severe in certain hotspots (Dakouo et al., 2005; Chantereau and Fliedel, 2013), resulting in panicles devoid of grain. Infestations are often severe where farmers cultivate varieties of contrasting flowering dates as in Northern Ghana, where early maturing-introduced varieties resulted in sorghum midge population buildups that decimated later flowering local varieties, and in eastern Burkina Faso, where midge build up on early flowering red sorghum types that can devastate later flowering susceptible white sorghums.

Head bugs lay eggs and feed on developing caryopsis, causing direct loss and, by providing entry points for fungal pathogens, contribute to grain mould in what can be called a head bug–grain mould complex. *Eurystylus oldi* Poppins is the dominant head bug species in WCA (Ratnadass et al., 2003). Particularly severe damage occurs on compact-panicle varieties bred with introduced germplasm (Ratnadass et al., 2003) and on panicles with poor exertion. The local varieties, with their lax panicles, typically suffer less damage as insects are more exposed to predation and desiccation.

Grain mould is caused by a complex of fungi (Marley et al., 2003; Rooney, 2004). Grain mould may severely discolour grain, reducing yield of usable grain, and may lead to mycotoxins (Marley et al., 2003) that could affect human and animal health. Grain mould also causes significant reduction of seed viability. Anthracnose also infests grains but is independent of head bug feeding.

Bird feeding can cause severe yield losses. Traditional lax panicles and tall plant height may be of some benefit as the swaying panicle is less favourable for birds to perch and feed. Grazing by cattle is also a frequent concern, with plant heights greater than 2.0 m desired to minimize damage to panicles by transhumant cattle.

4.5 Striga: overarching all growth phases

The parasitic weed *Striga* (*Striga hermonthica*) is present throughout WCA, and farmers need to monitor and control this potentially devastating pest. The biology and genetics of the parasite and the host–parasite interactions are now understood in more detail (Ejeta, 2007). Based on these insights, a wide array of crop management alternatives has been tested with farmers for control of *Striga* (Table 2) (Van Mourik, 2007). Farmers' experiences with integration of control techniques indicate the need to combine at least three or more control options. Approaches that contribute to *Striga* control include: (1) enhancement of sorghum growth, e.g. fertilization with optimum timing and placement; (2) reduction of the number of *Striga* seeds in the soil by a) cultivating a false host (trap crop) that induces suicidal *Striga* germination, b) applying compost that increases soil microbial activity which in turn reduces the viability of *Striga* seeds and c) cultivating a resistant variety; (3) suppression of *Striga* growth by practices such as sowing a spreading cowpea that smothers the *Striga* plants and (4) reduction of *Striga* seed production and dispersal by late weeding and particularly by hand-pulling of *Striga* plants. Early sowing of the sorghum crop may also help reduce *Striga* parasitism as *Striga* seeds require two to three weeks of moist 'conditioning' before they can germinate (Ejeta, 2007) and attach to sorghum roots.

Soil fertility and water are closely interlinked factors determining sorghum as well as *Striga* growth and reproduction. The level of soil fertility, particularly plant-available P, contributes to two important, mutually reinforcing dynamics: a) the amount of sorghum growth and extent of shading that limits *Striga* growth and b) the extent of sorghum root exudates that mobilize P as well as stimulate *Striga* germination (Jamil et al., 2013). Thus nutrient-deficient soil conditions contribute to both more *Striga* germination and enhanced *Striga* growth and seed production with reduced shading by poorly developed sorghum canopy. Compost applications showed advantages over inorganic fertilizer for integrated soil fertility and *Striga* management for both sorghum growth and reduction of *Striga* seed viability in on-farm experimentation (Van Mourik, 2007).

Farmers' understanding of *Striga* biology and its interactions with sorghum is essential for them to choose and adopt control practices that fit their specific labour and economic conditions. The fact that *Striga* multiplies by seed is a critical revelation, since the seeds are so small and barely visible that farmers often conclude that magic, rather than seed, is the source of *Striga* infestations. Videos have been produced and made available online for informing farmers (Agro-Insight et al. 2012). Farmers' adoption of *Striga* control practices also depends on obtaining some benefits in the first year of application, even if benefits accrue over many years, since sorghum growers operate under severe resource constraints. The use of spreading cowpea is an example of a particularly attractive practice, as it impedes *Striga* growth and produces valuable hay.

Table 2 Overview of *Striga* control techniques and their effects on sorghum growth, *Striga* growth and development, and *Striga* seed bank in the soil (adapted from van Mourik, 2007)

Control technique/tool	Effect on sorghum growth	Effect on *Striga* growth and development	Effect on *Striga* seed bank
Striga-tolerant sorghum variety	Good growth, yield is generally maintained	No hindrance to growth	Increased risk of post-harvest seed production
Resistant sorghum variety	Good growth, yield is maintained	Reduced growth and development	*Striga* seed production may be reduced
Sowing sorghum early	Good growth using early nitrogen flush	Crop shading reduces *Striga* growth	Risk of post-harvest seed production
Herbicide-coated seeds of a herbicide-resistant variety	Good early growth especially with metalaxyl	Death of early attaching *Striga*	Risk of post-harvest seed production
Intercrop sorghum with high-density cowpea	Possible competition from cowpea	*Striga* growth suppressed by shading	Usually reduced. Risk of post-harvest seed production
Rotation with trap crop, e.g. cotton or soybean	Sorghum benefits from fertilizer applied to cotton	*Striga* germinates but does not develop on trap crop	*Striga* seed bank can be reduced
Rotation with pearl millet	Possible negative preceding crop effect	Sorghum-specific *Striga* populations do not grow on pearl millet	May reduce seed bank due to suicidal germination and poor growth of surviving *Striga*
Rotation with maize	Sorghum benefits from residual fertilizer effects	Sorghum-specific *Striga* populations do not grow on maize	Minimal seed production
Application of fertilizer with placement in or near planting hole	Sorghum establishment and initial development stronger	More *Striga* may germinate but develops poorly due to shading	Risk of post-harvest *Striga* seed production if moisture available
Application of compost before sowing	Sorghum development is favoured	Good germination but growth is reduced by shading	Reduced by the greater microbial activity. Risk of post-harvest seed production
Late weeding, third weeding of sorghum	Limited benefit if previous weedings were effective	*Striga* growth and development is reduced significantly	Reduced addition to *Striga* seed bank
Pulling up *Striga* before they set seed	No direct benefit	Interruption of the growth cycle	Seed bank reduced

5 Genetic diversity and genetic enhancement of sorghum in WCA

5.1 Sorghum diversity and its uses in WCA

Guinea-race sorghums, with flattened, corneous grains, twisting relative to open glumes, are the dominant race in the western Savannah zone from Senegal across Mali and Burkina Faso, but also in Chad and South Sudan (Harlan and de Wet, 1972; Touré and Scheuring, 1982; Barro-Kondombo et al., 2010). The hard corneous grain, pendulous panicles and wide glume opening contribute to good weathering resistance under wet conditions (Harlan and de Wet, 1972; Haussmann et al., 2012). All three Guinea sub-races are cultivated in WCA: guineense, medium-sized grain and most widespread; conspicuum, with larger grain size, is cultivated in North West Nigeria, Cameroon; and margaretiferum, with very small grains, which is genetically and evolutionarily distinct from other Guinea sorghums (Folkertsma et al., 2005; Deu et al., 2006), and is the dominant type cultivated in highest rainfall environments of Sierra Leone and Guinea, but it is also grown in drier environments across WCA, primarily by women, frequently as intercrop in their groundnut fields. The margaretiferum sub-race of sorghum has very hard grains that can be prepared like rice.

Durra-race sorghums, with rounded grain apex and tapering to a wedge-shaped base, are well adapted to sandy soils found across the Sahel from Senegal to Chad and beyond. Sorghums grown on residual moisture as well as the late-flowering Dune-sorghums found in the Sahel are nearly all of the Durra race. They tend to have very specific adaptations of their crop cycle to the prevailing water availability and pest situations, and appear to harbour specific adaptations to high water vapour pressure deficits, that could be useful for future breeding targeting drought resistance (Vadez et al., 2011).

Caudatum-race sorghums are also cultivated in WCA, but are of relatively less importance traditionally than other races. Tall, photoperiod-sensitive and long-duration Caudatum landraces exist, as well as early maturing, photoperiod-insensitive landraces, primarily in the Sahelian zone. Caudatum-race materials, highly represented in sorghum breeding programmes in the United States and India (Harlan and de Wet, 1972), have been introduced and used in WCA sorghum breeding programmes. The direct contribution of these materials has been primarily limited to more niche environments. Serious adaptation difficulties occurred in the Savannah zone with direct use per se as pure-line varieties that risk yield reductions or total loss due to the short crop duration and the compact panicle form exposing crops to bird damage and to the head bug–grain mould complex. However, introgression of Caudatum materials at an appropriate dose into adapted, longer duration materials, has contributed valuable diversity.

Bicolor race sorghums are primarily represented by sweet sorghums (Nebie et al., 2013) which are grown in small areas in most villages, often by children, for their juicy sweet stems. They are genetically distinct from other groups of sorghum.

Intermediate races are of tremendous importance. For example the Kaura (classified as Durra-Caudatum) and the Fara-Fara sorghums (Guinea-Caudatum) are two major types of sorghums cultivated in Nigeria, the largest sorghum-producing country in WCA. In Chad six of the ten possible racial intermediates are cultivated, with Guinea-Caudatum, Durra-Caudatum and Guinea-Bicolor being the most frequent (Yagoua, 1997).

5.2 Breeding for improved sorghum productivity, yield stability and quality

Experience over the past four decades has shown the necessity for WCA sorghum breeding programmes to emphasize the use of germplasm from, and adapted to, the target production ecosystem. Parental material that provides the basic array of adaptive traits and grain qualities is required to breed for yield, stability and value. Therefore, the germplasm bases for breeding in Nigeria, where Kaura and Fara-Fara sorghums predominate, and in Burkina Faso and Mali, with mostly Guinea-race sorghums, are distinct and emphasize farmer-demanded traits. Even within country, distinct breeding streams are required to serve zones where earlier- or later-maturing varieties are demanded.

This approach, however, does not preclude enhancing genetic diversity by introgressing introduced germplasm. The creation of a random-mating Guinea-race population by intermating Guinea landrace varieties from across West Africa is one example (Rattunde et al., 1997). Introduction of novel diversity from unadapted introduced germplasm can also be valuable if it is done at an appropriate dose to retain required grain and adaptation traits at sufficient levels for agronomic utility and acceptability.

5.2.1 Breeding methods

Diverse methods for establishing breeding objectives through collaborative researcher–farmer priority setting have been developed based on experiences in WCA and elsewhere (Christinck et al., 2005). Conventional pedigree breeding, recurrent selection in broad based populations, including subsequent pedigree selection of derived lines, and backcross designs are methods applied in WCA sorghum breeding. The application of molecular tools in applied sorghum breeding has been rare to date, and mostly limited to individual projects. The development of markers for culling early generation progenies with unacceptable grain or glume traits before entering yield testing holds promise for enhancing the efficacy of applied breeding programmes, and research is ongoing to this end (personal communication, Chiaka Diallo, 25 May 2017). Farmer participatory breeding methods have long been employed in the region (Weltzien et al., 2008a), and their use includes on-farm testing of experimental lines (Rattunde et al., 2016) and determination of selection criteria (vom Brocke et al., 2010).

Variety development efforts have primarily targeted pure-line open-pollinated varieties over the past 40 years, although hybrid breeding has recently gained considerable momentum in the region. Both population improvement (discussed below) and bi-parental pedigree breeding methods are being used for pure-line and hybrid parent development in WCA.

Recurrent selection in random-mating populations is conducted to concentrate favourable genes through rapid cycles of selection so as to increase the frequency of progenies with superior performance for the targeted trait(s), while retaining variation for these selected as well as non-selected agronomic traits. This approach offers advantages for sorghum improvement in WCA for a) addressing the large number of plant, panicle and grain traits required for acceptability and adaptation and b) sharing improved germplasm with other breeders or farmers interested to further select for adaptive or quality traits required for their particular production system and maturity zone. The diversity retained for maturity, grain type or plant height, for example, has enabled diverging selection paths to address new target traits or plant types. For instance out of a tall (over 3m plant height)

Guinea population, based on West African Guinea landrace varieties (Rattunde et al., 1997) but containing dwarfing genes at low frequencies, a new population with shorter height was created that provides valuable source material for breeding novel dual-purpose grain/fodder varieties that combine Guinea-race panicle and grain traits with superior stover quality. The variety 'Lata' derived from this population is now both cultivated by Malian farmers as a variety per se and used as the male-parent for successful hybrids (Rattunde et al., 2013; Kante et al., 2017).

A modified backcross approach with only one or at most two backcrosses to an elite donor parent followed by inbreeding of a wide panel of F2-derived backcross lines has proven ideal for enhancing genetic diversity of sorghum in WCA. This approach involves introgression of unadapted germplasm, even of contrasting races, that may offer useful genetic variation for particular traits, combined with selection within backcross F2 families for segregants conforming to the required adaptation/quality norms. This approach is proving most useful in Mali for combining powerful genetic analysis with a progressive, stepwise long-term introgression programme as described by Jordan et al. (2011).

Hybrid breeding with West African sorghum germplasm has shown potential for increasing grain yields (Andrews, 1975; Touré and Scheuring, 1982). However, hybrids based on Caudatum seed parents and Guinea landrace varieties used as male parents lacked adaptation and grain quality traits required for production in the Guinea-race zone, indicating the need to develop both female and male parents that combine well for these traits (Touré and Scheuring, 1982). The first hybrids based entirely on West African Guinea-race germplasm are now available. These hybrids give 30% yield advantages over local varieties in farmers' fields (Rattunde et al., 2013; Kante et al., 2017) and adoption is starting (Smale et al., 2014). As these yield gains were obtained with only a limited number of parental lines and low selection intensity, even greater yield advantages should be attainable by implementing a full-scale hybrid breeding programme that uses the broad genetic diversity of West African sorghums. High levels of heterosis were shown to be achievable by crossing between distinct sub-pools of Guinea-race germplasm (e.g. guinense from Mali and Burkina Faso and conspicuum from Nigeria and Cameroon) (Dagnoko, 2008).

5.2.2 Collaborative multi-location and multi-environment testing

Achieving genetic gains for grain yield requires testing in multiple environments that represent the targeted cropping systems. Such large-scale yield testing, however, is challenging for sorghum breeding programmes in WCA due to resource and technical limitations. Whereas breeders' experimental station fields are typically managed at higher input levels and are few in number, farmers' field conditions and agronomic practices are highly diverse, even within a target production system, and are mostly at suboptimal fertility levels.

Farmer–breeder collaboration for on-farm yield testing provides one approach for increasing genetic gain for grain yield (Weltzien et al., 2008b; Rattunde et al., 2016). One breeding line identified via multi-environment on-farm testing showed a stable 11% yield superiority over the adapted local checks across the full range of production conditions. This line showed both superior phosphorus use efficiency and total P uptake (Leiser et al., 2014), a remarkable combination which may have contributed to its superior yielding ability in farmers' fields. Another approach for increasing gains for grain yield is breeders' collaboration for joint testing of breeding materials across more environments than can be managed by individual breeding programmes.

Large-scale testing of early generation materials would help increase genetic gains for yield by exploiting the high genetic diversity available in early generations and by enabling higher selection intensities than are feasible in later generations. Heritabilities for grain yield in large-scale on-farm testing systems can be sufficient for discriminating among progenies using various experimental designs (Rattunde et al., 2016). Furthermore, farmer collaboration with breeders can help focus efforts on materials that combine yield with farmer-preferred traits, and so improve breeding efficiency for the development of successful varieties.

5.2.3 Improving and combining grain- and stalk-quality

Discussing grain yield with farmers in WCA revealed that they often define yield in terms of food produced rather than weight of grain per area. One major determinant of food yield is the decortication yield, the yield of grits and flour from a given weight of grain after removing the seed coat. Decortication yields of farmer landrace and some bred varieties are typically about 80%, whereas they may be as low as 60% for varieties bred with introduced germplasm. Decortication yields therefore need to be routinely monitored. Farmers also observe varietal differences for the weight of grain needed to feed their family. Free threshing of grain is also an important determinant of food yield, as grain with clasping glumes would be discarded. Varieties that maintain open glumes and are free threshing even under stress conditions are therefore valued for giving more yield stability and food security.

The nutritional value of sorghum is also vital since sorghum, together with pearl millet, form the basis of rural diets in the zone with approximately 800 mm rainfall or less. In this zone the diets of children below five years of age are so highly based on these cereals that they provide approximately half of the iron and zinc intake (Tuinsma, 2006). The widespread and exceedingly high micronutrient deficiencies, especially among children and women of reproductive age, results in weakened immune systems and poor cognitive development, making the micronutrient content of sorghum grains a priority issue. WCA sorghums have significant genetic variability for micronutrient contents and breeding for improved nutritional value is ongoing. Monitoring for acceptable decortication yields will also help maintain micronutrient levels, since approximately half of iron and zinc content is lost by removing the bran (Hama et al., 2011). Various options for improved food preparation can also contribute to nutritional yield. For example, soaking, surface drying and grinding whole grain, where mechanical mills are available, help retain most of the iron zinc, and fibre, and reduce loss of the oil and protein-rich germ. Also, preparation of foods using fermentation or addition of fruit juices containing vitamin C can enhance the bioavailability of iron for improved nutrition.

The development of novel multi-use grain and fodder, or grain-, fodder- and sweet-sorghum types offers options for increasing total yield and crop value. The stems of the tall, highly lignified, landrace sorghum varieties of WCA are of low nutritional quality for ruminant livestock. Newly developed short (less than 2.5 m) sorghum varieties have stems with less lignification, higher digestibility and stover quality far superior to traditional varieties. Selecting in the opposite sense, for high stem lignification, might be taken up in areas where cooking fuel is not easily available or stemborers cause significant damage. The newly bred sweet sorghum varieties and lines combine quality grain with sweet juicy stems that can produce food, fodder and sweet sorghum syrup for human consumption or biofuel. Thus, these dual- or multipurpose sorghums can provide farmers with new cropping options for increasing total value and diversifying production, and thereby contribute to improved resilience.

6 Sorghum and seed system development

Local varieties of sorghum are widely used by farmers in the WCA region. These, mostly smallholder, farmers maintain and manage sorghum seed in a traditional manner on an individual or social network basis. Sorghum, as an indigenous staple crop, plays an important role in the culture for farmers of this region similar to maize in Mexico or wheat in the Middle East. This implies that sorghum seed in WCA is more than a commodity or an agricultural input, as it is associated with traditions and obligations of solidarity and risk sharing within communities.

Effective seed dissemination is vital for ensuring that farmers can access and use new varieties for enhancing value and productivity. The dissemination of newly bred varieties, however, has been problematic in many sorghum-growing countries of WCA, with both cultural aspects and varietal adaptation or quality requirements at play. Therefore all seed system actors, including sorghum breeders, need to understand farmers' seed management practices and culture such that variety development and dissemination efforts address farmers' priority needs and objectives.

6.1 Seed system analysis

Recognizing that the sorghum seed system in WCA, and elsewhere in Africa, is a 'human activity system' means that the first step in developing a seed dissemination approach is to understand the existing seed system and its main actors. This may include grain markets, traditional and commercial dissemination channels and farmers' practices and cultural norms for seed production and seed security in the target production zones. Comprehensive guidelines for seed security assessments have been developed (Sperling, 2008) and conducting such an analysis in non-emergency situations provides valuable information on the system's functioning for the targeted farmers (McGuire and Sperling, 2016).

Important pieces of information include under which circumstances farmers may acquire seed from outside their own farm and by which criteria they choose the variety, the supplier and the seed, as well as the preferred way for reimbursing the supplier of the seed. Understanding the variety and seed requirements for different types of farmers (i.e. gender, family and farm size, role and type of livestock) in a given area is another critical element.

Such a detailed analysis gives critical insights into the type of varieties that farmers seek, in terms of not only agronomic and adaptive traits but also quality traits for use and processing, and degree of uniformity or intra-varietal diversity that is accepted or has intrinsic value. This analysis can thus help guide both the organization of seed production, information exchange and seed dissemination in an effective and socially acceptable manner, and enable sorghum breeders to better orient their work and partnerships for increased impact (Christinck et al., 2005; Weltzien and Christinck, 2008).

Some findings from the Sudan Savannah in Mali illustrate the cultural issues related to seed that bear on efforts to develop effective seed dissemination options:

- 'A good farmer produces his own seed'. Most farmers choose good fields in which they select the best panicles of each variety they intend to sow the following year. The seed panicles are stored in a manner that everyone knows that they are for seed and will not be touched for food. The farmers' own seed is trusted to be adapted to

their conditions and needs, and taking seed 'from outside', from a field they have not seen or from 'someone else', may be seen as putting their family at risk (Siart, 2008). However, farmers who see a desirable variety grown by someone they know may 'place an order' to receive a quantity of seed for the following season.
- 'If someone asks me for seed during the sowing period, I am obliged to give it'. The person requesting seed may pay for the seed by exchanging an equal volume of grain either immediately, or if lacking grain, after the subsequent harvest. Taking money for seed is not practised (Siart, 2008).

Farmers in Mali are thus often reluctant to talk about their seed or new varieties: the more people ask for seed the more seed the farmer may have to give away with little or no compensation. Traditionally therefore seed has no commercial value and it is not culturally acceptable for an individual farmer to develop a seed enterprise and profit from others' needs. A group-based seed enterprise, where activities and benefits are shared in a transparent manner, is acceptable, however. A contrasting situation occurs in the Sudan Savanna of Kano and Katsina states in Nigeria where sorghum farmers with appreciable quantities of clean healthy seed or grain at planting time are highly respected in the society and they commonly share some with family and friends as well as sell or exchange with grain in a grain-to-seed ratio of 3:2.

6.2 Sorghum seed dissemination models

The first attempts of producing and disseminating newly bred sorghum and other crop varieties in WCA were directed through state agencies such as agricultural extension- and national seed-services. Although these efforts attained some success, especially with crops such as cotton, this approach was largely abandoned in the 1990s during economic adjustment programmes.

Considerable donor investment in WCA has gone into seed system development that could be called a linear supply chain model. This model operates primarily on the concept that there is a direct flow of creativity and production, initiating with the breeder, which is followed by seed producers and seed sellers, with input from control services, to arrive at the product to be purchased by the farmer. Donor support has primarily targeted public breeding programmes and private companies involved in seed processing and sale, and less so to seed producers, extension services or local retailers. The long-term success of this model for sorghum in WCA will hinge on its ability to add value and options for smallholder farmers that complement their varieties and traditional seed provisioning, establish farmers' trust and provide convenient and timely access to seed through local retailers.

The profitability of sorghum seed marketing enterprises appears to be better in Nigeria, where bakeries and malting or brewing enterprises provide farmers with seed. Sorghum processing industries that need to procure grain of uniform and specified qualities are now contracting growers for the production of specified varieties. These industries may support the dissemination of varieties, provide farmers with credit for seed and other inputs, as well as ensure the market for the product. This model is also being developed for sorghum in Ghana and Sierra Leone and has contributed to the dissemination of sorghum varieties.

A more decentralized seed system model is emerging with farmer cooperatives and associations conducting many seed system functions for sorghum and other staple cereals in Mali, Burkina Faso and Benin. Certified sorghum seed in Mali is now mostly produced by farmer cooperatives, and a great portion of the seed purchased directly by farmers is bought from nearby cooperatives. One factor contributing to this is that the cooperative model is a good fit to the sociocultural values and ecological context. This is particularly true for smallholder farmers for whom food security as well as market considerations are priorities. Critical elements are farmers' knowledge and trust, both of the people involved and of the varieties on offer.

The decentralized, cooperative seed production and dissemination model is a good match for public breeding programmes interested in collaborative large-scale variety testing with organized farmer groups. Such collaboration is ongoing. The long-term collaboration between researchers, development organizations and farmer seed cooperatives can and is helping these farmer groups strengthen their cooperative governance and finance and asset management as their seed production, processing and marketing activities grow.

Decentralized seed enterprises based on farmer groups are not new. They played a role in founding the modern seed systems of Europe (Harwood, 2012) and the United States, with certain seed companies today tracing back to farmer cooperatives and public sector–farmer collaboration. Investing in farmer seed cooperatives to advance sustainable seed system development for smallholder farmers of WCA, however, is currently rare. This will hopefully change when development donors gain awareness of the advantages that this approach can provide. While farmers, and particularly cooperative members, are very aware of the difficulties associated with cooperative management, they highly appreciate local farmer-managed seed supply and dissemination that responds to their variety and seed quality preferences and provides income for them and their communities.

7 Conclusion

Sorghum is the staple food for millions of people in WCA and, until recently, was mostly consumed by the farmers who produce it. Urbanization, however, is accelerating and the opportunities, and the necessity, for farmers to sell sorghum and other grains are increasing. This chapter presents a wide range of entry points for improving the productivity and sustainable intensification of sorghum production systems in WCA. These opportunities, identified through long-term engagement in the region, are tremendous for this resilient and resource-efficient cereal crop. The Savannah zone, where sorghum is predominant, would be the region of highest priority. The need for research to drive these developments is equally great and diverse, from biological and agronomical issues, to institutional arrangements for seed production dissemination and product marketing. Collaboration of a wide range of stakeholders, including farmers, focused on responding to the priorities and needs of farmers and their partners, is critical for sustainably improving livelihoods in sorghum-based agroecologies of WCA.

8 Where to look for further information

Two general tracks for guiding future research to understand and enhance the diverse sorghum production systems in WCA can be suggested: a) focus on options for creating greater value for farmers and their partners, and b) focus on the sorghum systems per se, with attention to actor relationships, communication flows and opportunities for collaboration.

Attention to environmental and social context and individual skills and differential user needs is vital for both tracks in the WCA region, and thus leads to nearly limitless possibilities for research and development. Yet, certain major themes such as the following can be identified for the 'creating-value track': 1) enhancement of food yield and nutritional value, including grain quality for storability, home food preparations and micronutrient content and bioavailability; 2) development of novel multipurpose varieties that combine specific grain and stover qualities for food and feed uses; 3) soil-fertility (especially P) and weed management options that are feasible and profitable for smallholder farmers; 4) hybrid development for the various regions with distinct agroecological conditions and grain quality requirements; and 5) the improvement and joint employment of agronomic practices and varietal resistances for limiting losses due to stem borers, sorghum midge and *Striga*.

Recognizing sorghum production and seed systems as human activity systems makes it necessary to look closely at the actors and their goals, activities and relationships (Long, 2001). Gatzweiler and von Braun (2016) suggested distinct innovation strategies for different groups of smallholder farmers along a gradient of human capability and agroecological potential, and noted that sustainable intensification by smallholder farmers is not just a technical optimization problem, but a 'task of creating value through innovations in the institutional, organizational and technological systems of societies'. Practical approaches for researcher–farmer collaboration in priority setting for breeding and seed system interventions are detailed in a handbook based on sorghum improvement work in WCA (Christinck et al., 2005). And a framework for seed system analysis, proposed by McGuire and Sperling (2016), can help guide organization of seed production, information exchange and seed dissemination in an effective and socially acceptable manner.

A community of collaborating research institutions and farmer organizations is central to the past and ongoing research on sorghum improvement in WCA. National institutes engaged in this research include l'Institute d'Economie Rurale (IER, Mali), Institut de l'Environnement et Recherches Agricoles (INERA, Burkina Faso), Institute of Agricultural Research of the Ahmadu Bello University, Zaria, Nigeria, Institut National de la Recherche Agronomique du Niger (INRAN, Niger), Savanna Agricultural Research Institute (SARI, Ghana), Institut Sénégalais de Recherches Agricoles (ISRA, Senegal), and Institut Tchadien de Recherche Agronomique pour la Développement (ITRAD, Chad). International research organizations include the International Crops Research Institute for the Semi-Arid Tropics (ICRISAT) and CIRAD (French Agricultural Research Centre for International Development). Universities, both within and outside of the region, contribute particularly through student research activities. Farmer organizations collaborating in sorghum research activities include the Union Locale des Producteurs de Cereales (ULPC) and the Association des Organisations des Producteurs Professionnelles (AOPP) in Mali, Union des Groupements pour la Commercialisation de Produits Agricoles de la Boucle du Mouhoun (UGCPA/BM) and l'Association Minim Song Panga (AMSP) in Burkina Faso and FUMA Gaskiya and Mooriben in Niger.

Several foundations and government agencies have provided important support for the sorghum improvement work in WCA, including the McKnight Foundation's Collaborative Crop Research Program for WCA, the Bill and Melinda Gates Foundation, the United States Agency for International Development and the German Federal Ministry for Economic Cooperation and Development.

9 Acknowledgements

We are most appreciative to all who have contributed to the joint learning that we attempt to synthesize from the sharing of farmers, collaborating in countless trials and discussions, the input of research partners and colleagues, as well as the challenges to address adoption, scaling and different types of impact raised by development partners and donor agencies. We also thank the reviewers, Benoit Clerget and anonymous, for their thoughtful and constructive suggestions and Willmar Leiser for preparation of Fig. 1.

10 References

Abdalla, E., A. Osman, M. Maki, F. Nur, S. Ali, and J. Aune. 2015. The response of sorghum, groundnut, sesame, and cowpea to seed priming and fertilizer micro-dosing in South Kordofan state, Sudan. *Agronomy* 5: 476–90. doi:10.3390/agronomy5040476.

Agro-Insight, ICRISAT, and UACT. 2012 Available at: https://www.accessagriculture.org/integrated-approach-against-striga (verified 24 May 2017).

Ajayi, O. 1989. Stem borers of sorghum in West Africa with emphasis on Nigeria. In *International Workshop on Sorghum Stem Borers*. ICRISAT, Patancheru, India, pp. 27–31.

Andrews, D. J. 1975. Sorghum grain hybrids in Nigeria. *Experimental Agriculture* 11: 119–27.

Aune, J. B., M. Doumbia and A. Berthe. 2007. Microfertilizing sorghum and pearl millet in Mali – agronomic, economic and social feasibility. *Outlook Agric* 36: 199–203.

Barnaud, A., M. Deu, E. Garine, D. McKey and H. I. Joly. 2006. Local genetic diversity of sorghum in a village in Northern Cameroon: structure and dynamics of landraces. *Theoretical and Applied Genetics*. 114: 237–48. doi:10.1007/s00122-006-0426-8.

Barro-Kondombo, C., F. Sagnard, J. Chantereau, M. Deu, K. vom Brocke, P. Durand, E. Gozé and J. D. Zongo. 2010. Genetic structure among sorghum landraces as revealed by morphological variation and microsatellite markers in three agroclimatic regions of Burkina Faso. *Theoretical and Applied Genetics* 120: 1511–23. doi:10.1007/s00122-010-1272-2.

Breman, H. 1998. Amerlioration de la fertilité des sols en Afrique de l'Ouest: Contraintes et perspectives. In *Soil Fertility Management in West African Land Use Systems*. Margraf Verlag, Niamey, Niger, pp. 7–20.

Buerkert, A., A. Bationo and H.-P. Piepho. 2001. Efficient phosphorus application strategies for increased crop production in sub-Saharan West Africa. *Field Crops Research* 72: 1–15. doi:10.1016/S0378-4290(01)00166-6.

Buerkert, A. and P. Hiernaux. 1998. Nutrients in the West African Sudano-Sahelian zone: Losses, transfers and role of external inputs. *Z. Pflanzenernaehrung Bodenkunde* 161: 365–83.

Buerkert, A., H.-P. Piepho and A. Bationo. 2002. Multi-site time-trend analysis of soil fertility management effects on crop production in Sub-Saharan West Africa. *Experimental Agriculture* 38. doi:10.1017/S0014479702000236.

Chantereau, J. and G. Fliedel. 2013. Le sorgho. Quae; CTA; Presses agronomiques de Gembloux, Versailles; Wageningen; Gembloux.

Christinck, A., E. Weltzien and V. Hoffmann. 2005. *Setting Breeding Objectives and Developing Seed Systems with Farmers. A Handbook for Practical Use in Participatory Plant Breeding Projects*. Magraf Publishers and CTA, Weikersheim, Germany and Wageningen, The Netherlands.

Clerget, B., M. Dingkuhn, E. Goze, H. F. W. Rattunde and B. Ney. 2008. Variability of Phyllochron, Plastochron and Rate of Increase in Height in Photoperiod-sensitive Sorghum Varieties. *Annals of Botany* 101: 579–94. doi:10.1093/aob/mcm327.

Comas, J. and H. Gomez MacPherson. 2002. La culture du sorgho de décrue en Afrique de l'Ouest et du Centre. Situation actuelle et définition d'un Plan d'Action Régional. In *La culture du sorgho de décrue en Afrique de l'Ouest et du Centre*. Situation actuelle et définition d'un Plan d'Action Régional. Nouakschott, Mauretania.

Cuevas, H. E., L. K. Prom, T. Isakeit and G. Radwan. 2016. Assessment of sorghum germplasm from Burkina Faso and South Africa to identify new sources of resistance to grain mold and anthracnose. *Crop Protection* 79: 43–50. doi:10.1016/j.cropro.2015.10.007.

Dagnoko, S. 2008. Dissecting heterosis in Guinea race sorghum: Simultaneous analysis by conventional breeding and molecular marker approaches.

Dakouo, D., G. Trouche, N. B. Malick, A. Neya and K. B. Kaboré. 2005. Lutte génétique contre la cécidomyie du sorgho, Stenodiplosis sorghicola : une contrainte majeure à la production du sorgho au Burkina Faso. *Cahiers d'Agriculture* 14: 201–8.

de Ridder, N. and H. van Keulen. 1990. Some aspects of the role of organic-matter in sustainable intensified arable farming systems in the West-African Semi-Arid-Tropics (sat). *Fertility Research* 26: 299–310.

Deu, M., F. Rattunde and J. Chantereau. 2006. A global view of genetic diversity in cultivated sorghums using a core collection. *Genome* 49: 168–80.

Doumbia, M. D., L. R. Hossner and A. B. Onken. 1993. Variable sorghum growth in acid soils of subhumid West Africa. *Arid Soil Research and Rehabilitation* 7: 335–46. doi:10.1080/15324989309381366.

Doumbia, M. D., L. R. Hossner and A. B. Onken. 1998. Sorghum growth in acid soils of West Africa: Variations in soil chemical properties. *Arid Land Research and Management* 12: 179–90. doi:10.1080/15324989809381507.

Ejeta, G. 2007. Breeding for resistance in sorghum: exploitation of an intricate host–parasite biology. *Crop Science* 47: S-216–S-227. doi:10.2135/cropsci2007.04.0011IPBS.

Falconnier, G. N., K. Descheemaeker, T. A. Van Mourik, O. M. Sanogo and K. E. Giller. 2015. Understanding farm trajectories and development pathways: Two decades of change in southern Mali. *Agricultural Systems* 139: 210–22. doi:10.1016/j.agsy.2015.07.005.

FAOSTAT. 2017. Available at: http://faostat.fao.org (verified 24 May 2017).

Fatondji, D., C. Martius, R. Zougmore, P. L. G. Vlek, C. L. Bielders and S. Koala. 2009. Decomposition of organic amendment and nutrient release under the zai technique in the Sahel. *Nutrient Cycling in Agroecosystems* 85: 225–39. doi:10.1007/s10705-009-9261-z.

Folkertsma, R. T., H. F. W. Rattunde, S. Chandra, G. S. Raju and C. T. Hash. 2005. The pattern of genetic diversity of Guinea-race *Sorghum bicolor* (L.) Moench landraces as revealed with SSR markers. *Theoretical and Applied Genetics* 111: 399–409. doi:10.1007/s00122-005-1949-0.

Gale, L.R. 2003. A population genetic approach to variation in colletotrichum graminicola, the causal agent of sorghum anthracnose. In Leslie, J. F. (ed.), *Sorghum and Millets Diseases*. Iowa State Press, Ames, Iowa, USA, pp. 191–9.

Gigou, J., K. Traore, F. Giraudy, H. Coulibaly and B. Sogoba. 2006. Farmer-led contour ridging can reduce water runoff in African savannahs. *Cahiers Agricultures* 15: 116–22.

Giller, K. E., P. Tittonell, M. C. Rufino, M. T. van Wijk, S. Zingore, P. Mapfumo, S. Adjei-Nsiah, M. Herrero, R. Chikowo, M. Corbeels, E.C. Rowe, F. Baijukya, A. Mwijage, J. Smith, E. Yeboah, W. J. van der Burg, O. M. Sanogo, M. Misiko, N. de Ridder, S. Karanja, C. Kaizzi, J. K'ungu, M. Mwale, D. Nwaga, C. Pacini and B. Vanlauwe. 2011. Communicating complexity: Integrated assessment of trade-offs concerning soil fertility management within African farming systems to support innovation and development. *Agricultural Systems* 104: 191–203. doi:10.1016/j.agsy.2010.07.002.

Goldsworthy, P. R. 1967. Responses of cereals to fertilizers in Northern Nigeria. I. Sorghum. *Experimental Agriculture* 3: 29. doi:10.1017/S0014479700004087.

Hama, F., C. Icard-Vernière, J.-P. Guyot, C. Picq, B. Diawara and C. Mouquet-Rivier. 2011. Changes in micro- and macronutrient composition of pearl millet and white sorghum during in field versus laboratory decortication. *Journal of Cereal Science* 54: 425–33. doi:10.1016/j.jcs.2011.08.007.

Harlan, J. R. and J. M. J. de Wet. 1972. A simplified classification of cultivated sorghum. *Crop Science* 12: 172–6. doi:10.2135/cropsci1972.0011183X001200020005x.

Harris, D. 2006. Development and testing of 'On-farm' seed priming. *Advances in Agronomy* 90: 129–78. doi:10.1016/S0065-2113(06)90004-2.

Harwood, J. 2012. *Europe's Green Revolution and Its Successors: The Rise and Fall of Peasant-friendly Plant Breeding*. Routledge. Available at: https://books-google-com.ezproxy.library.wisc.edu/books?hl=en&lr=&id=shnXQjmkaVoC&oi=fnd&pg=PP2&dq=The+European+Green+Revolution+and+Others+Since+Then&ots=VIzbeWIGnH&sig=kj2jrnBAPtC20oWQASMz1KStAXs (verified 9 November 2016).

Haussmann, B. I. G., H. Fred Rattunde, E. Weltzien-Rattunde, P. S. C. Traoré, K. vom Brocke and H. K. Parzies. 2012. Breeding Strategies for Adaptation of Pearl Millet and Sorghum to Climate Variability and Change in West Africa (F. Asch, Ed.). *Journal of Agronomy and Crop Science* 198: 327–39. doi:10.1111/j.1439-037X.2012.00526.x.

Hess, D. E., R. Bandyopadhyay and I. Sissoko. 2002. Pattern analysis of sorghum genotype x environment interaction for leaaf, panicle, and grain Anthracnose in Mali. *Plant Disease* 86: 1374–82.

Jamil, M., T. A. Van Mourik, T. Charnikhova and H. J. Bouwmeester. 2013. Effect of diammonium phosphate application on strigolactone production and *Striga hermonthica* infection in three sorghum cultivars: Diammonium phosphate application on *Striga hermonthica*. *Weed Research* 53: 121–30. doi:10.1111/wre.12003.

Jordan, D. R., E. S. Mace, A. W. Cruickshank, C. H. Hunt and R. G. Henzell. 2011. Exploring and Exploiting Genetic Variation from Unadapted Sorghum Germplasm in a Breeding Program. *Crop Science* 51: 1444. doi:10.2135/cropsci2010.06.0326.

Kante, M., H. F. W. Rattunde, W. L. Leiser, B. Nebié, B. Diallo, A. Diallo, A. O. Touré, E. Weltzien and B. I. G. Haussmann. 2017. Can tall guinea-race sorghum hybrids deliver yield advantage to smallholder farmers in West and Central Africa? *Crop Science* 57: 833. doi:10.2135/cropsci2016.09.0765.

Kenga, R., M. M'Biandoun, A. Njoya, M. Havard and E. Vall. 2003. Analysis of constraints to agricultural production in the Sudan-Sahelian zone of Cameroon using a diagnostic survey. In *Actes du Colloque*. CIRAD, Montpellier, Garoua, Cameroon.

Kihara, J., G. Nziguheba, S. Zingore, A. Coulibaly, A. Esilaba, V. Kabambe, S. Njoroge, C. Palm and J. Huising. 2016. Understanding variability in crop response to fertilizer and amendments in sub-Saharan Africa. *Agriculture, Ecosystems & Environment* 229: 1–12. doi:10.1016/j.agee.2016.05.012.

Leiser, W. L., H. F. W. Rattunde, H.-P. Piepho, E. Weltzien, A. Diallo, A. E. Melchinger, H. K. Parzies and B. I. G. Haussmann. 2012. Selection strategy for sorghum targeting phosphorus-limited environments in West Africa: analysis of multi-environment experiments. *Crop Science* 52: 2517–27. doi:10.2135/cropsci2012.02.0139.

Leiser, W. L., H. F. W. Rattunde, E. Weltzien and B. I. G. Haussmann. 2014. Phosphorus uptake and use efficiency of diverse West and Central African sorghum genotypes under field conditions in Mali. *Plant and Soil* 377: 383–94. doi:10.1007/s11104-013-1978-4.

Marley, P. S., M. Diourt, A. Neya, S. K. Nutsugah, P. Srm, S. O. Katil, D. E. Hess, D. F. Mbaye and Z. Ngoko. 2003. Sorghum and pearl millet diseases in West and Central Africa. In Leslie, J. F. (ed.), *Sorghum and Millets Diseases*. Iowa State Press, Ames, Iowa, USA, pp. 419–25.

McIntire, J., J. Powell, S. Fernandez-Rivera and T. Williams. 1993. African semi-arid tropical agriculture cannot grow without external inputs. In Addis Ababa, Ethiopia.

McGuire, S. and L. Sperling. 2016. Seed systems smallholder farmers use. *Food Security* 8: 179–95. doi:10.1007/s12571-015-0528-8.

Mohamed, E. M. I., C. Jordi, C. David, M. Luciano and G. oacute mez, M. Helena. 2016. Irrigated sorghum and cowpea after wet-season rice as a pathway out of subsistence agriculture in the Senegal River Valley in Mauritania. *African Journal of Agricultural Research*. 11: 1824–35. doi:10.5897/AJAR2016.10876.

Muturi, P. W., P. Rubaihayo, M. Mgonja, S. Kyamanywa, H. C. Sharma and C. T. Hash. 2012. Novel source of sorghum tolerance to the African stem borer, Busseola fusca. *African Journal of Plant Science* 6 doi:10.5897/AJPS12.051.

Nebie, B., R. Nanema, P. Bationo-Kando, E. Traore, V. Labeyrie, N. Sawadogo, M. Sawodogo and J. Zongo. 2013. Variation de caractères agromorphologiques et du Brix d'une collection de sorghos à tige sucrée du Burkina Faso. *International Journal of Biological and Chemical Sciences* 7: 1919–28. doi:10.4314/ijbcs.v7i5.12.

Nicholson, S. E. 2013. The West African Sahel: A review of recent studies on the rainfall regime and its interannual variability. *International Scholarly Research Notices* 2013: e453521. doi:10.1155/2013/453521.

Office of the Honorable Minister of Agriculture. 2012. Agricultural transformation action plan, sorghum transformation value chain, Abjuja, NIgeria.

Ollenburger, M. H., K. Descheemaeker, T. A. Crane, O. M. Sanogo and K. E. Giller. 2016. Waking the sleeping Giant: Agricultural intensification, extensification or stagnation in Mali's Guinea Savannah. *Agricultural Systems* 148: 58–70. doi:10.1016/j.agsy.2016.07.003.

Ouédraogo, E., L. Brussaard, A. Mando and L. Stroosnijder. 2005. Organic resources and earthworms affect phosphorus availability to sorghum after phosphate rock addition in semi-arid West Africa. *Biology and Fertility of Soils* 41: 458–65. doi:10.1007/s00374-005-0840-0.

Ratnadass, A., P. Marley, M. Hamada, O. Ajayi, B. Cissé, F. Assamoi, I. D. Atokple, J. Beyo, O. Cisse, D. Dakouo, M. Diakite, S. Dossou-Yovo, B. Le Diambo, M. Vopeyande, I. Sissoko and A. Tenkouano. 2003. Sorghum head-bugs and grain molds in West and Central Africa: I. Host plant resistance and bug–mold interactions on sorghum grains. *Crop Protection* 22: 837–51. doi:10.1016/S0261-2194(03)00066-8.

Rattunde, H. F. W., S. Michel, W. L. Leiser, H. P. Piepho, C. Diallo, K. vom Brocke, B. I. G. Haussmann and E. Weltzien. 2016. Farmer participatory early-generation yield testing of sorghum in West Africa: Possibilities to optimize genetic gains for yield in farmers' fields. *Crop Science* 56: 1–13. doi:10.2135/cropsci2015.12.0758.

Rattunde, H. F. W., E. Weltzien, P. J. Bramel-Cox, K. Kofoid, C. T. Hash, W. Schipprack, J. W. Stenhouse and T. Presterl. 1997. Population improvement of pearl millet and sorghum: Current research, impact and issues for implementation. In *Proceedings of the International Conference on Genetic Improvement of Sorghum and Pearl Millet*. Lubbock, Texas USA, pp. 188–212.

Rattunde, H. F. W., E. Weltzien, B. Diallo, A. G. Diallo, M. Sidibe, A. O. Touré, A. Rathore, R. R. Das, W. L. Leiser and A. Touré. 2013. Yield of photoperiod-sensitive sorghum hybrids based on guinea-race germplasm under farmers' field conditions in Mali. *Crop Science* 53: 2454. doi:10.2135/cropsci2013.03.0182.

Rooney, W. 2004. Sorghum improvement – integrating traditional and new technology to produce improved genotypes. In *Advances in Agronomy*. Elsevier, pp. 37–109.

Sanogo, O. M., N. de Ridder and H. van Keulen. 2010. Diversity and dynamics of mixed crop-livestock agricultural households in Southern Mali. *Cahiers Agricultures* 19: 185–93. doi:10.1684/agr.2010.0401.

Siart, S. 2008. Strengthening local seed systems: Options for enhancing diffusion of varietal diversity of sorghum in Southern Mali, In H. Boland, V. Hoffmann and U. J. Nagel (Eds.), *Kummunikation und Begratung, Sozialwissenschaftliche Schriften zur Landnutzung und Ländlichen Entwicklung (series # 85)*, Magraf Publishers GmbH, Weikersheim, Germany, 185p. ISBN 978-3-8236-1525-5; ISSN 0947-0352.

Smale, M., Alpha Kernga, Amidou Assima, Eva Weltzien and Fred Rattunde. 2014. An overview and economic assessment of sorghum improvement in mali. *Michigan State University International Development Working Paper*. Working Paper 137: 40.

Sperling, L. 2008. When disaster strikes: a guide for assessing seed security. Centro International de Agricultura Tropical (CIAT), Cali, Colombia. Available at: https://cgspace.cgiar.org/handle/10568/69690 (verified 24 June 2017).

Tonitto, C. and J. E. Ricker-Gilbert. 2016. Nutrient management in African sorghum cropping systems: Applying meta-analysis to assess yield and profitability. *Agronomy for Sustainable Development*. 36. doi:10.1007/s13593-015-0336-8.

Touré, A. B. and J. F. Scheuring. 1982. Presence de genes mainteneurs de l'androsterilité cytoplsamique parmi les varietiés locales de sorgho au Mali. *Agronomie Tropicale* 37: 362–5.

Tuinsma, M. 2006. La consommation et la contribution au régime alimentaire du mil et du sorgho au Mali, Niger et Burkina Faso: L'apport en (micro) nutriments, International Crops Research Institute for the Semi-Arid Tropics, Bamako, Mali, 43p .

Vadez, V., L. Krishnamurthy, C. T. Hash, H. D. Upadhyaya and A. K. Borrell. 2011. Yield, transpiration efficiency, and water-use variations and their interrelationships in the sorghum reference collection. *Crop and Pasture Science* 62: 645–55. doi:10.1071/CP11007.

van Mourik , T. A. 2007. Striga hermonthica seed bank dynamics: process quantification and modelling, Wageningen University, Wageningen, The Netherlands, 123p. ISBN 978-90-8504-692-9.

vom Brocke, K., G. Trouche, E. Weltzien, C. P. Barro-Kondombo, E. Gozé and J. Chantereau. 2010. Participatory variety development for sorghum in Burkina Faso: Farmers' selection and farmers' criteria. *Field Crops Research* 119: 183–94. doi:10.1016/j.fcr.2010.07.005.

Weltzien, E., K. vom Brocke, A. Touré, F. Rattunde and J. Chantereau. 2008a. Revue et tendances pour la recherche en sélection paticipative en Afrique de l'Ouest. *Cahiers Agricultures* 17: 165–71.

Weltzien, E. and A. Christinck. 2008. Participatory plant breeding: Developing improved and relevant crop varieties with farmers. In *Agricultural Systems: Agroecology and Rural Innovation for Development*. Academic Press, Burlingont, MA, USA and London UK, pp. 211–51.

Weltzien, E., A. Christinck, A. Touré, F. Rattunde, M. Diarra, A. Sangare and M. Coulibaly. 2006. Enhancing farmers' access to sorghum varieties through scaling-up participatory plant breeding in Mali, West Africa. In Almekinders, C. and Hardon, J. (eds), *Bringing Farmers Back into Breeding. Experiences with Participatory Plant Breding and Challenges for Institutionalisation*. Agromisa Foundation, Wageningen, Netherlands, pp. 58–69.

Weltzien, E., M. Kanouté, A. Toure, F. Rattunde, B. Diallo, I. Sissoko, A. Sangaré and S. Siart. 2008b. Sélection participative des variétés de sorgho à l'aide d'essais multilocaux dans deux zones cibles. *Cahiers Agricultures* 17: 134–9.

Yagoua, N. D. 1997. Caracterisation du sorgho pluvial, (Sorghum bicolor (L.) Moench.), de la zone soudanienne du Tchad. In Actes du Colloque 'Gestion des Ressources Génétiques des plantes en Afrique des Savanes'. Brochage Imprimerie Chirat, St-Just-La Pendue, France, Bamako, Mali, pp. 111–20.

Zougmoré, R., A. Mando and L. Stroosnijder. 2009. Soil nutrient and sediment loss as affected by erosion barriers and nutrient source in semi-Arid Burkina Faso. *Arid Land Research and Management* 23: 85–101. doi:10.1080/15324980802599142.

Zougmoré, R., A. Mando and L. Stroosnijder. 2010. Benefits of integrated soil fertility and water management in semi-arid West Africa: An example study in Burkina Faso. *Nutrient Cycling in Agroecosystems* 88: 17–27. doi:10.1007/s10705-008-9191-1.

Zougmoré, R., Z. Zida and N. F. Kambou. 2003. Role of nutrient amendments in the success of half-moon soil and water conservation practice in semiarid Burkina Faso. *Soil and Tillage Research* 71: 143–9. doi:10.1016/S0167-1987(03)00050-3.

Index

Lightning Source UK Ltd.
Milton Keynes UK
UKHW020921070819
347547UK00001B/18/P

9 781786 761248